SPRINGER SERIES IN SURFACE SCIENCES 41

Springer
Berlin
Heidelberg
New York
Barcelona
Hong Kong
London
Milan
Paris
Tokyo

Physics and Astronomy ONLINE LIBRARY

http://www.springer.de/phys/

SPRINGER SERIES IN SURFACE SCIENCES

Series Editors: G. Ertl, R. Gomer, H. Lüth and D.L. Mills

This series covers the whole spectrum of surface sciences, including structure and dynamics of clean and adsorbate-covered surfaces, thin films, basic surface effects, analytical methods and also the physics and chemistry of interfaces. Written by leading researchers in the field, the books are intended primarily for researchers in academia and industry and for graduate students.

38 **Progress in Transmission Electron Microscopy 1**
Concepts and Techniques
Editors: X.-F. Zhang, Z. Zhang

39 **Progress in Transmission Electron Microscopy 2**
Applications in Materials Science
Editors: X.-F. Zhang, Z. Zhang

40 **Giant Magneto-Resistance Devices**
By E. Hirota, H. Sakakima, and K. Inomata

41 **The Physics of Ultra-High-Density Magnetic Recording**
Editors: M.L. Plumer, J. van Ek, and D. Weller

Series homepage – http://www.springer.de/phys/books/ssss/

Volumes 1–37 are listed at the end of the book

M.L. Plumer J. van Ek D. Weller
(Eds.)

The Physics of Ultra-High-Density Magnetic Recording

With 163 Figures

Springer

Martin L. Plumer
Johannes van Ek
Seagate Technology, 7801 Computer Ave., S., Bloomington, MN 55435, USA
e-mail: martin_l_plumer@notes.seagate.com, johannes_van_ek@notes.seagate.com

Dieter Weller
Seagate Technology, River Park Commons, Suite 550, 2403 Sidney Street, Pittsburgh, PA 15203, USA
e-mail: dieter_weller@notes.seagate.com

Library of Congress Cataloging-in-Publication Data

The physics of ultra-high-density magnetic recording / M.L. Plumer, J. van Ek, D. Weller. p. cm. - - (Springer series in surface sciences, ISSN 0931-5195 ; 41) Includes bibliographical references and index.
ISBN 3540423702 (alk. paper)
1. Magnetic recorders and recording. I. Plumer, M. L. (Martin L.), 1954- II. Van Ek, J. (Johannes), 1964- III. Weller, D. (Dieter), 1957- IV. Series.
TK7881.6 P56 2001
623.389'3- -dc21 2001049029

ISSN 0931-5195
ISBN 3-540-42370-2 Springer-Verlag Berlin Heidelberg New York

Springer-Verlag Berlin Heidelberg New York
a member of BertelsmannSpringer Science+Business Media GmbH

http://www.springer.de

Typesetting: PTP, Heidelberg Berlin
Cover concept: eStudio Calamar Steinen
Cover production: *design & production* GmbH, Heidelberg

Printed on acid-free paper 57/3111 - 5 4 3 2 1

Preface

Over the past few years, the storage capacity in magnetic recording has approximately doubled every year. This unprecedented growth rate has in part been fueled by discoveries in basic physics research, such as the Giant Magnetoresistive (GMR) effect, which has made possible the use of much narrower recording tracks on disc drives. As the size of a stored bit on a disc continues to shrink, there are formidable fundamental and technical challenges that emerge. For example, smaller recorded bits require smaller grains in the magnetic recording medium in order to reduce transition noise. On the other hand, the magnetization of smaller grains is more susceptible to thermal fluctuations and the grains may become superparamagnetic, in which case the information stored on the medium will degrade. Issues such as this have forced the hard-disc drive industry to re-examine some fundamental physics and engineering associated with the recording process. The number of grains in a recorded di-bit at an information storage density of 100×109 bits per square inch (100 Gb/in^2) is nearly five times smaller than for today's commercially available 20 Gb/in^2 disc drives. The range of possibilities for quick engineering solutions to difficult problems, which served well in the past, are diminishing. The need is for a more thorough understanding of "recording physics" cannot be over emphasized. This applies especially to the centerpiece of the magnetic recording system, namely the recording medium.

This book is an attempt to highlight some of the basic understanding of magnetic recording at very high areal densities which has evolved over the past few years and which will serve as part of its foundation for continued growth. The following 11 chapters were written by 18 experts from academia and industry. Five general topics are addressed: a micromagnetic view of the recording process in Chap. 1; media microstructure and processing issues in Chap. 2; switching dynamics and thermal fluctuations (mostly media) in Chaps. 3–6; alternate media in Chaps. 7–9; magneto-transport and head design in Chaps. 10 and 11.

Early models of the recording process, which treat the medium as a continuum and which do not account for finite track widths or writer pole-tip saturation effects, must be used with caution in high-areal-density applications. Chapter 1 presents a view of recording physics based on the thin films modeled as an assembly of microscopic bar magnets and the effects arising from their stray fields and interactions. Micromagnetics is revealed as a powerful tool in the examination of writer head field profiles, the effects of medium microstructure on the recording process, as well as track width effects on magneto-resistive playback.

Understanding and control of multilayer medium microstructure has become increasingly important in recent years. Chapter 2 demonstrates the impact of the layer-by-layer growth and composition of the multitude of thin films used in a hard disc on grain size, grain size distribution, grain orientation, chemical segregation, interlayer diffusion, etc. The impact of these properties on superparamagnetism is emphasized.

The characterization of thermally agitated magnetization reversal over a large time interval is a formidable challenge. Fundamental understanding of granular "spin" dynamics and damping mechanisms has led to the new model of switching discussed in Chap. 3, which exhibits good quantitative agreement with experiments at short (sub-nanosecond) as well as long timescales. The experimental measurement of dynamics in thin films requires as wide a range of techniques as there are reversal mechanisms. Waveguide, optical and transport apparatus, used to investigate the variety of domain-wall, coherent and incoherent motion that can arise, are summarized in Chap. 4. Thermal and dynamic effects specific to high-density recording media are addressed in Chaps. 5 and 6. Measurement of di-bit decay, transition demagnetization fields, pulse-length-dependent coercivities and medium noise are discussed. Possibilities for partial relief from the onset of superparamagnetism by using antiferromagnetically coupled media or perpendicular recording media are summarized.

One effect of the superparamagnetic problem on industry road map planning has been an energetic renewal of interest in alternatives (to the traditional longitudinal) recording schemes. The current favorite, namely the use of media with anisotropy oriented perpendicular to the disc surface, is reviewed in Chap. 8. Media requirements, such as squareness and grain size are addressed. Compounds and alloys under investigation, soft under-layer noise issues and the advantages of super-lattices are reviewed. Patterned media offer the possibility of ultra-high-density recording. A thorough review of problem areas in photolithography techniques, materials, recording physics, tribology, detection channels, servotracking, etc., as well as some suggestions for solutions, are covered in Chap. 7. Self-assembling FePt nanoparticle arrays offer yet another exciting alternative medium to record data, as discussed in Chap. 9. Such natural patterning can be made with either longitudinal or perpendicular orientation with grain sizes as small as 2–3 nm.

Reductions in track width and bit length have put significant pressure on head performance criteria. Fundamental to the performance of read-back sensors for present and foreseeable technologies is spin-dependent electron transport. A discussion of two-current models, semiclassical theories (based on the inhomogeneous Boltzmann equation), multilayer systems, and specular and diffuse scattering mechanisms applied to current-in-plane and current-perpendicular to the plane modes, are discussed in Chap. 10. Finally, a general discussion of recording head design issues, with a focus on the impact of bit aspect ratio requirements from media thermal stability are reviewed in Chap. 11. Flux delivery schemes for the writer, as well as reader sensitivity and stability, are discussed. Design solutions for both longitudinal and perpendicular recording are put forth.

This volume is clearly focused on magnetism-related issues. Space limitations have forced us to omit many important topics such as servotracking, the detection channel, and mechanical and tribology requirements at ultra-high areal density. We hope, however, that this book will serve as a guide to an understanding of many of the basic physical principles involved in ultra-high-density magnetic recording.

We thank all of the contributors for their excellent manuscripts and especially Springer-Verlag for expediting the publication of their work.

Bloomington, MN, and Pittsburgh, PA,
June 2001

Martin Plumer,
Johannes van Ek,
Dieter Weller

Contents

List of Contributors

Martin Plumer
Segate Technology
7801 Computer Ave. S.
Bloomington, MN 55435

Johannes van Ek
Segate Technology
7801 Computer Ave. S.
Bloomington, MN 55435

Dieter Weller
Seagate Technology
2403 Sidney St. Ste. 550
Pittsburgh, PA 15203-2116

Bin Lu
Seagate Technology
2403 Sidney St. Ste. 550
Pittsburgh, PA 15203-2116

David E. Laughlin
ALCOA Professor of Physical Metallurgy
Materials Science and Engineering Department
Carnegie Mellon University
5000 Forbes Avenue
Pittsburgh, PA 15213

Vladimir Safonov
Center for Magnetic Recording Research,
University of California - San Diego,
9500 Gilman Drive
La Jolla, CA 92093-0401

Neal H. Bertram
Center for Magnetic Recording Research,
University of California - San Diego,
9500 Gilman Drive
La Jolla, CA 92093-0401

Tom Silva
National Institute of Standards and Technology (NIST)
m/s 814.05
325 Broadway
Boulder, CO 80303

Andreas Moser
K63/E3
IBM Almaden Research Center
650 Harry Road
San Jose, CA 95120

Hans Jürgen Richter
Seagate Recording Media
47010 Kato Road
Fremont, CA 94538

Gordon Hughes
Center for Magnetic Recording Research,
University of California - San Diego,
9500 Gilman Drive
La Jolla, CA 92093-0401

Randall Victora
4-174 EE/CSci
University of Minnesota
200 Union Street SE
Minneapolis, MN 55455-0154

Shouheng Sun
IBM T. J. Watson Research Center
P. O. Box 218 / Route 134
Yorktown Heights, NY 10598
USA

William H. Butler
Metals and Ceramics Division
Oak Ridge National Laboratory
Mail Stop 6114
Oak Ridge, TN 37831-6114

Olle Heinonen
Segate Technology
7801 Computer Ave. S.
Bloomington, MN 55435

Dr. Michael Mallary
Maxtor Corp.
SHR1-3/e23
333 South St.
Shrewsbury, MA 01545-4112

1 Introduction to Micromagnetic Recording Physics

Jan van Ek and Martin Plumer

Extremely effective reduction of physical dimensions in the recording head through photolithography techniques, in combination with tremendous development in thin film recording media, has led to a spectacular increase in areal density in the magnetic recording industry. Since the fundamental length scale (exchange length) in ferromagnetic materials is on the order of 10^{-8} meter, the small physical dimensions have made more detailed micromagnetic analysis of the recording head and the recorded data track feasible: The number of discretisation elements required in the numerical simulation of objects with small dimensions is now of a manageable size. The intent of this Chapter is to provide an elementary introduction to aspects of magnetostatics, leading up to popular implementations of the micromagnetic method. This also serves as a background to many of the concepts discussed in the subsequent chapters in this book. Examples of the application of simple magnetostatic equations and of the micromagnetic method to the recording head and medium will serve to illustrate the utility of the micromagnetic method, as well as some of its shortcomings.

Emphasis is on the fact that the micromagnetic method relies on the description of interactions that are well understood, and hence provides a simple tool for the understanding of the behaviour of magnetic objects of finite dimensions. In particular, it should be emphasised that one of the primary attributes of the micromagnetic method is that it accounts for the exchange interaction and so-called shape anisotropy effects. This facilitates explanations for the observation that the response of ferromagnetic devices to external fields can be very different from what one might expect on the basis of the intrinsic (bulk) properties of the materials.

Methods that solve Maxwell's equations on the basis of auxiliary initial permeability curves, and that are often implemented as a finite element method, suffer from the shortcoming that the material essentially is assumed to be paramagnetic. At vanishing magnetic field the magnetisation in the sample identically vanishes. The micromagnetic method assumes locally conserved magnetic moments that are short-range coupled through the exchange interaction, and thus respects elementary aspects of ferromagnetic materials.

Nevertheless, the finite element method routinely is used in the development of write heads, and hence deserves a brief discussion.

This chapter is organised as follows. Section 1.1 gives a concise overview of the main ingredients of the micromagnetic method. Sections 1.2 and 1.3 describe the application of magnetostatics and the micromagnetic method to write heads and magnetoresistive sensors for play back. Section 1.4 reviews the application of a micromagnetic recording model to aspects of longitudinal recording. The 5th section is on the magnetostatic implications of oriented media for vital recording parameters. Some challenges in recording at ultra high areal density are discussed.

1.1 The Micromagnetic Method in Recording

1.1.1 Elementary Ingredients of Recording Physics

The fundamental premise in magnetic recording is to bring the magnetically active elements of a recording head in close proximity with the magnetic recording medium, such that inductively generated magnetic stray fields in the vicinity of the write gap are strong enough to record transitions (reversal of write current polarity) on the medium. At different points in the 'history of magnetic recording' excellent publications on the topic have become available [1–8]. Chapter 11 of this book reviews some fundamentals of the magnetic recording head. In the recording process, the magnetisation direction of individual grains in the medium is affected such that, on average, the magnetisation along the down-track direction changes direction at a transition. This implies a relation between the magneto crystalline anisotropy of the medium grains and the write field required in the medium. On a coarser scale, the coercivity of the medium and the saturation magnetisation of the materials around the write gap must be related. Thermal stability requirements of the medium at high areal density translate into high medium coercivities, and hence impact the choice of materials for the write head. In addition, since the easy axes of planar isotropic media are randomly distributed, fluctuations in the medium magnetisation form a significant source of noise (see Chaps. 2, 5 and 6 of this book). At the medium level, the ratio between the intended stray field from the transition and superimposed random fluctuations must guarantee an adequate signal-to-noise ratio.

In contemporary disk drives, play back of the recorded signal is realised with magnetoresistive sensors. The intrinsic magnetoresistive properties combined with the magnetostatic response to the stray field from the recorded transition must be adequate in terms of the signal-to-noise ratio at the recording head level ([3,8] and Chap. 10).

These basic considerations quickly branch out into a complex set of 'engineering' requirements that is governed by an even larger set of variables. Nevertheless, it is safe to assume that as a starting point elementary, magnetostatic requirements need to be fulfilled. It is here that magnetostatics and the micromagnetic method can assist in understanding those needs.

1.1.2 Magnetostatic Expressions for Stray Fields – Bar Magnets

From elementary magnetostatics the expression for the magnetic field anywhere in space from a magnetised object, is given by (cgs units) [4,9]

$$\vec{H}(\vec{r}) = -\int_V d\tau \frac{\vec{\nabla} \cdot \vec{M}(\vec{r}')(\vec{r}-\vec{r}')}{|\vec{r}-\vec{r}'|^3} + \int_S dA \frac{\hat{n}(\vec{r}') \cdot \vec{M}(\vec{r}')(\vec{r}-\vec{r}')}{|\vec{r}-\vec{r}'|^3} \tag{1.1}$$

Here $\vec{M}(\vec{r}')$ is the magnetisation inside the object of volume V, or on the surface R with normal vector $\hat{n}(\vec{r}')$. The integrals are over the interior and the surface of the object, respectively. The first term relates the volume magnetic charge density, $-\vec{\nabla} \cdot \vec{M}(\vec{r}')$, to the magnetic field while the second term yields the contribution from charge density at the surface of the object, $\hat{n}(\vec{r}') \cdot \vec{M}(\vec{r}')$.

For bar magnets, or uniformly magnetised rectangular prisms, the first term vanishes $\vec{\nabla} \cdot \vec{M}(\vec{r}') \equiv 0$, while the second term reduces to

$$\vec{H}(\vec{r}) = \left[\int_S dA \frac{\hat{n}(\vec{r}') \otimes (\vec{r}-\vec{r}')}{|\vec{r}-\vec{r}'|^3} \right] \cdot \vec{M} = \overleftrightarrow{N}(\vec{r}-\vec{r}') \cdot \vec{M} \tag{1.2}$$

where the 3×3 demagnetisation tensor $\overleftrightarrow{N}(\vec{r}-\vec{r}')$ is a purely geometrical factor that depends on the relative distance to the bar magnet. It consists of elements of the form

$$N_{\alpha\beta}(\vec{r}-\vec{r}') = \sum_{i=1}^{6} \int_{S_i} dA \frac{n_{\alpha,i}(\vec{r}')(r_\beta - r'_\beta)}{|\vec{r}-\vec{r}'|^3} \tag{1.3}$$

with $\alpha, \beta = x, y, z$, and the index i runs over the six faces of a rectangular prism. Analytical expressions for the integrals over the faces of the prism can be derived. Hence, the stray field from a uniformly magnetised bar magnet is known everywhere in space [10–12].

This expression for the stray field arising from a uniformly magnetised prism is one essential ingredient of the micromagnetic method, where large numbers of small cells are magnetostatically coupled. However, with the form above one can already build a heuristic model of a pair of writer poles and examine the stray field in the recording medium as a function of the magnetisation direction and track width. Figure 1.1 illustrates the configuration of the two pole tips, and the graph shows the field in the medium along the down track direction as a function of the magnetisation angle θ and pole width W. The write gap, G, is 50 nm and the head to centre-of-medium spacing, y, is 15 nm. The two poles have a track width varying from 0.05

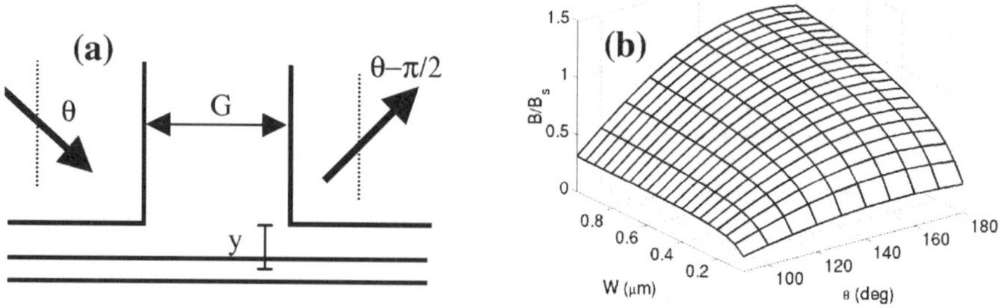

Fig. 1.1. Bar magnet model of a writer. **a** Cross section showing write gap, G, head-to-medium spacing, y, and the magnetisation vectors at polar angles θ and $\theta - \pi/2$. **b** Normalised field along the down-track direction, at the centre of the track as a function of θ and track width W

to 1 μm, and each extend 100 μm in the down-track and perpendicular directions. Figure 1.1b shows the dependence of the field in the medium on track width: for fixed head to medium spacing a narrower track implies loss of magnetic field in the medium. This can be understood easily in view of Eq. (1.1), where for a narrower track less surface charge density contributes to the field. The angle θ at which the field in the medium is largest was found to be between 150 and 165 degrees, depending on track width. Note that the field in the medium can exceed the saturation magnetization of the material. This is a consequence of the proximity of the medium to the divergence in the stray field from uniformly magnetized objects along edges. This unphysical artefact is not observed in a proper micromagnetic calculation, in which demagnetisation effects at surfaces and edges prevent complete saturation. Nevertheless, the trends shown in Fig. 1.1b still hold.

1.1.3 Exchange Interaction – From Atoms to Exchange Length

At the microscopic quantum mechanical level, either parallel or anti-parallel alignment of electron spin angular momentum, and hence the alignment of spin magnetic moment vectors, can be energetically favourable through the exchange interaction. In the transition metals Fe, Co and Ni this phenomenon leads to the formation of a ferromagnetic ground state. A simple description of the exchange interaction is provided by the Heisenberg energy expression

$$E_{ex} = -2J \sum_{(i,j)} \vec{s}_i, \vec{s}_j \ , \tag{1.4}$$

where $\vec{s}_i$ and $\vec{s}_j$ are electron spin vectors on neighbouring atomic sites. The exchange energy, J, between the pair of spins arises from the Pauli exclusion principle and the Coulomb interaction between the atomic electron wave functions. For ferromagnetic systems J is greater than zero (antiferromagnetic $J < 0$). In micromagnetic calculations one typically deals with systems that do not allow for a treatment at the atomic level. The micromagnetic method involves interactions between 'grains' or cells, composed of many atoms or molecules [13,14] – typically the size is on the order of the exchange length, which for Permalloy ($Ni_{80}Fe_{20}$) is about 12.5 nm. A coarse grained version of Eq. (1.4) is given by the exchange energy density expression

$$\epsilon_{ex} = \frac{A}{M_s^2} \left| \nabla \vec{M} \right|^2 , \tag{1.5}$$

where A is the exchange constant (about 1 μerg/cm for Permalloy) and M_s is the saturation magnetisation. The exchange field is given by the functional derivative of the energy density with respect to the magnetisation

$$\vec{H}_{ex} = -\frac{\delta \epsilon_{ex}}{\delta M} = \frac{2A}{M_s^2} \nabla^2 \vec{M} \approx \frac{2A}{a^2 M_s^2} \sum_{i=nn} \vec{M}_i , \tag{1.6}$$

where a is the lattice constant of the discretisation, and the summation is over nearest neighbour sites.

Contrary to the long-range magnetostatic interaction, the exchange field in a micromagnetic calculation has a short range. Both for relatively soft magnetic materials such as writer poles and the magnetoresistive sensor, as well as for hard magnetic thin film recording media, the exchange interaction can play an important role.

1.1.4 The Micromagnetic Method

For systems with only uniaxial anisotropy, in terms of the first three anisotropy constants, K_1, K_2 and K_3, the effective uniaxial magneto crystalline anisotropy field can be written as

$$\vec{H}_k = (\hat{k} \cdot \hat{m}) \hat{k} [H_{K_1} + 2H_{K_2}(1 - (\hat{k} \cdot \hat{m})^2) + 3H_{K_3}(1 - (\hat{k} \cdot \hat{m})^2)^2] \tag{1.7}$$

with $H_{K_1} = \frac{2k_i}{M_s}$ and with $\hat{k}$ and $\hat{m}$ unit vectors along the anisotropy and the magnetisation axes, respectively. For a collection of uniformly magnetised particles the effective field on each is given by the sum of the magnetostatic, exchange, anisotropy, and external fields

$$\vec{H}_{eff} = \vec{H}_{magn.} + \vec{H}_{ex} + \vec{H}_k + \vec{H}_{ext} . \tag{1.8}$$

The Landau–Lifschitz–Gilbert (LLG) equation describes the motion of the magnetisation vectors of the particles. The equation consists of a precession term and a phenomenological damping term.

$$(1+\alpha)^2 \frac{d\vec{M}}{dt} = -\gamma \vec{M} \times \vec{H}_{eff} - \frac{\alpha\gamma}{M_s} \vec{M} \times \left(\vec{M} \times \vec{H}_{eff} \right) \tag{1.9}$$

It is possible to derive the precession term from fundamental principles (see Chap. 3 of this book), and it states that the magnetic moment will gyrate around the effective field at a rate proportional to H_{eff} and γ, the gyromagnetic constant. The second term is the damping term and α is the dimensionless damping constant, typically in the range of 0.05 to 0.2. This term forces the magnetisation toward the effective field, and combined with the precession results in a spiralling trajectory of the magnetic moment with time [15].

Note that the multiple grains in a micromagnetic system are short-range coupled through the exchange interaction while the magnetostatic field couples all the grains over a long range, typically the entire simulation volume. In practice the determination of the magnetostatic field acting on each element in the sample consumes the main fraction of the total processing time. Efficient algorithms are available for the computation of the magnetostatic field on a regular three-dimensional grid (Fast Fourier Transform method [10,12] and on irregular grids (Fast Multipole Summation method [16]). Equation (1.9) must be integrated with a robust ordinary differential equation solver, and at each (relatively small) time step the effective field on each element must be recomputed.

In a micromagnetic simulation of a soft magnetic material the exchange interaction and the magnetostatic interaction are competing on different length scales. The LLG equations must then be solved carefully since converge problems are to be expected. In particular, finding an equilibrium solution (t $\rightarrow \infty$) for a finely meshed (i.e., $a \sim 10$ nm) system of many particles is nearly impossible without the short-range exchange interaction. In absence of the exchange interaction, dipole-dipole interactions can favour a head-to-tail configuration for neighbouring magnetisation vectors, rather than the characteristic slowly varying, nearly parallel alignment. An interesting collection of problems along with possible solutions generated by the micromagnetic community can be found in reference [17].

1.1.5 Landau–Lifschitz–Gilbert Equation and Excitations in Sub-micron sized Ferromagnets

Even though the damping term in the LLG equation is of a phenomenological nature, encouraging fits of dynamic phenomena observed in soft magnetic samples suggest that the LLG equation can provide at least a qualitatively correct description of dynamic phenomena in ferromagnetic materials ([18–22] and Chapter 4 of this book). In this section the impact of shape anisotropy on the characteristic precession frequencies of small ferromagnetic platelets is examined with the dynamic solutions of LLG equation. Focus will be on 2.5 nm thick platelets with $M_s = 1440$ emu/cc ($Co_{80}Fe_{20}$). These samples were excited with short pulses (200 ps) of a weak external field (10 Oe), or they are subject to a harmonic (frequencies from 0.5 to 2 GHz) strong driving force (200 Oe amplitude). The calculations with the weak pulses resemble the typical experimental set-up, with the exception that now significant effects from shape anisotropy can be expected since the samples are very small. The strongly driven systems are of interest since they resemble the situation in a sensor for play back in magnetic recording.

Systems with varying aspect ratio W/L, but constant $W = 100$ nm, were excited with weak, short field pulses along the x-direction (along W). A constant 10 Oe bias field was applied along the y-direction. The temporal evolution of M_x is monitored for samples ranging in size from $W \times L = 100 \times 80$ nm^2 to 100×800 nm^2. The cell size in the LLG calculation was $10 \times 10 \times 2.5$ nm^3 and the damping parameter was $\alpha = 0.01$. Figure 1.2a shows the temporal evolution of the x-component of the magnetisation. At the 10:8 aspect ratio ($W \times L = 100 \times 80$ nm^2) the remnant state ($t = 0$) is largely magnetised along the x-direction, constrained by shape anisotropy (upper curve). After a short field pulse the system is dissipating energy through viscous damping, but the magnetisation never deviates substantially from the remnant state. The upper right panel in Fig. 1.2b shows the power spectrum of the temporal evolution of M_x. A relatively narrow mode at 6.3 GHz dominates the spectrum.

The oscillatory mode softens and broadens when the aspect ratio approaches unity. At this aspect ratio the combined bias and the pulse fields make the state with diagonal magnetisation ($M_x \sim 0.71\ M_s$) energetically favourable, and the system strives toward this state. Once the field pulse vanishes, the system relaxes with roughly half the precessional frequency observed at the 10:8 aspect ratio.

When the aspect ratio decreases, shape anisotropy poses well-defined constraints on the magnetisation of the sample. The excitation narrows (Fig. 1.2b) and has shifted to slightly lower frequency (5.7 GHz). At very small aspect ratio of 10:40 and 10:80, spectral power develops at even higher frequency (8.6 GHz). The origin of this beating phenomenon is not clear.

Similar work on larger $Ni_{60}Fe_{40}$ thin film samples yielded characteristic frequencies much closer to the oscillation frequency for infinite thin films [19]. The results presented here demonstrate how shape anisotropy effects in small ferromagnetic objects enhance the characteristic frequencies.

In an attempt to make this type of calculation more relevant to the play back process in magnetic recording, a model system of a sensing layer with permanent magnet abutted junction style stabilisation was constructed. The permanent magnet stabilisation forces the magnetization of the sensing layer nominally along the x-direction. The dimensions of the sensing structure (e.g., a CoFe sensing layer in a GMR spin valve) were chosen as $100 \times 100 \times 2.5$ nm^3. This device was exposed to an external field that varied linearly from 300 Oe at the front ('air bearing surface'), to 10 Oe at the back of the device to mimic the spatial decay of a field arising from a transition in the recording medium. This field was then varied in time as a sinusoid at frequencies ranging from 0.5 and 1 GHz. The damping constant was chosen to be $\alpha = 0.01$. Figure 1.3

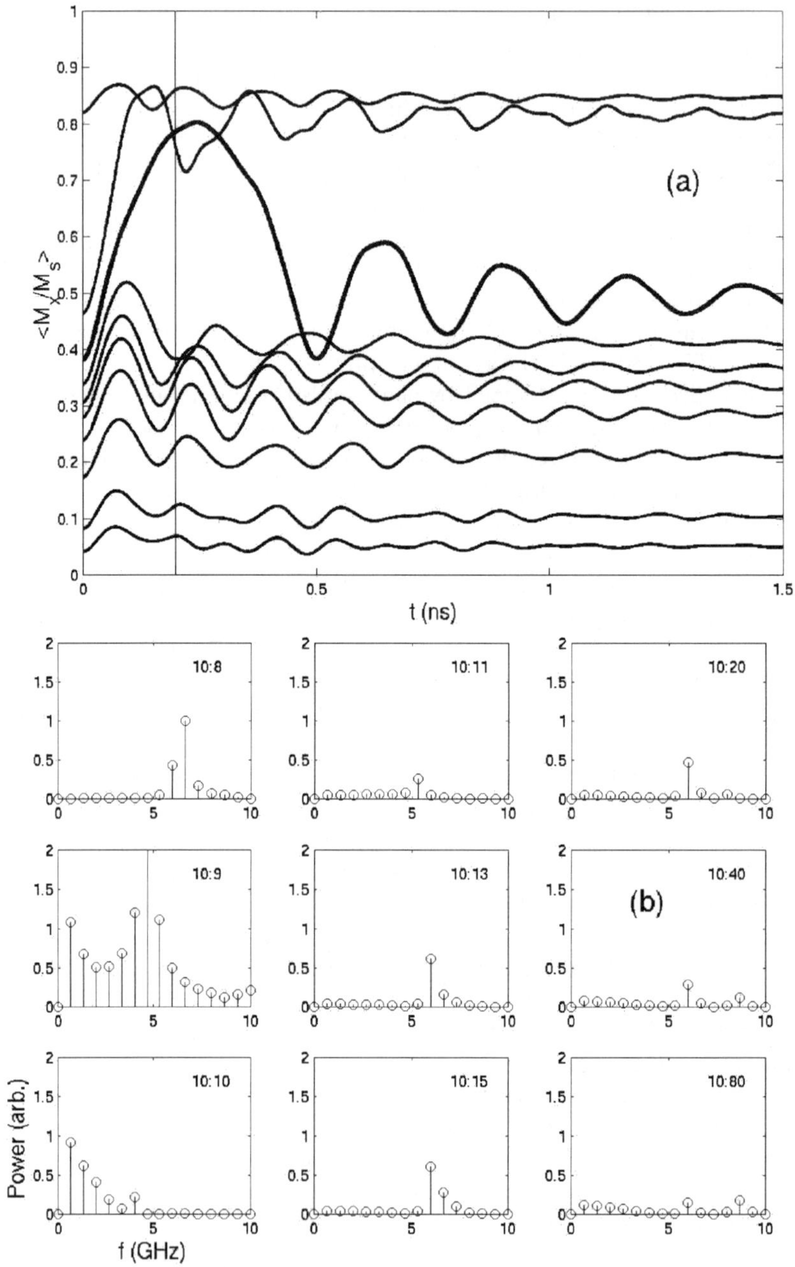

Fig. 1.2. Temporal response of 2.5 nm thick CoFe platelets (M_s = 1440 emu/cc) with different aspect ratios. **a** Time evolution of the magnetisation component along the pulse direction for platelets with aspect ratio ranging from 10:8 (*upper curve*) to 10:80 (*lower curve*). The vertical line indicates the end of the 200 ps field pulse. **b** Power spectral content of the temporal responses in **a**

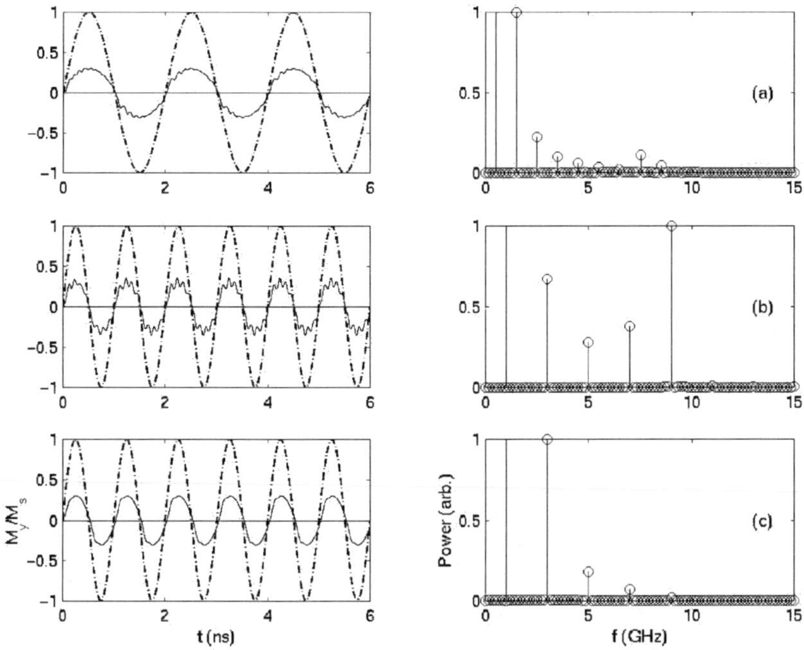

Fig. 1.3. Temporal response of the magnetisation in an abutted junction style sensor at different excitation frequencies (*left column*) and the spectral content of $M_y(t)$. **a** Driving field oscillates at 500 MHz with $\alpha = 0.01$. **b** Driving field oscillates at 1 GHz with $\alpha = 0.01$. **c** Driving field oscillates at 1GHz with $\alpha = 0.1$

shows the temporal response of the y-component of the magnetisation in the free layer over a number of cycles for different frequencies, along with the spectral content of $M_y(t)$. In Figs. 1.3a and b the driving frequencies are 0.5 and 1 GHz, respectively.

Interestingly, only a very small phase shift of $\delta = 0.053$ rad can be detected in the response of the free layer magnetisation. The response to the external field essentially is instantaneous, and this can be understood in terms of the large effective fields governing the system. The permanent magnets in this model produce a field of 1500 Oe at the edges, and about 200 Oe at the centre of the sensor (parabolic interpolation). In addition, the sinusoidal driving field is strong, on average about 150 Oe. Estimated characteristic frequencies, f, derived from just the precession term in Eq. (1.10), i.e., $f = \gamma H_{eff}/2\pi$ are on the order of 0.7 to 4 GHz. The spectral power exhibits sharp resonance at $f = (2n+1)f_0$, where f_0 is the driving frequency. Presumably this can be explicitly demonstrated by decomposition into Eigenmodes, but this has not yet been attempted.

It turns out that the impact of varying α is limited. Increasing α from 0.01 to 0.1 at 1GHz driving force results in a smoother response to the driving field by the sensor (see Fig. 1.3c), albeit with a slightly larger phase shift of $\delta = 0.22$ radians. Apparently, the intrinsic ability of the system to respond to strong driving forces at high frequencies is largely determined by the magnetostatic boundary conditions, and is relatively insensitive to the particular choice of α. A conjecture related to this apparent insensitivity to the choice of is that the fundamental mechanisms associated with this kind of damping are somewhat irrelevant for sensor response.

As a consequence it is concluded that for properly stabilized magnetic sensors only a small phase lag between the driving force and the response of the sensing layer is to be expected for frequencies up to 1 GHz.

1.2 Write Head Fields

1.2.1 Karlqvist, Lindholm, Bar Magnet, and FEM Write Fields

Section 1.1.2 alluded to changes in the write head field as a function of the magnetisation configuration in a simple model of the pole tips. These considerations are further illustrated by comparing write head fields derived within finite element method (FEM), Karlqvist, Lindholm, and bar magnet approximations. The Karlqvist head field corresponds to infinitely wide and long symmetric pole pieces and can be shown to be equivalent to the stray field emanating from two infinitely wide planes of magnetic poles separated by the write gap length. This corresponds with two bar magnets as in Fig. 1.1a and a magnetisation angle of $\theta = 90°$ [1,4]. The Lindholm head field [23] is a finite track width version of the infinitely wide Karlqvist head [1]. It was derived under the assumption of uniform scalar potential on the surfaces of each infinitely permeable pole tip.

Shown in Fig. 1.4a below are FEM results from a 9-turn writer with a notched lower pole [24], a pole width of 0.25 µm and a write gap length $G = 0.1$ µm. The down-track component of the field is depicted as a function of down-track distance relative to the gap centre, and at a distance of

25 nm from the air-bearing surface (in the centre of the medium). Upper (trailing edge) and lower poles are composed of a 2.1 T and 1.8 T materials, respectively. A moderately high write current was used that gave a field value at the air bearing surface of $H_0 = 10500$ Oe. The corresponding 'deep-gap' field parameter in the Karlqvist and Lindholm expressions was *set* at this value without a-priori justification. At track centre, the comparison between the analytic approximations and the FEM results reveals the differences due to the asymmetric pole structure as well as the fact that the simple approximations underestimate the spatial variation of the field (i.e., the field gradient [24]). Reduction of the write field at narrow pole widths is also evident in the graphs of the head field after averaging across the pole width (Fig. 1.4a), where again the FEM results show a larger spatial extent.

Figure 1.4b shows a comparison between FEM and results from the two bar magnet model. Each of the bar magnets has dimensions 0.25, 0.5, and

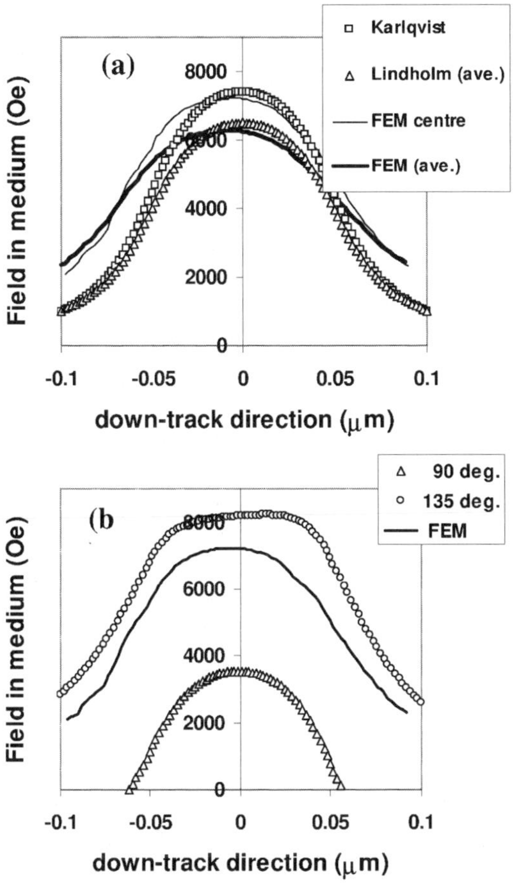

Fig. 1.4. Down-track component of the write head field as a function of down-track direction (relative to gap centre). Writer poles are 0.25 μm wide (Lindholm, FEM) with a 0.1 μm long write gap. **a** Comparison of FEM with two analytic approximations where Karlqvist and Lindholm are nearly equivalent at track centre. **b** Comparison at track-centre of FEM results with a bar magnet model of the writer with magnetisation directed at $\theta = 90°$ or at an angle $\theta = 135°$ (as in Fig. 1.1)

1.0 μm along the cross-track, down-track and perpendicular directions, respectively. Upper and lower magnets have the same saturation magnetisation values as in the FEM model. Two different moment directions are chosen, namely $\theta = 90°$ and $135°$, as defined in Fig. 1.1. The 'Karlqvist' result for $\theta = 90°$ severely underestimates the overall magnitude of the head field by about a factor of two. Better agreement is achieved with the moments at an angle of 135 degrees. Examination of the magnetisation profile obtained from the FEM calculation reveals that the magnetisation direction near the gap is definitely not equal to $\theta = 90°$ (Karlqvist and Lindholm), but closer to 135 degrees, as would be expected on the basis of shape-anisotropy effects.

1.2.2 Landau–Lifschitz–Gilbert Model

Until recently, no micromagnetic results for write fields were available. Micromagnetic simulations for a pair of pole tips, subject to boundary conditions that supply flux density, indicate that the head field is able to follow the driving fields in time as a result of fast vortex motion in the pole tip magnetisation [25]. Another study concludes that the head fields resemble those derived from a Lindholm model, and that magnetisation reversal takes place through domain wall creation and annihilation [26]. Both studies suggest that the dynamic response of the system, as measured from the head field dynamics, is not very sensitive to the choice of the damping parameter α. This is similar to the dynamic results for sensor structures, discussed in Sect. 1.1.5.

The studies mentioned in the previous paragraph have in common that they reveal the presence of magnetisation vortices in the pole tips. In this chapter, a micromagnetic calculation for a pair of pole tips was performed for the purpose of illustration. The writer poles are 100 nm wide and extend 400 nm along the down track direction as well as into the head (see Fig. 1.5). The write gap is 50 nm and on each side of the gap the first 200 nm of the pole tips consists of 1920 emu/cc (2.4 T) material. The rest of the pole tip consists of 1440 emu/cc (1.8 T) material. The cell size is $10 \times 10 \times 10\ \text{nm}^3$ and

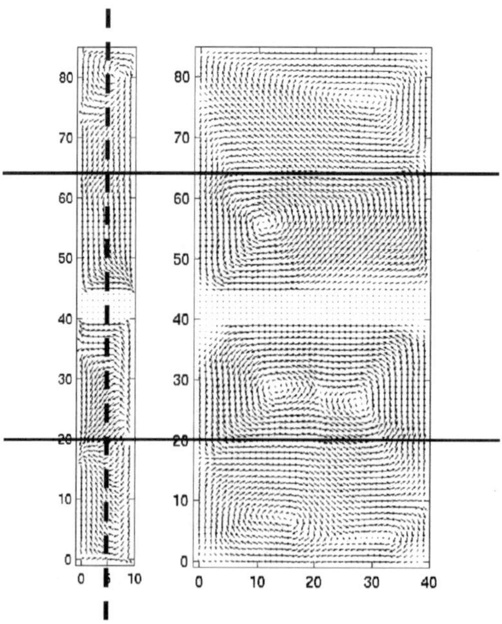

Fig. 1.5. Remnant magnetisation in simple micromagnetic pole tip model. A front view of the pole tips is shown on the left. The magnetisation in the cross sectional area along the *dashed line*, into the head, is shown on the right. The horizontal lines represent the boundary between B_s = 1.8 T and B_s = 2.4 T material. The 50 nm write gap starts at cell number 40 and ends at cell 44 (*vertical axis*)

the exchange constant was assumed to be 1 μerg/cm. The computed remnant state, starting from a random initial configuration, is shown in Fig. 1.5. Note the vortex structure in the magnetisation profile, consistent with observations made in [25,26]. Compliance with shape anisotropy is evident in Fig. 1.5. Also notice the energetically favourable head-to-tail alignment of the magnetisation around the write gap. The remnant magnetisation was negligible, and the magnitude of the stray field from the pole tips, at the centre of the gap and about 15 nm away from the pole tips, is on the order of a few hundred Oersteds. Hence, there is no evidence that pole tips in their remnant state can affect a high coercivity-recording medium (remnant erasure).

Upon application of a uniform field along the vertical direction it was found that saturation sets in at about 2000 Oe. Clearly the assumption of infinite permeability in simple models of pole tips cannot be justified. For the structure shown in Fig. 1.5 the permeability estimated from the computed initial permeability curve is on the order of $\mu = 25$.

One could propose a combination of a micromagnetic model of the pole tips, subject to a boundary condition derived with the FEM method. However, the difficulty with such an approach is that it does not preserve the self-consistency that is contained in either LLG or FEM separately.

1.2.3 Finite Element Method versus Landau–Lifschitz–Gilbert

Obviously the micromagnetic treatment of the entire ring head is impractical. Even for a small writer of, say, roughly $20 \times 20 \times 4$ μm^3, this would imply a grid containing 1.6 million cells, 10 nm on the side. Typically, inductive write heads are examined with Finite Element Modelling (FEM) software, in which the writer materials are characterised by a relation between the local magnetic field, H, and the magnetisation, M, along the direction of the field. Just as for paramagnetic materials, at vanishing field there is no magnetisation. The auxiliary supplied M–H curves for bulk writer materials start with some initial permeability, and saturate at the saturation magnetisation. Maxwell's equations are iteratively solved with the finite element method. In the final solution the H-fields and magnetisation have been made self-consistent on a point-to-point-basis. As will be discussed in Sect. 1.4, an FEM model of the magnetisation of writer pole tips qualitatively reproduces 'pole tip saturation effects', i.e., the degradation of the recording characteristics as a function of current through the writer coil.

Since solving Maxwell's equations with the finite element method is a popular technique for the analysis of aspects of the performance of inductive write heads, an attempt to justify the assumption of paramagnetic materials ($M = 0$ if $H = 0$) in a model of the write head is appropriate. To this end FEM and LLG calculations were performed for a small ferromagnetic object. Stray fields were computed in a plane 0.02 μm from the 0.25×0.5 μm^2 face of a $0.25 \times 0.5 \times 0.75$ μm^3 piece of material with $M_s = 800$ emu/cc and $A = 1$ μ erg/cm ('Permalloy'). Figure 1.6 shows this geometry. The

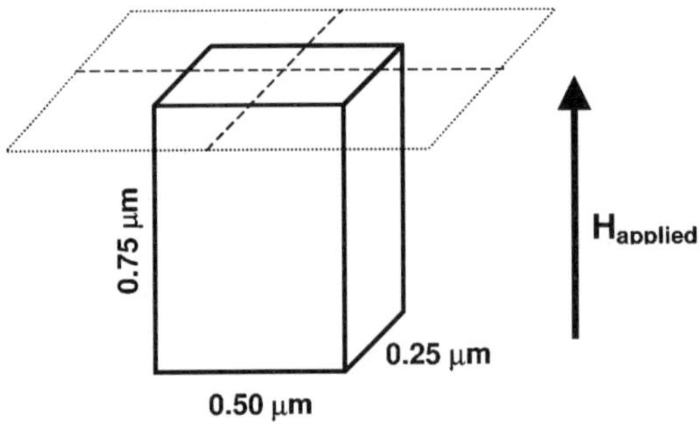

Fig. 1.6. Geometry of the Permalloy rectangular prism used in the FEM versus LLG comparison. The dashed plane is located 0.02 μm above the top face of the prism. The crossed dashed lines correspond to the horizontal axes in Fig. 1.7. The external field is applied along the vertical direction

Table 1.1. Comparison of the x, y, and z components of FEM and LLG-derived stray fields from a small Permalloy sample in an externally applied field

$H_{applied}$ (Oe)	*Mean error* (%)		
	x	y	z
0	100	100	100
100	154	96	141
300	29	102	35
500	12	84	15
1000	7	56	6
2000	3	20	3
3000	4	13	4
4000	2	6	2
5000	2	7	2

applied field is normal to the plane, along the long axis of the Permalloy object. The mean error in stray field can be defined as the mean value of $|H_{FEM} - H_{LLG}|$ divided by the mean value of $|H_{LLG}|$ over the area of the top face (0.25×0.5 μm^2), on a component-by-component basis. Table 1.1 shows that reasonable quantitative agreement is obtained for applied fields greater than 1000 Oe. It is noted that pole tip extremities typically experience driving H-fields that are on the order of 2000 Oe (inferred from FEM calculations). Hence, a reasonable description of the response of writer pole tips in an FEM model can be expected.

At lower applied field, the stray fields compare poorly because the FEM model does not describe ferromagnetic structure in the sample. In particular the effects of shape anisotropy are not properly accounted for since locally M is reduced from its full value if H is small. An interesting corollary is that it will be impossible to obtain a correct description of the dynamic behaviour of micron-sized ferromagnetic systems assuming paramagnetic properties since the magnetic configuration at small driving field is described very poorly.

Figure 1.7 compares the components of H_{FEM} and H_{LLG} along lines of symmetry in the sampling plane at applied fields of 300 and 5000 Oe. At low applied field there is structure in the stray field from the LLG calculation since the magnetisation of the prism has an asymmetric complex vortex structure. The FEM result always is smooth and symmetric. Note that, at strong applied fields, the FEM solution underestimates the z-component (perpendicular to the plane) at the centre of plane. This is due to the difference of the approach of saturation in bulk materials versus saturation subject to strong three dimensional demagnetisation effects.

Demagnetising effects are present in the FEM solution of Maxwell's equation since they are implicit in the magnetostatic equations. However, since in FEM the local magnitude of the magnetisation is proportional to the local H-field (paramagnetic response), demagnetising fields at low field are inaccurate. In a micromagnetic treatment the local magnetisation always is conserved.

1.2.4 Eddy Current Fields

One might be tempted to argue that for conductive ferromagnetic materials used in the fabrication of an inductive writer, it is necessary to include the effect of eddy currents as part of the effective field that enters the LLG equation. This would render the LLG equation implicit since the eddy current field, and hence the effective field, will depend on dM/dt itself through Faraday's law. Therefore, eddy current fields are typically omitted from LLG calculations, although attempts have been made to include them [27,28].

There is a more compelling physical argument for neglecting eddy current effects in micron sized inductive writers. This insight is based on inspection of Maxwell's equations in integral form, and recognizing that the magnitude of eddy current fields scales as a characteristic cross sectional area. Ampere's law reads

$$\oint_C d\boldsymbol{l}\cdot\boldsymbol{H}_{Eddy} = \int_S \vec{j}\cdot\hat{n}dA\,, \tag{1.10}$$

where the contour C encloses surface S. For the magnitude of the eddy current field, H_{eddy}, this approximately implies

$$H_{eddy} \propto \frac{jA}{l} \tag{1.11}$$

with A the surface area of S and l the length of C. From Faraday's law

$$-\int_{S'} \frac{\partial\vec{B}}{\partial t}\cdot\hat{n}dA = \oint_{C'} \vec{E}\cdot d\vec{l} \tag{1.12}$$

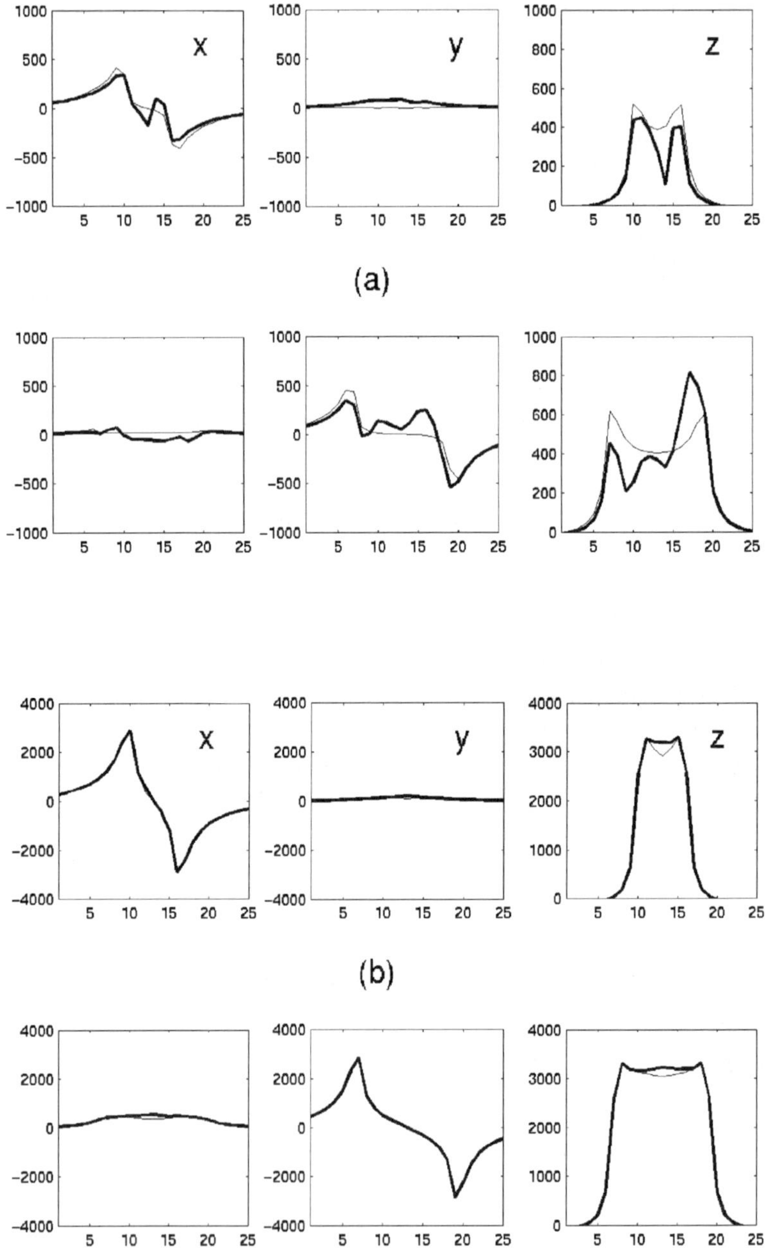

Fig. 1.7. Comparison of LLG (thick) and FEM (thin) derived stray fields outside a small ferromagnetic object. **a** The applied field is 300 Oe (top panel). Shown are x, y, and z component of the stray field along the short axis (*top row*) and the long axis (*bottom*). **b** The applied field is 5000 Oe (*bottom panel*)

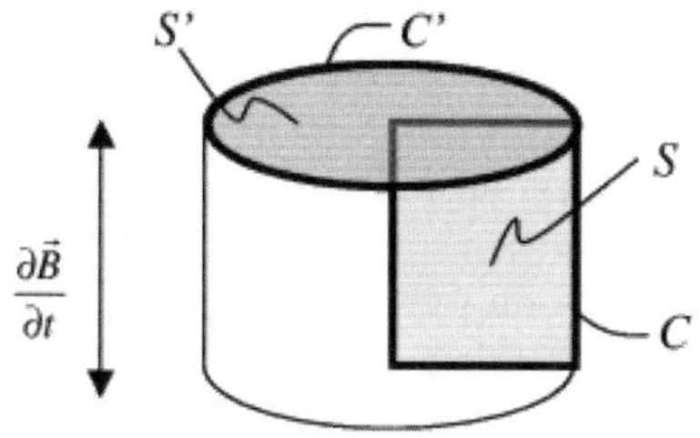

Fig. 1.8. Cartoon of the relevant surfaces and contours in the discussion of eddy current effects

one obtains an order of magnitude estimate for the current density, $j = \sigma E$,

$$j \propto \sigma \frac{\partial B}{\partial t} \frac{A'}{l'} . \tag{1.13}$$

Substitution of (13) in (11) yields

$$H_{eddy} \propto \sigma \frac{\partial B}{\partial t} \frac{A'}{l'} \frac{A}{l} . \tag{1.14}$$

Since A and A' as well as l and l' are of the same order of magnitude, it has now become manifest that eddy current fields scale as a characteristic area. Hence, eddy current fields that can be an important loss mechanism in transformer coils (centimeter length scale) are no longer important in write heads (micrometer length scale) since such effects are suppressed by eight orders of magnitude. These considerations should hold for frequencies well into the Giga-Hertz range. Figure 1.8 shows an example of a cylinder with the surfaces S and S', contours C and C', and $\vec{B}(t)$ indicated.

An analytic expression that describes the penetration of a harmonically oscillating external magnetic field into a paramagnetic cylinder [29] corroborates these findings. A strong dependence of the penetration depth on the diameter of the cylinder is observed. Also this expression exhibits a strong dependence on the choice for the permeability of the linear material. Note that for micron-sized ferromagnetic objects the assumption of a linear ferromagnetic material with isotropic permeability is quite inappropriate in the micromagnetic sense.

1.3 Micromagnetic Model of Magnetoresistive Sensors

Since dynamic effects in spin valve sensors are not of primary concern, the outcome of LLG and torque (or energy minimisation) calculations should be equivalent. Most of these models of spin valve sensors have in common that the films in the sensor multilayer are of uniform thickness and quality. Often the permanent magnets in the abutted junction are represented by bar magnets. Also, the shields that define the read gap are assumed to consist of an isotropic permeable material, i.e., not treated micromagnetically. Magnetic perturbations after a writing event in a real recording head are invariably ignored. Many of the model assumptions are becoming even more questionable for sensors aimed at ultra-high areal density recording. However, general trends induced by fundamental physical concepts, as well as the coarse exploration of magnetic stability of magnetoresistive sensors can be addressed. Impacts of variance in key sensor characteristics on overall parametric performance can be assessed with probabilistic simulations in order to develop a sense of the robustness of the sensor to variance in extrinsic and intrinsic properties [30]. After all, if feasibility and/or parametric performance of a novel proposal for a GMR sensor cannot be demonstrated with favourable assumptions, in an ideal and controlled computer simulation, there cannot be hope for a miracle in the real world.

In this section a stochastic evaluation of a spin valve sensor at a 6 Gb/in^2 areal density target is presented. The useful concept of sensor efficiency is introduced.

1.3.1 Stochastic Evaluation of Sensor Performance

Convergence of micromagnetic calculations is most easily obtained for systems that are subject to strong magnetic fields that suppress the formation magnetic structure (e.g., vortices and edge effects). The abutted junction stabilisation scheme combined with exchange stabilisation of the pinned layer in spin valve sensors gives precisely this result. As a consequence, micromagnetic calculations for such systems are very feasible. It turns out that for a two dimensional treatment of a spin valve sensor (only in-plane magnetisation), quadratic convergence can be obtained using a multi-dimensional Newton-Raphson method for energy minimisation.

At the minimum energy, E, all the derivatives of the energy with respect to the local magnetisation angle, $\partial E/\partial \varphi_i$, must vanish, while the diagonal elements of the Hessian matrix, $\partial^2 E/\partial \varphi_i \partial \varphi_j$, must be greater than zero. The Newton-Raphson scheme to obtain the next iteration on the magnetisation configuration, $\varphi + \delta\varphi$, is

$$\sum_{j=1}^{N} \frac{\partial^2 E}{\partial \varphi_i \partial \varphi_j} \delta \varphi_j = -\frac{\partial E}{\partial \varphi_i} \ . \tag{1.15}$$

Hence, the $N \times N$ Hessian matrix must be inverted at each iteration. The beauty of the method is that all the derivatives with respect to the angles φ_i can be evaluated analytically. Starting from the characteristic crossed magnetisation pattern for spin valves, convergence typically is obtained within 15 iterations. Since convergence is so fast, it is feasible to analyse variance in parametric sensor performance through stochastic variation of input parameters for the micromagnetic calculations.

The example concerns a NiMn spin valve sensor intended for a 6.2 Gb/in^2 areal density target. Table 1.2 summarises the mean values and standard de-

Table 1.2. Gaussian distributions for input parameters for simulations of a 6.2 Gb/in^2 NiMn based spin valve sensor

Variable	Mean value	Standard deviation	Unit
Free layer $M_s t$	760	15.2	memu/cm^2
Pinned layer $M_s t$	440	8.8	memu/cm^2
Free layer sheet resistance	50	0.98	Ω/square
Cu sheet resistance	70	1.4	Ω/square
Pinned layer sheet resistance	100	2.0	Ω/square
NiMn sheet resistance	75	1.5	Ω/square
Sensor width	0.9	0.023	µm
Sensor height	0.5	0.05	µm
Coupling across Cu	15	6	Oe
ΔR	0.75	0.077	Ω/square
Pinning angle	80	3	degree
First half gap	30	5	nm
Second half gap	48	6	nm
Permanent magnet $M_r t$	760	30	memu/cm^2

Table 1.3. Impact of bias scheme on variance of the peak-to-peak signal amplitude and the pulse asymmetry

Bias scheme	V_{p-p} (mV)	σ_V (mV)	V^+/V^-	$\sigma_{V+/V-}$
Current	2.0	0.21	1.01	0.10
Power	2.0	0.15	1.02	0.11
Voltage	2.0	0.12	1.02	0.11

viations used for the input parameters. Gaussian distributions are assumed, and the values in Table 1.2 represent an estimate based on statistical process control data early in the development cycle. From these stochastic variables 1000 sets of input parameters are generated and a corresponding 1000 micromagnetic calculations are performed. In each calculation the sensor is exposed to the stray field from arctangent transitions of opposite polarity ($B_r t = 0.28$ µ$''T$), such that peak-to-peak isolated pulse amplitude and peak amplitude asymmetry, can be collected and analysed.

Figure 1.9 shows distributions for isolated pulse amplitude and pulse asymmetry (defined as the positive pulse divided by the magnitude of the negative pulse) under three bias conditions of the sensor: constant current, constant power and constant voltage. Table 1.3 summarises the mean and standard deviation obtained from the calculations for each of the three bias schemes. From Fig. 1.9 and Table 1.3 it is clear that a constant voltage bias scheme considerably narrows the pulse amplitude distribution. The reason is that the primary origin of amplitude variance in the constant current case is rooted in the relatively poor sensor height control. A constant voltage bias

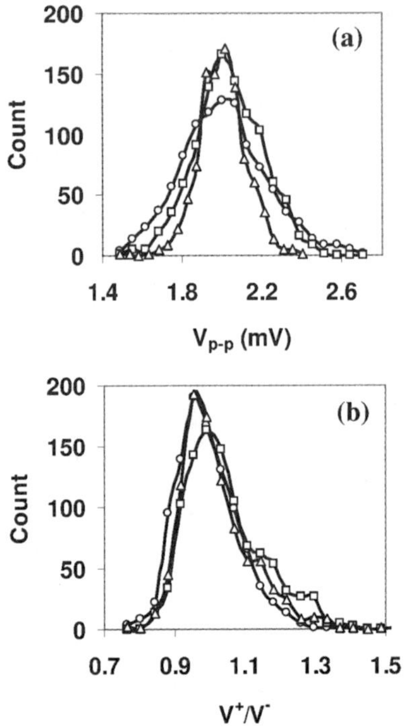

Fig. 1.9. Distributions for **a** isolated pulse peak-to-peak amplitude and **b** pulse asymmetry under three bias schemes: Constant current (*circles*), constant power (*squares*), constant voltage (*triangles*)

scheme will supply more sense current to a sensor with large stripe height and less to a short sensor. This narrows the distribution (Fig. 1.9c), and in principle the entire distribution can be moved closer to the reliability limit (higher mean sense current), thereby potentially improving performance of the sensors in the low amplitude tail of the distribution. The results on pulse asymmetry indicate no specific preference for one of the three bias schemes. This is due to the fact that asymmetry variance mainly is induced by variance in the inter layer coupling, rather than sensor height.

Williams carried out a similar analysis [8]. In this study emphasis is on thermal effects in IrMn versus PdPtMn-based spin valves. The analysis is not based on a micromagnetic model of a sensor, but also clearly shows the overwhelming impact of poor sensor height control on signal amplitude variance under constant current bias.

1.3.2 Efficiency of the Magnetoresistive Sensor

In magnetic recording at high areal density several trends regarding the magnetoresistive sensor are observed. As track widths become smaller, the sensor height is reduced in order to minimise shunting of current through the back of the sensor. With decreasing $B_r t$ of the recording medium, the thickness of the ferromagnetic films in the sensor are reduced not only in order to enhance sensitivity through improvement of the magnetoresistive effect, but also through enhancement of the rotational response of the sensor magnetisation.

The sensitivity of a magnetoresistive sensor is conveniently described in terms of an efficiency number. The efficiency, ε, of a magnetoresistive sensor is a measure of the amount of the magnetoresistive effect that is used in an actual application, i.e., a finished recording head. More directly, it measures the amount of rotation of the free layer magnetisation, i.e. $\varepsilon = 1$ for the full $180°$ rotation. Efficiency combines the magnetostatic interactions between the transition in the medium, the magnetoresistive sensor, and the shields. In principle ε is a function of medium $B_r t$, head-to-medium spacing, sensor height and width, the sensor stabilisation scheme, the $M_s t$ products of the ferromagnetic films in the sensor (t is film thickness), and possibly also a function of coupling between the films in the sensor and the coupling to the antiferromagnetic material in case of a spin valve. However, the phenomenological expression for the efficiency simply reads

$$\varepsilon = \frac{V_{p-p}}{\frac{w}{h}\Delta R_s I_s}\,, \tag{1.16}$$

where V_{p-p} is the peak-to-peak isolated pulse amplitude, w is the sensor width, h the sensor height, and I_s is the sense current. The maximum GMR effect, i.e., from parallel to anti-parallel alignment of the free and pinned layer magnetisation is denoted by ΔR_s in units of Ω/square (sheet film value). At the coupon or sheet film level demagnetisation effects can be entirely ignored and the comparison of the sheet resistance for parallel and anti-parallel alignment is through application of homogeneous fields is relatively straightforward. In a real sensor under operating conditions the situation is very different. The field from the transition is non-uniform and decays into the read gap. Moreover, strong demagnetisation effects that limit the rotational response are induced by the top and bottom edge of the sensor. The limitation of rotational response by shape anisotropy effects is an important argument in favour of the reduction of the $M_s t$ product of the free layer through reduction of t. One may wonder how this is possible, since for a given applied field, reduction of $M_s t$ in principle implies a reduced torque on the free layer. However, the demagnetisation field in the free layer, H_{dem}, also strongly depends on $M_s t$. The torque arising from the mean transition field, H_{trans}, depends on $M_s t$ as $\tau_{trans} \propto M_s t H_{trans}$, assuming fixed sensor width and height. For the uniformly magnetised case the torque arising from the demagnetisation field varies as $\tau_{dem} \propto M_s t H_{dem} \propto M_s^2 t N \propto (C M_s t)^2$, with C a constant characterised by fixed sensor width and height. Hence the ratio of the transition and demagnetisation torque is of the order $\frac{\tau_{trans}}{\tau_{dem}} \propto \frac{H_{trans}}{C M_s t}$. Thus, for a given transition field the demagnetisation field generated upon rotational response of the free layer will dominate and limit the response. From this magnetostatic point of view it is beneficial to reduce $M_s t$. In addi-

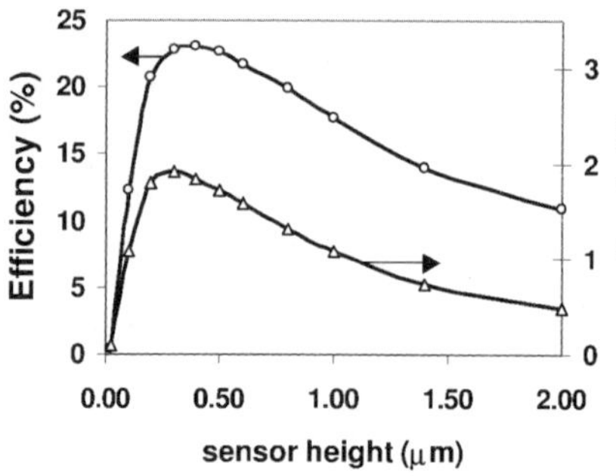

Fig. 1.10. Computed efficiency and signal amplitude as a function of sensor height for a 0.9 µm wide spin valve sensor operated under constant voltage conditions

tion, changes in the magnetoresistive effect can occur as a result of changes in film thickness. These must be taken into account in the optimisation process.

Figure 1.10 shows the impact of stripe height on efficiency and signal amplitude for a sensor operated at constant voltage (approximately constant current per unit stripe height). These are results from micromagnetic calculations for a 0.9 µm wide spin valve sensor while the sensor height varies from 0.05 to 2 µm. At large sensor height the field from the transition does not impact the magnetisation of the upper part of the sensor very much. The sense current that passes through the upper part of the sensor does not contribute to the signal and must be considered a shunting loss. As a result, the efficiency and signal amplitude are reduced. For small sensor heights shape-anisotropy effects do not allow for a large rotation of magnetisation in the free layer magnetisation in response to the field from the transition in the medium. A pronounced drop in efficiency is the result. At an intermediate sensor height the efficiency reaches a maximum. Typical maximum efficiencies (empirical and theoretical) range from 15 to 25%.

Note that the sensor heights corresponding to maximum efficiency and to maximum amplitude do not necessarily coincide. Two causes come to mind. First, variation in efficiency is purely determined by the sensor height variation, while the applied sense current is a function of sensor resistance and changes in the contact resistance as a function of sensor height. Second, alterations of stripe height, and concomitantly in sense current, can induce pulse amplitude asymmetry. Typically the sensor heights for maximum ε and V_{p-p} are very close.

Efficiency is also severely impacted by the physical sensor width. Figure 1.11 shows this effect through a micromagnetic calculation for a generic abutted junction spin valve sensor with a constant sensor width-to-stripe height ratio equal to 2 and 3.5 nm thick free layer of an 800 emu/cc material.

Unless the head-to-medium spacing and free thickness layer can shrink as fast as the sensor width (which is impossible), efficiency must approach zero at vanishing sensor width. This is due to two effects. One is the fact

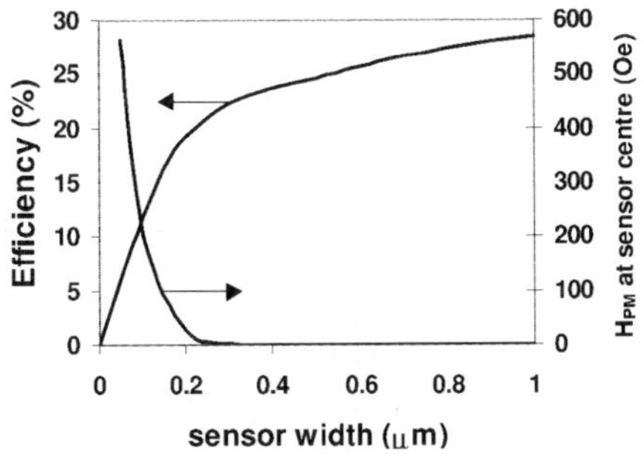

Fig. 1.11. Computed efficiency versus sensor width for constant free layer thickness and aspect ratio equal to two. The stray field from the permanent magnets, H_{PM}, at the centre of the sensor versus sensor width is also shown (secondary axis)

that for an abutted junction design, the magnitude of the stray field from the permanent magnets at the centre region of the free layer becomes very large as the sensor width diminishes, as shown in Fig. 1.11. Reducing the permanent magnet field strength is one option but this would significantly reduce its effectiveness in stabilising the edges of the free layer. The other effect is due to the self-demagnetisation field of the free layer. As the sensor width and stripe height decrease, shape anisotropy effects are increased unless the film thickness is reduced. The latter option can be realised in practice only to a limited extent since the GMR effect itself as well as inter-layer diffusion and film deposition issues limit this method of increasing efficiency.

1.4 Micromagnetic Recording Model

A comprehensive recording model can readily be constructed from a combination of FEM solutions for the write head field, a micromagnetic medium model, and a micromagnetic model of a magnetoresistive sensor for play-back [24]. Detailed examinations of the recording process in the static limit are now possible within the aproximations of the Micromagnetic Recording Model (MRM). As an example, the impact of write current on the recording and play-back process will be discussed in this section. It is emphasised that classical, simple recording models fail to describe important features of the recording process at high areal density, e.g., at 10 Gb/in^2.

1.4.1 Magnetisation Profiles of Recorded Transitions

The characteristics of recorded transitions were investigated as a function of write current or, equivalently, as a function of both write field magnitude and field gradient in the medium. The writer pole width was 0.6 μm, the write gap was 0.2 μm, and the pole material had a saturation magnetisation of 1440 emu/cc (1.8 T). A micromagnetic model of a 0.6×0.3 μm^2 generic spin valve sensor with abutted junction edge stabilisation was used for playback. The medium consists of a regular array of $13 \times 13 \times 20$ nm^3 grains with in-plane random uniaxial anisotropy of $H_k = 6700$ and $M_s = 350$ emu/cc. No exchange coupling between the grains was present. The head-to-centre-of-medium separation was 30 nm (see [24] for details). These are roughly 10–15 Gb/in^2 dimensions.

Increasing the current through the writer coil serves not only to increase the field reaching the medium but also enhances pole-tip saturation. FEM calculations support the notion that such saturation degrades the field gradient in the medium. It is suspected that this can lead to degradation of the quality of the recorded tracks, especially at the edges of the track. Figure 1.12 illustrates this point. Dibits were recorded at a spacing corresponding to a linear density of 200 kbpi, using the nominal FEM 10 Gb/in^2 writer and LLG medium model. The images each correspond to an average over ten dibits recorded on media samples with e different anisotropy axes distribution. Four different write currents were chosen to give the values of H_m indicated. Here H_m is the down-track component of the head field at the center of the write gap and at the point mid-way in the medium thickness. The field in the medium increases with increasing write current. In Fig. 1.12 the images of the LLG solution for the down-track component of the magnetization vector (M_x) clearly show track curvature, track edge distortion, and an increased level of percolation as the pole-tip saturation increases. Also shown are profiles of the down-track magnetization M_x (averaged over the pole width). The remnant magnetisation inside the dibit is reduced with write current and least-squares fitting of the magnetization profiles to $\tanh(2x/\pi \mathrm{a})$ indicates that the transition parameter increases with write current. Figure 1.12 shows fitted magnetization profiles (dashed lines) as well as the effective transition parameter extracted from the fit.

Figure 1.13 quantitatively shows how track-edge and on-track gradient degradation serve to increase the effective transition parameter, a. The fitted transition parameter increases with the percentage of the track width used for the average down-track magnetization profile. Higher write currents (larger H_m) increase the transition parameter even further. Note that even in the centre portion of the track an increase in write current serves to increase the transition length due to degradation of the on-track field gradient with write current. These results show that pole-tip saturation impacts the gradient, and hence the quality of the recorded transitions, across the entire track width.

In general, it is seen that for the FEM head field, 'a' increases with increasing H_m, from about 13 to 18 nm over the range of H_m (field in the medium) considered. As an illustration of the effect of head field gradient, the simple Williams-Comstock expression is used to evaluate the transition parameter for two values of H_m. The transition parameter decreases with increasing record field in contrast to the simulation results. The reason is that this simple

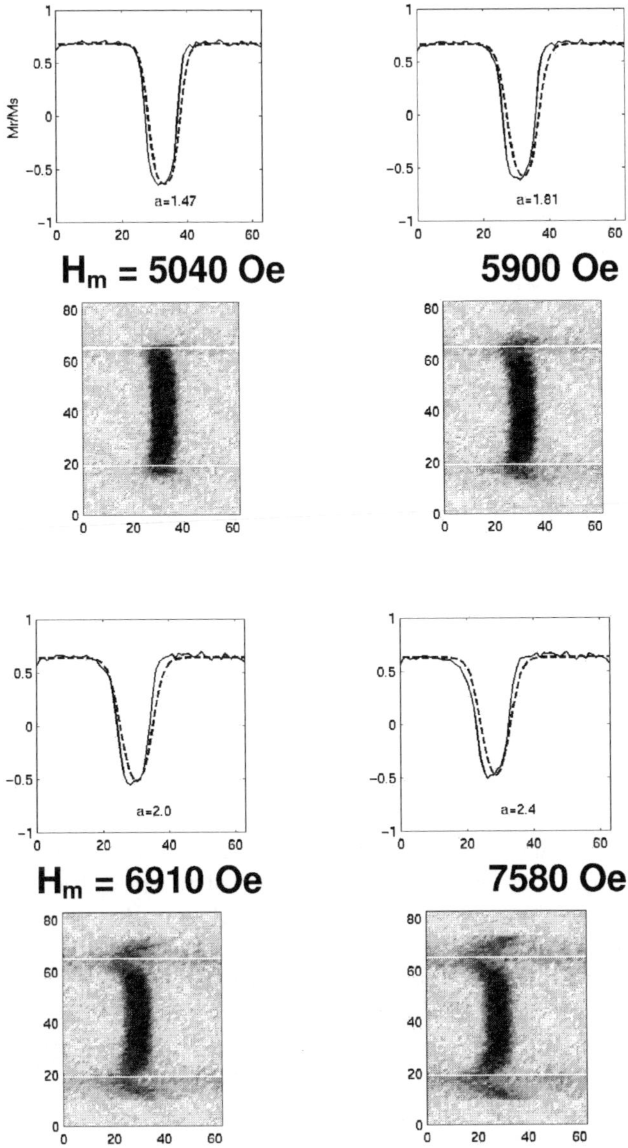

Fig. 1.12. Track averaged down-track magnetization profiles (solid curves) and images of the down-track magnetization for four different write currents. The FEM head field at the center of the media is indicated. Also shown is the result of tanh fits (dashed) and fitted transition parameters (in units of grain size) are indicated. White lines in the images outline the writer pole width

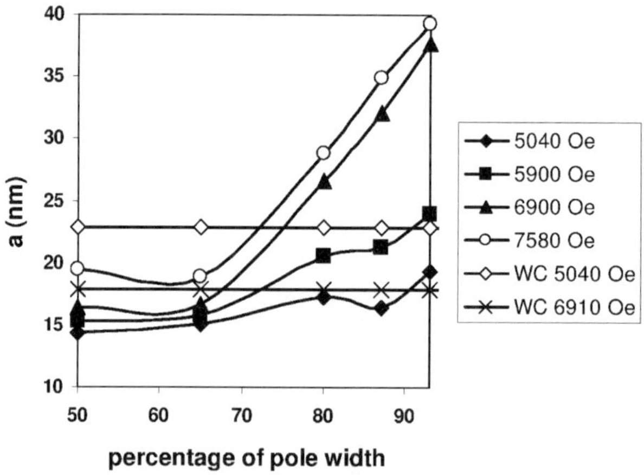

Fig. 1.13. Fitted transition parameter versus percentage of the track width used for averaging. Inset shows the field in the medium, H_m, indicative of the level of pole-tip saturation. Typical results from the Williams-Comstock model (WC) are shown

model does not include pole tip saturation and the head-medium parameters utilized are such that increasing the current increases the field gradient.

1.4.2 Signal Pulse Width

Measurement of the playback voltage pulse width for isolated transitions provides another means for characterisation of the recorded transitions. If the signal pulse-width-at-half-maximum (PW_{50}) is large, intersymbol interference will negatively impact the playback performance of the recording system. In classical recording theory any dependence of PW_{50} on, for instance, write current enters through the transition length.

Figure 1.14 shows PW_{50} as a function of write current, derived with the 10 Gb/in^2 FEM writer head field and a generic spin valve magnetoresistive sensor (from [24]). Pulse widths are estimated from the average of ten runs, where the in-plane easy axis distribution of the medium is re-generated for each run. The error bars in Fig. 1.14 indicate the range of obtained PW_{50} values. Also shown are results from the Williams-Comstock approximation for 'a' for the pulse width [1]. In addition, results are included where FEM values for the field gradient were used in the Williams-Comstock relation for the transition parameter (where the Karlqvist field H_0 was set to the FEM result as in Sect. 1.2.1). Results in Fig. 1.14, derived with the MRM, recover the very typical and pronounced minimum in PW_{50} as function of write current seen experimentally. On the other hand, at medium to high write current the Williams-Comstock values for PW_{50} depend only very weakly on the write head field, and PW_{50} decreases monotonically with write current.

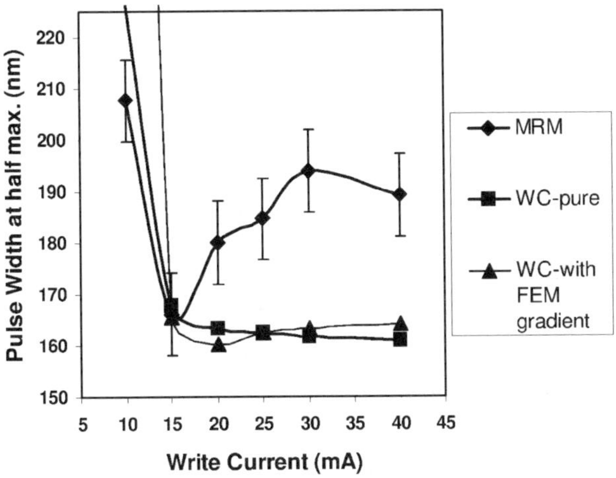

Fig. 1.14. PW_{50} versus write current as derived from FEM write head fields using the recording model. Also shown are PW_{50} for pure Williams-Comstock and Williams-Comstock with FEM field gradients. Error bars on the MRM curve indicate the range of pulse widths obtained in ten runs for each write current

These results further illustrate how analytic recording models fail to capture essential features of high-density magnetic recording that are a consequence of narrow track widths, pole-tip saturation effects, and the granular nature of real media.

1.5 Recording on Oriented Longitudinal Media and Challenges in Ultra High Density Recording

The MRM was also used to examine the interplay between pole-tip saturation and the orientation ratio of the medium at 10–15 Gb/in^2 as well as 100–150 Gb/in^2 dimensions [31,32]. Orientation ratio can be defined either in terms of the measured medium remnant magnetisation ratio, circumferential to radial, or in terms of the ratio of coercivities. We use the latter definition in the present description: $OR = H_{cx}/H_{cz}$, where x and z are circumferential and radial directions, respectively. In the case of 100–150 Gb/in^2 recording, the writer pole width was 0.13 µm with a gap of 0.04 µm, and the pole materials had a M_s of 1920 emu/cc (2.4 T). The model GMR spin valve had dimensions 0.1×0.05 µm^2. The medium was composed of 8 nm grains and was 10 nm thick grains with in-plane uniaxial anisotropy of $H_k = 11.5$ kOe and with $M_s = 430$ emu/cc. The medium coercivity in the isotropic case was about 5500 Oe. Orientation effects were accounted for by assigning a Gaussian distribution to the grain anisotropy axes orientation. No exchange coupling between the grains was included. The head-to-centre-of-medium separation was 12 nm.

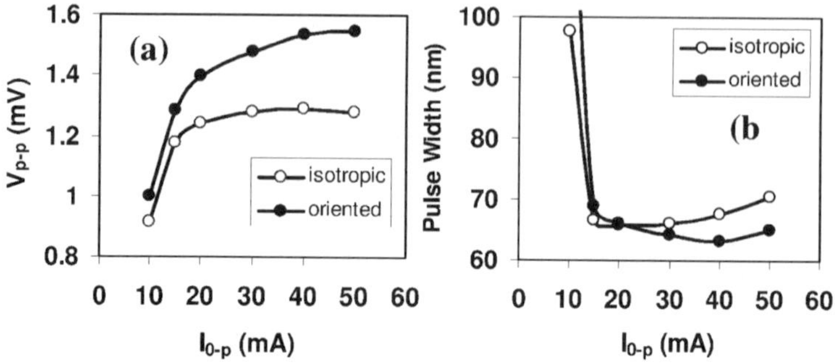

Fig. 1.15. MRM results from [32] for playback amplitude and pulse width as functions of write current for isotropic and oriented ($OR = 1.6$) media

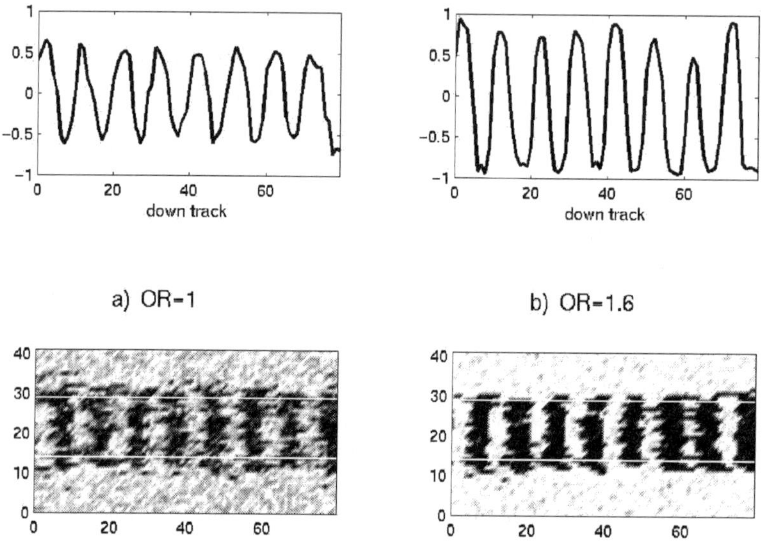

Fig. 1.16. Normalized down-track component of the magnetization vector M_x/M_s for multiple transitions written on **a** isotropic and **b** oriented media ($OR = 1.6$) at 640 kbpi

The effect of oriented media on playback amplitude and pulse-width for isolated transitions, using a model writer with a 0.13 µm wide pole, are shown in Fig. 1.15. Two effects are illustrated: The use of oriented media (in this case $OR = 1.6$) can significantly improve these two performance metrics and pole-tip saturation effects are reduced.

Recording on oriented media also results in tracks that are better defined at higher linear density. Figure 1.16 compares images of transitions at 640 kbpi in the case of isotropic and oriented media ($OR = 1.6$). The significant percolation seen in the isotropic case is reduced to a large extent with the use

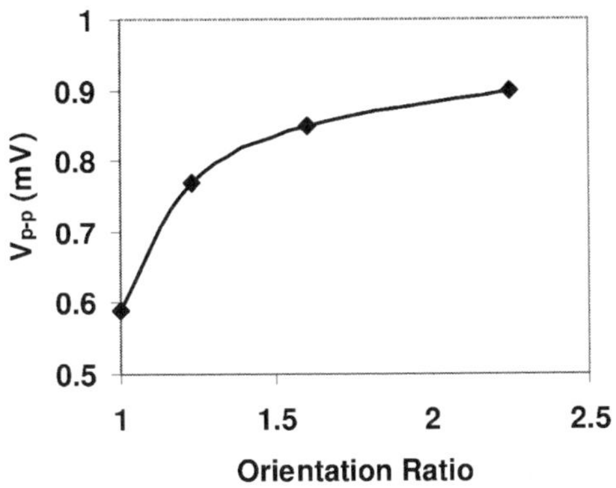

Fig. 1.17. Playback amplitude as a function of orientation ratio at 540 kbpi

of oriented media. This translates directly into higher playback amplitude at 540 kbpi as shown in Fig. 1.17.

Oriented media has also been shown, using the MRM, to result in reduced transition noise [32] at 10 as well as 100 Gb/in^2 dimensions. The signal-to-noise ratio (SNR) was calculated using

$$SNR = 10 \log \left[\frac{\int_0^L \langle V(x) \rangle^2 dx}{\int_0^L \langle V^2(x) \rangle - \langle V(x) \rangle^2 dx} \right] , \tag{1.17}$$

where L is the length of the sample, $V(x)$ is the play-back signal as a function of down-track position x, and $\langle \ldots \rangle$ indicates an average over the fifty independently recorded tracks. The sample length included about 10 transitions at each areal density. This is similar to the approach taken by Zhou and Bertram [33,34], with the exception of the playback, which is done micromagnetically in this study. The MRM was used to record transitions at mid-frequency (200 kbpi at 10 Gb/in^2 and 390 kbpi at 100 Gb/in^2 areal density), and perform playback with a micromagnetic model of a spin valve. Each SNR number originates from a statistical average over fifty individual record-and-playback passes. The SNR values are computed as the ratio of total signal and total noise power at mid frequency. For playback, the sensor was centred either over the track centre or over the track edge.

Results are presented in the table below. At both areal densities, the oriented medium significantly improves the signal-to-noise ratio, even when the sensor is centred over the track edge rather than over the centre of the track. At 10 Gb/in^2, positioning of the sensor over the track edge has a much larger impact on SNR than at 100 Gb/in^2. This is consistent with the notion that a 0.13 µm pole width produces two track edges with very little 'good' track centre region. Also consistent with previous results from micromagnetic calculations (not shown here), was the observation that weak intergranular exchange coupling further degrades the SNR [33,34].

Table 1.4. Calculated signal-to-noise ratios as a function of medium coercivity orientation ratio (OR) for 10 and 100 Gb/in^2 areal densities. The sensor is positioned over the centre of the track and 50% off-track, over the track edge

10 Gb/in^2			100 Gb/in^2		
OR	SNR (dB)	SNR (dB) Track edge	OR	SNR (dB)	SNR (dB) Track edge
1	20	14	1	14	12
1.6	26	18	1.5	19	17

In Sect. 4.1 it was found that the quality of the transitions degrades toward the track edges as a result of a degradation of the head field. It is interesting to note that this degradation in principle could have been avoided through scaling of the *entire* head-medium interface (i.e., head-to-medium spacing, pole width, medium grain size, materials properties, etc.). However, in practice this has not been the case. The head-to-medium spacing is not following the reduction in writer pole width. Thermal stability requirements for the medium impose lower limits on the grain size and the uniaxial anisotropy. The result is that there are relatively fewer grains across the track, while the ratio between the medium coercivity and the saturation moment of available pole tip materials increases to problematic values.

Finally, two interesting observations regarding the magnetisation in models of high coercivity thin film recording media are made. First, the magnetisation in each of the medium grains assumes only the two configurations along the unidirectional axis. Deviations from these states due to magnetostatic interactions are negligible. Consider the quantity $\frac{1}{N}\sum_{i=1}^{N}\left|\hat{k}_i \cdot \hat{m}_i\right|$, where N is the number of grains and $\hat{k}_i$ and $\hat{m}_i$ are unit vectors along the easy axes and the magnetisation vectors, respectively. If deviations of the magnetisation from the easy axes are small, the sum should be close to unity. Calculations for high coercivity media (H_k = 10500 Oe, M_s = 430 emu/cc) at zero remnant magnetisation yielded a value of 0.99. After a track was recorded at 400 kbpi with a 0.13 μm wide writer pole, this value did not change.

Second, alignment of the magnetisation with the easy axis for essentially in-plane isotropic media implies significant variance of magnetisation in the cross-track direction. If the recorded track is wide relative to the grain size, the cross-track component of the medium magnetisation averages to zero, suppressing generation of noise during playback. In the example in Table 1.4, for 10 Gb/in^2 the spin valve sensor averages the magnetisation of about 46 grains in the cross track direction. At 100 Gb/in^2 this number was reduced to 16 grains, and is accompanied by an inherently noisier playback waveform due to enhanced fluctuations in the medium magnetisation.

1.6 Summary

Understanding of fundamental aspects of the magnetic recording process can be greatly enhanced through an examination of the interaction of small ferromagnetic objects with applied fields. A micromagnetic treatment of recording physics can reveal detailed information on write the medium magnetisation process during recording, the write head field, and the response of magnetoresistive sensors. Such knowledge is becoming increasingly important as the rush to even higher areal density applications demands a deeper understanding of the physics of the devices that need to be volume manufactured to meet the expectations of the storage industry.

It is hoped that this first chapter will serve as an introduction to some of the basic concepts discussed in the following contributions to this book, where key elements will be explored in much more detail. Here basic concepts in magnetism and the micromagnetic method were concisely reviewed, and the utility of the micromagnetic method was illustrated with examples that are relevant to magnetic recording.

Acknowledgement. The authors would like to express gratitude for valuable discussions and assistance to Harry S. Edelman, Frank E. Stageberg, Vee S. Kong, Hong Zhou and H. Neal Bertram.

References

1. R. M. White, *Introduction to Magnetic Recording*, 1st ed. (IEEE Press, New York, 1985)
2. C. D. Mee, E. D. Daniel, *Magnetic Recording, Volume 1: Technology*, 1st ed. (McGraw-Hill, New York, 1987)
3. J. C. Mallinson, *The Foundations of Magnetic Recording*, 2nd ed. (Academic Press, San Diego, 1993)
4. H. N. Bertram, *Theory of Magnetic Recording*, 1st ed. (Cambridge Univ. Press, Cambridge, 1994)
5. J. C. Mallinson, *Magneto-resistive Recording Heads*, 1st ed. (Academic Press, San Diego, 1996)
6. K. G. Ashar, *Magnetic Disk Drive Technology*, 1st ed. (IEEE Press, New York, 1997)
7. R. L. Comstock, *Magnetism and Magnetic Recording*, 1st ed. (Wiley, New York, 1999)
8. E. M. Williams, *Design and Analysis of Magnetoresistive Recording Heads*, 1st ed. (Wiley, New York 2001)
9. J. D. Jackson, *Classical Electrodynamics*, 2nd ed. (Wiley, New York 1975)
10. M. Mansuripur, R. Giles, IEEE Trans. Magn. **24**, 2326 (1988)
11. M. R. Scheinfein, J. Unguris, J. L. Blue, K. J. Coakley, D. T. Pierce, R. J. Celotta, P. Ryan, Micromagnetics of Domain Walls at Surfaces, Phys. Rev. B **43**, 3395 (1991)

12. S. Yuan, Micromagnetics of Domains and Walls in Soft Ferromagnetic Materials, Ph.D. Dissertation, University of California, San Diego (1992)
13. W. F. Brown: *Micromagnetics*, Reprint (Krieger Publishing Company, Hintington 1978)
14. G. F. Hughes, Magnetisation Reversal in Cobalt-phosphorous films, J. Appl. Phys. **54**, 5306 (1983)
15. C. Kittel, *Introduction to Solid State Physics*, 5th ed. (Wiley, New York 1976)
16. J. L. Blue, M. R. Scheinfein, Using Multipoles Decreases Computation Time for Magnetostatic Self-Energy, IEEE Trans. Magn. **27**, 4778 (1991)
17. The µMAG site at `www.ctcms.nist.gov/~rdm/mumag.org.html`
18. W. K. Hiebert, A. Stankiewicz, M. R. Freeman, Direct observation of Magnetic Relaxation in a Small Permalloy Disk by Time-Resolved Scanning Kerr Microscopy, Phys. Rev. Lett. **79**, 1134 (1997)
19. R. H. Koch et al., Magnetization Reversal in Micron-Sized Magnetic Thin Films, Phys. Rev. Lett. **81**, 4512 (1998)
20. G. M. Sandler, H. N. Bertram, T. J. Silva, T. M. Crawford, Determination of the Magnetic Damping Constant in NiFe Films, J. Appl. Phys. **85**, 5080 (1999)
21. T. J. Silva, T. M. Crawford, Methods for Determination of Response Times of Magnetic Head Materials, IEEE Trans. Magn. **2**, 671 (1999)
22. T. J. Silva, C. S. Lee, T. M. Crawford, C. T. Rogers, Inductive Measurement of Ultrafast Magnetisation Dynamics in Thin-film Permalloy, J. Appl, Phys. **85**, 7849 (1999)
23. D. A. Lindholm, Magnetic Fields of Finite Track Width Heads, IEEE Trans. Magn. **MAG-13**, 1460 (1977)
24. J. van Ek, M. L. Plumer, H. Zhou, H. N. Bertram, Micromagnetic Recording Model of Pole Tip Saturation Effects, IEEE Trans. Magn. **36**, 3975 (2000)
25. C. Y. Mao, J.-G. Zhu, R. M. White, T. Min, Effect of Damping on the Switching Limit of a Thin Film Recording Head, J. Appl. Phys. **85**, 5870 (1999)
26. K. Gao, H. N. Bertram: Three Dimensional Micromagnetic Analysis of Write Dynamics and Field Patterns, to be published in IEEE Trans. Magn. (2001)
27. G. M. Sandler, H. N. Bertram, Micromagnetic Simulations with Eddy Currents of Rise Time in Thin Film Heads, J. Appl. Phys. **81**, 4513 (1997)
28. G. M. Sandler, Computational Analysis of Magnetisation Dynamics in Thin Film Write Heads, Ph.D. Dissertation, University of California, San Diego (1998)
29. R. M. Bozorth: *Ferromagnetism*, (IEEE Press, New York, 1993)
30. B. A. Gurney et al., Can Spin Valves be Reliably Deposited for Magnetic Recording Applications?, J. Appl. Phys. **81**, 3998 (1997)
31. M. L. Plumer and J. van Ek, Micromagnetic Study of Track Edge and Medium Orientation Effects in High Areal Density Recording, IEEE Trans. Magn. (Intermag 2000)
32. M. L. Plumer, J. van Ek, Micromagnetic Recording Model of Pole-Tip Saturation and Medium Orientation Effects at 100 Gb/in^2, IEEE Trans. Magn. (Intermag 2001)
33. H. Zhou and H. N. Bertram, Micromagnetic Study of Longitudinal Thin Film Media: Effect of Grain Size Distribution, IEEE Trans. Magn. **35**, 2712 (1999)
34. H. Zhou and H. N. Bertram, Micromagnetic Analysis of Anisotropy Orientation on Recording Performance in Longitudinal Thin Film Media, J. Appl. Phys **87**, 5675 (2000)

2 Microstructure of Longitudinal Media

Bin Lu and David E. Laughlin

2.1 Introduction

In this chapter we give an overview of the effects of various microstructural features on the resulting magnetic properties of magnetic thin films. The role of microstructure on certain properties of materials has a long history of investigation, starting in the discipline of physical metallurgy where the role of microstructure on properties was investigated in great detail [1]. It became clear that some properties were relatively insensitive to changes in microstructure and others were strongly sensitive to changes in microstructure. The former were called *intrinsic properties* or sometimes primary properties, since they were *intrinsic* to the material. The latter were called *extrinsic properties* or secondary properties, since they were dependent on features that were not inherent to the material [2]. With the advent of the discipline of materials science the role of material processing was included in the way extrinsic properties were understood. Changing the method of producing a material often changed some of the resulting physical properties of the material. This change was shown in many cases to be mediated through the change in microstructure of the material. This lead to the so called materials science paradigm: processing determines microstructure which in turn determines the extrinsic properties of a material.

Magnetic thin films are a good example of this paradigm. Small changes in the way a thin film is produced often give rise to large changes in some of the magnetic properties of the thin film. This is best understood by observing how the microstructure of the film changes with processing and then correlating the microstructure directly with the properties of the thin film.

A typical structure of a thin film used for magnetic recording is shown in Fig. 2.1. It can be seen that the magnetic layer is in contact with an intermediate layer, which in turn is in contact with an underlayer. The seedlayer and the underlayer set up the grain size as well as the crystallographic orientations in the layers. The intermediate layer enhances the epitaxy growth of the magnetic layer. Furthermore, atoms from the intermediate layer may diffuse up the grain boundaries, giving rise to magnetic isolation of the grains within the magnetic layer. All these processes affect the magnetic properties of the film.

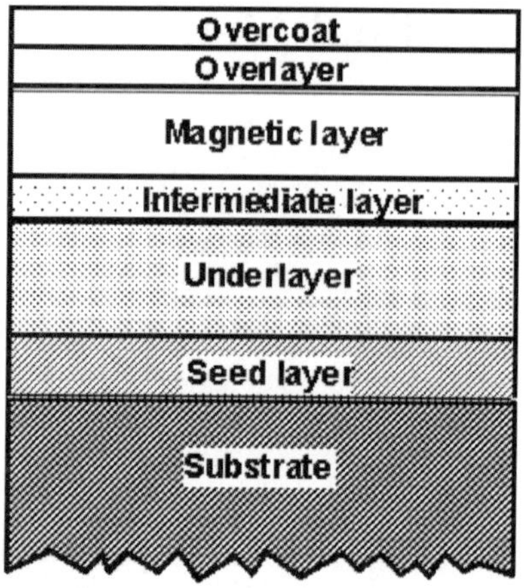

Fig. 2.1. Schematic of the layer structure of thin films used for longitudinal magnetic recording

In this chapter we will give an overview of the role that some microstructural features of thin films play in determining the properties of the magnetic layer in films used for longitudinal recording. These features include grain size and grain size distribution, crystallographic orientation, defects and chemical segregation. We will discuss various methods used to control these features during the deposition process and show how with careful control the magnetic properties and hence the recording performance of the films can be optimized.

2.2 The Size and Distribution of Thin Film Grains

2.2.1 Introduction

High areal density has been achieved in the longitudinal magnetic recording media by continuously shrinking the size of the recorded bits. Since the noise power of magnetic recording media is strongly related to the number of magnetic particles (or grains) per bit, in order to keep the signal to noise ratio (SNR) at a readable level, the number of grains per bit must be maintained above a certain statistical minimum. For this reason the average grain size in the magnetic layer has been scaled continuously with the bit length and the track width. This can not continue indefinitely, as the magnetic energy stored in each grain (K_uV) is becoming so small that the thermal energy (k_BT) is approaching the same magnitude. When this happens the superparamagnetic limit is reached and the media is no longer thermally stable [3] (also see Chap. 5).

Therefore the effect of grain size is two-fold. On the one hand the grains should be small enough to keep high SNR, on the other hand the grain size should be large enough to keep the recorded information thermally stable. For Co-alloy media, assuming a $K_u = 2.5 \times 10^6$ erg/cc, and a cubic grain shape, and using the media stability criterion of $K_uV/k_BT \geq 60$ (thermally stable for 10 years) [3], the smallest grain size that can be used at room temperature (300 K) is calculated to be about 9.7 nm. In fact, in the 10 Gbits/in^2 media design announced in 1998 [4], the grain size was about $D = 12 \pm 4$ nm. It is obviously that there is little room left for grain size reduction with the current Co-alloy media.

Another important feature is grain size distribution. The sizes of the grains in media are not uniform but are spread about an average value. The information stored in the smaller grains decay more rapidly, while the larger grains give rise to a higher noise level. It has been reported by Weller and Moser [5] that narrowing the grain size distribution can alleviate about 25% of the thermal decay problem. Likewise, it will also significantly enhance the signal to noise ratio (SNR).

When discussing grain size and grain size distribution, it is usually assumed that the Co-alloy grains are magnetically well separated by nonmagnetic grain boundaries, which either consist of nonmagnetic materials or simply voids. If this assumption is not true, magnetic clusters of grains must be taken into account [6,7]. These magnetic clusters consist of several grains magnetically coupled by inter-granular exchange coupling. The grains inside a magnetic cluster behave in concert under magnetic fields. Therefore for media with strong exchange coupling, the effective magnetic particle size will be much larger than the grain size, causing a higher media noise [8–12]. On the other hand, there is also a possibility of forming microstructural clusters, inside which the grains are of similar crystallographic orientation. The grain boundaries inside such clusters are less distinguishable than the cluster boundaries. In such a case, the media noise depends strongly on the microstructural cluster size [13].

For these reasons the control of the grain size and grain size distribution of the underlayer and magnetic layer is of critical importance to control and to optimize the magnetic properties and recording performance of the magnetic media.

2.2.2 Grain Size Distribution

Three kinds of distributions can be used to describe grain size distribution.

1. Gaussion (normal) distribution:

$$y = \frac{1}{\sigma\sqrt{2\pi}} \exp\left(-\frac{(x-A)^2}{2\sigma^2}\right) , \tag{2.1}$$

   where A is the arithmetic mean and σ is the standard deviation.

2. Log normal distribution:

$$y = \frac{1}{log\sigma_g\sqrt{2\pi}} \exp\left[-\frac{(logx - logx_g)^2}{2log^2\sigma_g}\right] \tag{2.2}$$

It can be seen that the log normal distribution is formed from normal distribution by substituting $\log x$ in Eq. (2.2) for x. As a result, x_g is the geometric mean, and σ_g the geometric standard deviation, whose logarithm is the standard deviation of the variable $\log x$.

3. Rosin-Rammler distribution:

$$y = \frac{n}{d^n}x^{(n-1)} \exp\left(-\frac{x^n}{d^n}\right), \quad (n > 1)\,. \tag{2.3}$$

The parameter n primarily depends on the skew of the distribution and the parameter d is primarily determined by the breadth of the distribution.

Since the Rosin-Rammlar distribution has two parameters, it can be easily fit to the grain size distribution in Co-alloy films [14]. However, both the normal and Rosin-Rammlar distribution allow for negative variable values, which are not physically applicable to grain size distributions.

The log normal distribution allows only positive variable values. Moreover, it has been found by Soderlund et al. [15] that in thin film deposition, if the adatoms diffuse or drift within a small but finite region, the time of growth of the nuclei before impingement can be accurately described by a log normal distribution. This will cause the grain size distribution to be a log normal one. It has been shown that the grain size distributions of CoCrPt, CoCrPtTa and CoCrPtB alloy thin films are log normal [16]. Hence, many calculations of magnetic properties of the media use the log normal grain size distribution [5].

It should be noted that when we consider the signal read from each recorded bit, it is the grain area distribution by area fraction that should be counted rather than the grain area distribution by count fraction. This is because the signal contributed by an individual grain is proportional to its moment, $M_r tS$, where M_r is the remnant magnetization of the magnetic layer and t is the layer thickness. Due to columnar growth of the grains, $M_r t$ is constant through out the entire media, the distribution of the signal from each grain should be expressed as grain area distribution by area fraction. Figure 2.2 shows the difference between the grain area distribution by count fraction and by area fraction for CoCrPt media.

At the initial nucleation stage of film deposition, a minimum number of atoms are required to form a critical nucleus. A sub-critical nucleus is unstable and will dissolve. Critical nuclei are small. For example, the size of critical nuclei for Zn on glass at room temperature (RT) is about 0.3 nm [18], it is 1.2 nm for Mg [18] on glass at RT, 1–2 nm for Cr on NiP [19], 3–5 nm for CoCrTa on Cr at 250°C [20]. Since the grains cannot be smaller than the

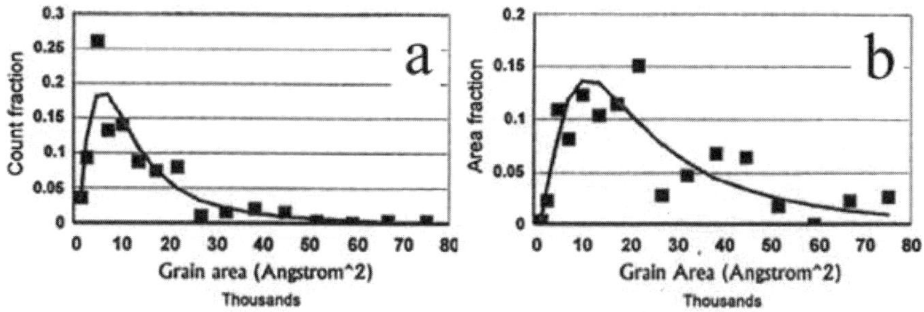

Fig. 2.2. Grain size distribution by number and by area in a longitudinal media [17] (© 2000 IEEE)

critical nucleus, there is a characteristic truncation on the left-hand side of the grain size distribution. Therefore, when the mean grain size approaches the size of the critical nucleus, the grain size distribution will tend to be narrower. Based on this reasoning, most of the current research focuses only on how to reduce the average grain size, expecting a corresponding decrease in the distribution breadth [17].

2.2.3 Control of Grain Size

In longitudinal media the Co-alloy grains are epitaxially grown on underlayers or intermediate layers, mostly consisting of Cr-alloy films, while adatoms of the Cr-alloy (CrX) films nucleate randomly and grow on the smooth surface of amorphous NiP or glass. Therefore, the control of the grain size and grain size distribution of the magnetic layer is realized through controlling the nucleation and growth of the CrX underlayer on the substrate.

The initial grain structure is formed following models of nucleation and growth-to-impingement. The heterogeneous nucleation of CrX on a smooth amorphous substrate can be described by capillarity theory [21]. According to the theory the nucleation rate is a strong function of the nucleation critical energy ΔG^*, which is the critical free energy of forming the nuclei. ΔG^* varies with deposition rate, substrate temperature, surface roughness and other factors.

Generally, ΔG^* increases with substrate temperature and decreases with deposition rate [22]. Consequently, the number of stable nucleus decreases rapidly with substrate temperature increase and a continuous film will take longer to develop, which results in larger grains. On the other hand, a larger number of stable nuclei form at higher deposition rates, and a continuous film is produced at lower average film thickness with smaller grain size.

After impingement, the growth of the grains is also affected by deposition conditions, such as, substrate temperature, the sputtering power, the argon pressure, etc. Thornton's zone model [23–25] best classifies the final

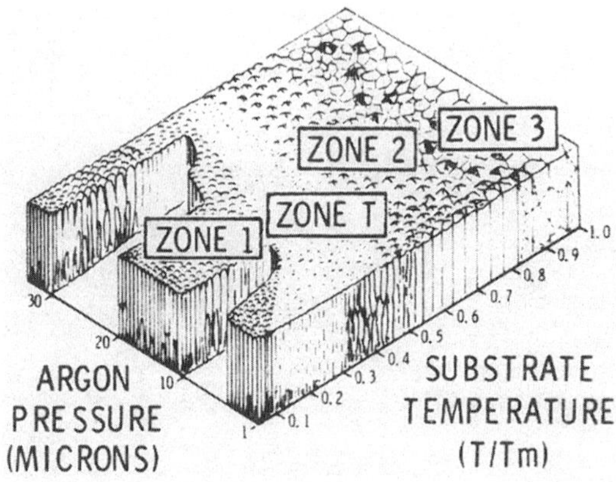

Fig. 2.3. Thornton's Zone model for sputter deposited metal films [23] (with permission, from the *Annual Review of Materials Science*, Vol. 7, © 1997 by Annual Reviews, `www.AnnualReviews.org`)

morphology of the sputter deposited metal films. Figure 2.3 shows the morphology of the film varying with the Ar pressure and substrate temperature. According to this model, at low temperatures ($T_{sub}/T_m < 0.3$, where T_{sub} is the substrate temperature and T_m is melting point of the material) and high Ar pressure, the film consists of tapered crystals with domed tops which are separated by voided boundaries. The internal structure is poorly defined, with a high dislocation density. The grain size increases with T_{sub}/T_m. The morphology of the films falls into Zone 1 of the model. This kind of deposition condition used to be utilized to grow very thick underlayers in order to enhance the voided grains boundaries, which decouple the magnetic grains and consequently give rise to lower media noise [26–31].

At higher substrate temperatures ($0.3 < T/T_m < 0.5$) and lower Ar pressures, the films consist of columnar grains separated by distinct dense grain boundaries. These grains are highly faceted at the surface of the films. Dislocations are primarily in the boundary regions. The nucleation energy is about equal to that for surface diffusion. This is defined as Zone 2. Zone T is a transition zone with processing parameters between those of Zones 1 and 2. A film in Zone T consists of a dense array of poorly defined fibrous grains without voided boundaries. The columnar grain diameter also increases with T_{sub}/T_m.

The influence of the substrate temperature, Ar pressure and substrate roughness has been discussed in detail in Thornton's review paper [23]. Further description of both thick and thin films has been reported by Grovenor et al. [25].

DC magnetron sputtering deposition at moderate substrate temperatures (300–550 K) is normally used in the manufacturing of magnetic recording media. The deposition rate is very high in the process (about 1000 nm/min).

Table 2.1. Dependence of grain size on O_2 exposure [34] (© 2000 IEEE)

Media	O_2 (L)	Grain size (nm)	σ (nm)
A	0	10.8	3.9
B	1.5	9.8	3.3
C	3.5	9.3	2.9

From the start, the nucleation rate is high but quickly decreases to zero as the number of nuclei saturates on the substrate. The substrate temperature for the Cr layer (T_{sub}/Tm) is about 0.14–0.25 and the Ar pressure is about 3–10 mtorr. For room temperature deposition of Cr (110) films, many defects and voids have been observed [29]. In this case the nucleation and growth conditions belong to Zone 1 of the Thornton's model. For Cr (200) textured layers deposited at 250°C, the deposition condition is in Zone T. Therefore, well defined columnar grains are commonly observed in the CrX thin films.

Columnar growth is primarily controlled by the competition between discrete atomic deposition and surface diffusion [32]. Assuming no coalescence, the average grain diameter, D, is related to the grain growth rate, G, and the nucleation rate, I, by [33]

$$D = 1.448 \times \left(\frac{G}{I}\right)^{\frac{2}{3}} \tag{2.4}$$

Therefore, any conditions which affect the nucleation or growth rates have a corresponding influence on the grain size. Many methods of grain size control have been studied based on the above theories. Depending on the layer structure of a media, those methods can be classified into the following categories: a) substrate surface modification, b) seedlayer studies, c) underlayer studies.

a) Surface modification of the substrate has not been considered as a method for grain size reduction until recently. Lightly oxidized NiP or NiP-Co plated Al-Mg substrates can be used to prepare smaller Cr grains [34]. Table 2.1 shows that exposing the substrate to oxygen after dry etching helps to reduce the Cr grain size and also narrows its distribution (σ is the standard deviation).

When Co is added to the NiP coating and oxidized, it is believed that Cr (5 nm thick) nucleates much easier at the Ni–O or Co–O sites than on NiP substrate without oxidation. Therefore, the nucleation rate will be high and the grain size will be smaller. Table 2.2 shows that the average grain size and the standard deviation of the distribution can be reduced by this method.

Surface roughness is also one of the factors that affects the nucleation rate by affecting the wetting factor of the critical nucleation energy. Chang et al. [19] have demonstrated that the nano-roughness of the seedlayer will vary the wetting of the Cr nuclei, increasing the nucleation rate of the Cr film, thus, decreasing the grain size of the Cr layer. Using HRTEM, the authors

Table 2.2. Dependence of grain size on Co concentration in Co-NiP coating [34] (© 2000 IEEE

Media	Co (ppm)	Grain size (nm)	σ (nm)
B	0	9.8	3.3
E	120	9.6	3.1
F	400	8.8	2.9
G	1200	8.6	2.7

also found fewer nuclei with size of 2–3 nm on the smooth NiP layer, while the nuclei were 1–2 nm on surface treated NiP layer (with nano-roughness). The smaller nuclei are speculated to be due to the rougher surface and to be located in the valleys of the rough NiP layer. As the Cr layer gets thicker (10 nm), the Cr grains on smoother NiP layers coalesce into larger grains with size of 30 nm, while the Cr grains on the nano-rougher NiP layers maintain their orientation and particle coalescence does not occur. This results in a smaller Cr grain size and a more uniform distribution.

b) Seedlayer is another important factor in controlling the grain size. Matsuda et al. [35] have found that a seedlayer of oxidized CoCrZr not only enhances the CrTi (200) texture but also reduces the grain size. The authors believed that Zr was selectively oxidized. As a result, ZrO nucleation sites can be formed at the seedlayer surface controlling the nucleation rate and growth of the subsequent CrTi layer. Likewise, the use of Ta_2O_5 [36] or NiP [37] seedlayers has also been reported to reduce the media grain size.

c) The underlayer is the layer that has been extensively studied in controlling the grain size of magnetic layer. The gain size of the NiAl underlayers are reported to be smaller than that of the Cr [38]. This is due to the stronger intermetallic bonds between the Ni and Al atoms than between the Cr bonds, which results in smaller nucleation energy Therefore NiAl films have a higher nucleation rate and subsequently smaller grain size [39].

The grain size of the underlayer can also be controlled by it own thickness. However, if the underlayer (CrX) is too thin (less than 10 nm), the film does not have a strong crystallographic texture. This will result in poor orientation and an increase of defect density in the magnetic layer. It is commonly observed that H_c increases with Cr thickness and levels off after a certain thickness [30]. To overcome this problem, ultra clean (UC) processes were adopted to obtain good microstructural and magnetic properties with thinner Cr underlayer thickness and lower $M_r t$ of the magnetic layer [40,41]. The reason for the effect of the UC process is not clear, but removal of oxygen, water vapor and nitrogen in the sputtering chamber is helpful in obtaining better surface mobility of the adatoms and less defects, during both the Cr and Co-alloy deposition.

Choosing alloys with higher melting temperatures, like CrW [42], to reduce T_{sub}/T_m is another practical way of decreasing the grain size. Moreover,

Table 2.3. Dependence of grain size on underlayers [45] (© 1999 IEEE)

Seedlayer	underlayer	⟨d⟩ (nm)	$\sigma/\langle d\rangle$
Ta/Zr	CrTi	15.9	0.31
	CrTiB	10.5	0.27
CoCrZr	CrTi	9.1	0.24
	CrTiB	7.1	0.18

the underlayer composition also affects the nucleation rate and growth kinetics, which eventually affects the grain size. Hosoe et al. [43] reported that CrTiB has a larger grain size (11 nm) than that of CrTi (9.5 nm) when DC magnetron sputter deposited onto NiP-plated Al–Mg substrates.

The same group has shown that a CoCrZr seedlayer has an effective influence on grain size reduction of the CrTi underlayer [44]. By combining the CoCrZr seedlayer and CrTiB underlayer the authors obtained the grain size reduction. Table 2.3 shows that the underlayer grain size is significantly influenced by boron additions and the type of seedlayer. The smallest grain size is 7.1 nm. The table also shows that when the average grain size of the underlayer is reduced, the dispersion of the grain size distribution is reduced as well.

In summary, the grain size of the magnetic Co-alloy thin films can be controlled through modifications of the substrate, the seedlayer and the underlayer. However, there is currently no known method to narrow the grain size distribution other than reducing the average grain size and expecting that the log normal distribution will narrow itself accordingly. To significantly narrow the grain size distribution, regular arrays of nucleation sites must be prepared on the substrate or seedlayer. There are ways of producing mono-dispersed magnetic particles by chemical synthesis [46,47] (also see Chap. 9); however, this method has not yet been implemented in the production of recording media.

2.3 Crystallographic Orientation

2.3.1 Introduction

For materials with the hcp structure a common texture is (00.1) when sputter deposited onto an amorphous substrate, i.e., the (00.1) planes are parallel to the plane of the film. In this case, there is often a mixture of hcp and fcc phases present, which causes poor magnetic properties for magnetic recording [48]. To make the c-axis of Co-alloy grain lie in or close to the film plane for use as longitudinal media, underlayers must be sputtered before the Co-alloy layer. It has been nearly a quarter of century since the discovery of bcc Cr underlayers [49,50]. Due to the similar atomic size of Cr and Co, there are

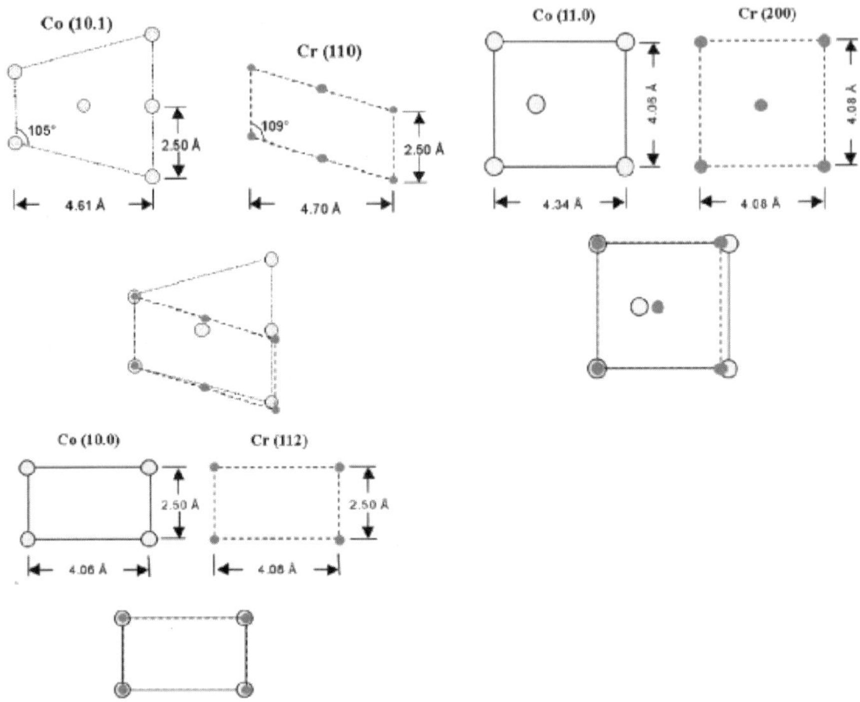

Fig. 2.4. Three common orientation relationships between bcc Cr and hcp Co films

several possible orientation relationships between the bcc Cr grains and the hcp Co grains. Three relationships that have been used in longitudinal media are shown in Fig. 2.4.

It can be seen that Co (10.0) and Cr (112) planes have the best lattice match. Actually the ratio of the two sides of the Cr (112) lattice plane is $\sqrt{8/3}$, which is exactly same as that of the Co (10.0) plane. For this reason, a perfect lattice match can be realized between Co-alloy (10.0) plane and a Cr-alloy (112) plane by varying the composition of the Cr-alloy.

When the Co thin film is oriented in a (10.1) texture, its c-axes are about 28.1° away from the thin film plane, while in the (11.0) and (10.0) texture, the c-axes are in the film plane. Much research has been performed to make single crystal Co-alloy thin films with these three kinds of orientations, not only to demonstrate the lattice matching but also to obtain the intrinsic magnetic properties (such as the magnetocrystalline anisotropy) of Co-alloy thin films. When Ag/Si templates are used, the epitaxial relationships between the substrate and thin films are:

Co(10.1)/Cr(110)/Ag(111)/Si (111) [52]
Co(11.0)/Cr(200)/Ag(200)/Si (200) [53,54]
Co(10.0)/Cr(112)/Ag(110)/Si (110) [55]

When MgO single crystal substrate is used the epitaxial relationships between the substrate and thin films are:

Co-alloy (11.0)/Cr (200)/MgO (200) [56,57]
Co-alloy (10.0)/Cr (112)/MgO (110) [58]

Other single crystal substrates such as, Cr (100) and Cr(110) [59], GaAs (100) [60], LiF and NaCl (100) [61] have also been studied. Reviews of the growth of single crystal Co-alloy thin films and their basic magnetic properties can be found in Ref. [62] and [63].

It is not practical to use single crystalline thin films for longitudinal magnetic recording, because the angular variation of H_c, M_rt and S^* in different circumferential directions will cause a modulation of output signals. Therefore it is better to allow the Co-alloy films to have their c-axes randomly oriented in the film plane. From the results of epitaxial growth of single crystal films, one can see that the availability of the (10.1), (11.0) and (10.0) textured Co-alloy thin films depends on the availability of the Cr (110), Cr (200) and Cr (112) textured underlayers, respectively. Reviews on the control of the texture orientation of polycrystalline Co-alloy thin films for longitudinal magnetic recording can be found in [51] and [64]. A detailed electron diffraction study of polycrystalline Co/Cr bilayers can be found in [65] where some of the less frequently occurring orientation relationships, such as Co (11.0)/Cr (111) and Co (10.0)/Cr (113) were also observed by electron micro-diffraction.

When Cr is deposited onto an amorphous substrate at room temperature, it yields films with (110) texture. Early longitudinal magnetic recording media had Co-alloy grains either randomly oriented [66] or with (10.1) preferred orientation [67]. Consequently the magnetic easy axis (c-axis) of the grains were either 3D randomly oriented or were oriented 28.1° from the film surface. In the latter case, the Co (00.2) texture can also be developed because there is certain degree of lattice matching between the plane Co (00.2) and Cr (110). As a result, there is a good portion of Co-alloy grains oriented with their c-axes perpendicular to the film, which is undesirable for the longitudinal recording. Therefore, the media with Co (10.1) texture were replaced by ones with the Co (11.0) texture.

Several research groups discovered in the late-1980's that films with Cr (200) texture could be grown when the films were deposited at elevated temperatures [68–70]. This texture of the underlayer lead to Co-alloy films with (11.0) texture, as shown by Daval and Randet [71] in 1970. For (11.0) textured Co-alloy films, the c axes are 2D-randomly oriented in the film plane. Moreover, when combined with circumferential mechanical texturing the films can give rise to in-plane circumferential anisotropy [72,73], that renders excellent recording properties [74,75]. However, the roughness introduced by mechanical texturing [76] increases the head to disk spacing. High-density recording requires that the heads fly as close to the media as possible. Hence there is a trade off between obtaining a smooth media surface and obtaining circumferential anisotropy. On the other hand, glass substrates not only have

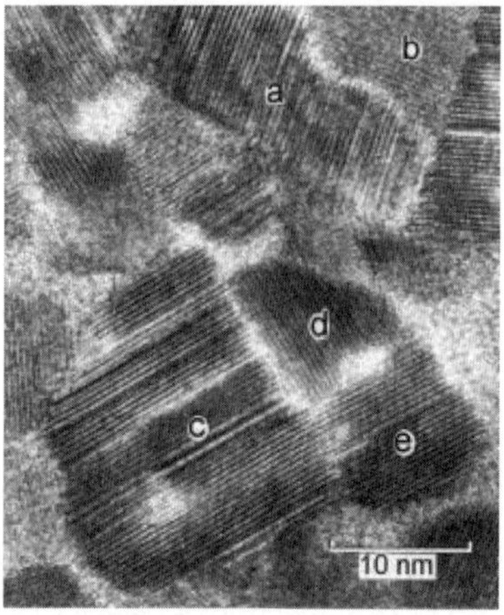

Fig. 2.5. Bicrystals of a CoCrTa film on Cr underlayer [77] (© 1995 IEEE, courtesy of Y. Hosoe)

smoother surfaces but also have mechanical advantages over NiP/Al-Mg substrates. High recording density media on glass substrates was made possible [4] after the introduction of the NiAl (112) underlayer as described below.

A distinctive feature of media with the Co-alloy (11.0) on the Cr (200) texture is its bicrystal structure [77,78]. Since the c-axis of Co-alloys can align with either of the $\langle 110 \rangle$ directions in the Cr (200) planes, the Co-alloy can form bicrystals with c-axis 90° to each other (Fig. 2.5).

The advantage of bicrystal grains is that they refine the grain size of the media and may reduce recording noise [79,80]. The disadvantage is that they are very likely to be exchange coupled with each other. The compositional studies have shown that Cr segregation at the bicrystal grain boundaries is much less than that at the regular grain boundaries [81]. This indicates that Cr segregation at the bicrystal boundaries is not sufficient to magnetically decouple the grains. When two bicrystals are exchange coupled, the diagonal directions between the orthogonal oriented c-axes are effectively easy axes, while either of the c-axes directions becomes effective hard axes [82]. In this case, the anisotropy energy (K_u) of an exchange coupled bicrystal grain is significantly decreased [83]. Obviously, the chances of having several bicrystals nucleate on one underlayer grain decrease with the decrease of underlayer grain size, viz. there will be a small amount of bicrystals in high density recording media with smaller grain size.

Co-alloy (10.0) on Cr (112) are the most desirable textures for longitudinal media. Theoretically it can have a perfect lattice match. It does not produce bicrystals. However, due to its higher Miller indices, the (112) textured Cr films are very hard to obtain. In 1994 Lee et al. [38] reported that NiAl can develop (112) texture when deposited onto amorphous substrates. NiAl has B2 (CsCl) structure, which is a bcc derivative structure. It also has nearly the same lattice parameter as Cr. Therefore, Co (10.0) textured film can be obtained when deposited either directly onto NiAl or on a Cr (112) intermediate layer. The magnetic properties of Co (10.0)/NiAl (112) films are inferior to the bicrystal Co (11.0)/Cr (200) media. However, once a thin Cr intermediate layer is added the magnetic properties improve greatly, exceeding the (11.0) oriented media [84].

It was later found and suggested by Lu et al. that the function of Cr intermediate is two-fold. First it inherits and continues to enhance the (112) texture epitaxial growth [85]. Secondly, it has metallic bonds, which are non-directional, compared to NiAl, whose bond types are a mixture of metallic, ionic and covalent with directional character [86,87]. The Co-alloy has hcp structure, which is different from the cubic B2 structure of NiAl, the directional bonds of NiAl make it difficult for the Co-alloy to epitaxially grow on it. Adding a layer of Cr enhances the epitaxy between the hcp Co alloy and its underlayer [88].

The weaker (112) texture of NiAl causes the Co-alloy (10.0) texture to be weaker than the Co-alloy (11.0). Thus, a portion of the Co-alloy grains are randomly oriented [85]. In this case, fcc regions and stacking faults cannot be controlled by the Cr to Co-alloy epitaxy. As a result, the randomly oriented grains not only have undesirable c-axis orientation but also have lower anisotropy energy. On the other hand, the epitaxially grown grains have perfect lattice match with the underlayers. These grains contain less stacking faults compared with the grains in (11.0) textured Co-alloy media

It should also be noted that in-plane circumferential anisotropy has not yet been achieved in (10.0) oriented media.

2.3.2 Preferred Orientation

Whenever a crystalline material is deposited onto an amorphous substrate, a certain degree of preferred orientation (PO) is obtained (PO is also referred to as a crystallographic texture). The specific texture of the film depends mainly on the crystal structure and substrate-adatom interaction.

The driving force behind the PO is the minimization of the surface (interface) free energy either by developing the lowest energy planes at the film-substrate interface or by growing the lowest energy planes at the surface. The lowest energy planes vary with crystal structure. Based on single crystal growth observations [89], some of the known facets for crystals grown under equilibrium states are: a) (111) planes for fcc Cu, diamond, ZnS and CaF_2 structure; b) (110) planes for bcc metals and B2 (CsCl) structure. The latter can also develop (100) and (112) planes when the crystals grow larger; c) (100) planes for NaCl structure; d) (00.2), (10.0) and (10.1) planes for hcp structure.

Unlike the single crystals, which are grown close to equilibrium conditions with all the surfaces being free, the texture of deposited films is governed by both the surface and interface free energy. Therefore a crystallographic texture can develop either during the nucleation stage before nuclei impingement or during the growth stage after such impingement. Accordingly, the textures can be classified as either nucleation textures or growth textures.

The following are the textures that are usually developed in deposited films: Co-alloy films usually have the (00.2) texture for the hcp phase and the (111) texture for the fcc phase. MgO (NaCl structure) films usually develop the (200) texture [90], Cr (bcc) films usually have the (110) texture when grown at room temperature, and the (200) texture when grown at higher temperatures (250°C) [91–94]. NiAl (CsCl structure) develops either the (110) or the (112) texture [38].

Nucleation textures are formed during the initial stage of film deposition. When the deposition conditions enable the adatoms to have enough mobility on the substrate surface, the nuclei can more or less approach the equilibrium shape before they impinge with each other. The orientation of the nuclei depends on the relative values of the surface energy and interface energy. Such deposition conditions include [89]:

1. The interaction between substrate and adatoms forming the nucleus is weak compared to the interaction amongst nucleus atoms.
2. The supersaturation is small.
3. The application of a substrate bias, which gives the adatoms extra kinematic energy from the bombardment of the sputtering gas and therefore emits loosely bonded atoms from the surface.
4. The sputtering gas pressure is low.
5. The nuclei must be large enough so that their surface can be described in terms of crystallographic planes.

In deposition processes, the supersaturation may be high in the chamber. The Ar pressure may also be high. To reduce the grain size, the substrate surface is sometimes modified to enhance the interaction between the substrate and adatoms. When these conditions are fulfilled, the nuclei are randomly oriented. In the growth stage after the impingement of nuclei, certain grains with lowest surface energy will grow faster than others. The film will be gradually dominated by the grains with this favorable orientation. This kind of PO is called growth texture.

It was found [95] that the Cr (200) texture can be initiated directly on the substrate surface while the Cr (110) texture appears not to nucleate on the substrate surface, but rather to form as a result of growth. Figure 2.6 shows that the PO of Cr films changes from (110) to (200) as the substrate temperature increases. This is because the development of texture changes from being dominated by growth to being dominated by nucleation.

At elevated substrate temperatures, the mobility of atoms is high, the nucleation rate is low, and the nuclei can reach the equilibrium shape before impingement. Because of the relative values of surface and interface free energy, the Cr nuclei tend to take (200) orientations. However, as can be seen

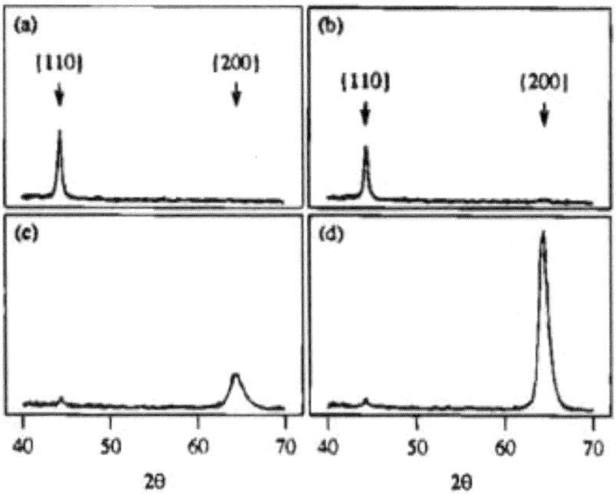

Fig. 2.6. The X-ray diffraction patterns of Cr (175 nm)/Glass films deposited at **a** 45°C, **b** 105°C, **c** 200°C, **d** 260°C [69] (© 1990 IEEE)

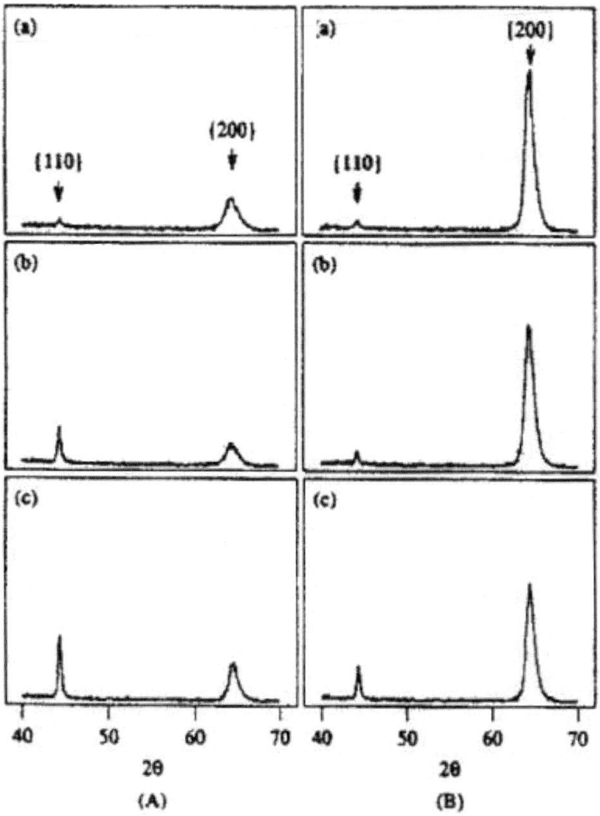

Fig. 2.7. X-ray diffraction patterns of Cr/glass films of **a** 170 nm, **b** 250 nm, **c** 400 nm thickness at A) 200°C B) 260°C [69] (© 1990 IEEE)

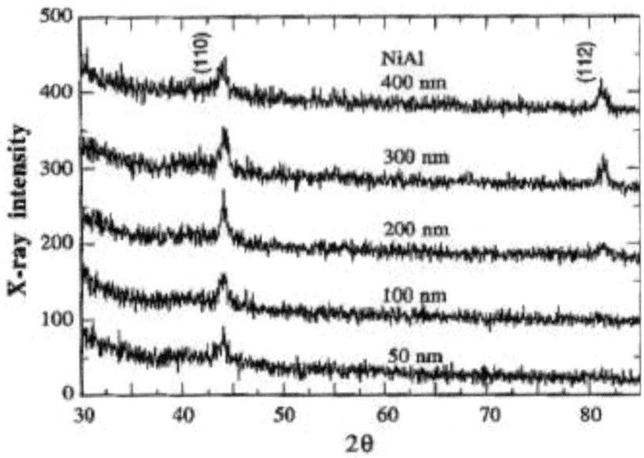

Fig. 2.8. NiAl films at varies thickness [98]

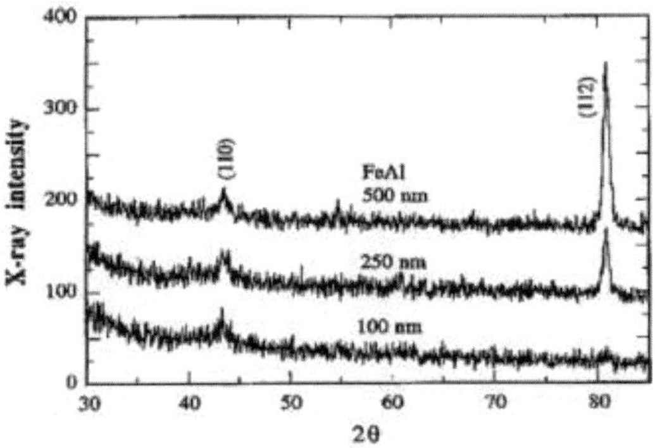

Fig. 2.9. FeAl films at varies thickness [98]

in Figs. 2.6 and 2.7, some of (110) oriented grains remain in the film because the ideal equilibrium conditions are not perfectly realized during deposition. As the film grows thicker there will be a greater volume of (110) oriented grains because of their faster growth. The x-ray diffraction curves in Fig. 2.7 show an increase of (110) peak intensity and a corresponding decrease of the (200) peak intensity as the film thickness increases.

It is interesting to note that materials with the B2 structure [96] develop different textures than those with the bcc structure. NiAl [38] and FeAl [97] tend to have the (112) texture mixed with the (110) texture when they are deposited at room temperature. The (112) texture gets stronger as the film grows thicker, suggesting a growth texture (Figs. 2.8 and 2.9). Figure 2.10 is a TEM cross-section dark-field image [39] of a 400-nm thick NiAl film, which shows that most of the (110) grains are near the film-substrate interface.

Fig. 2.10. (110) dark-field TEM image of a 400 nm thick NiAl film on glass substrate [39] (© 2000 IEEE, courtesy of Yu-nu Hsu)

2.3.3 Degree of Texture

To achieve a high degree of PO of the magnetic layer there are two main considerations. First, how to obtain a good texture of the underlayer; Second, how to obtain epitaxial growth of the magnetic layer on the underlayer.

2.3.3.1 Texture of Underlayer

Many underlayer materials have been studied for the magnetic recording media, such as, β-W [99], Ti [100], Cr, Mo, W [101,102], Ga [103] Ni_xP [31,104], Cr/seedlayer (NiP_3, $TiSi_2$, C, CrSi, etc.) [105,106], Cr/CrNi [107], Cr/pre-coating (Cr, Ta, W, and Zr) [108], NiCuP [109], Cr/amorphous carbon substrate [110], $Cr_{75}Ti_{25}$/CoTi (B2 structure) [111,112], Cr/NiAl [84], etc. Among these materials, CrX [113] and Cr/NiAl [4] are the most frequently used underlayers for longitudinal Co-alloy media.

To obtain a high degree of Cr (200) texture, the adatoms should have a high degree of surface mobility. A preferred way of doing this is through a heated substrate, where the T_{sub} is between 200°C and 300°C, and the Ar pressure is kept low [114]. By monitoring the texture quality using XRD rocking curves, Shan *et al.* [115] found that the degree of texture of both CrRu underlayers and CoCrPtTa magnetic layers improved monotonically as the substrate temperature increased.

Using a Ta seedlayer [108,116] can also improve Cr (200) texture. Ta serves as an excellent wetting layer and provides enough mobility for the Cr adatoms to nucleate in the (200) orientation.

Another way of obtaining Cr (200) textures is to deposit a seedlayer of MgO [90,117]. Because of its NaCl structure, MgO thin films usually obtain the (200) texture when deposited at room temperature. In this approach, substrate preheating is avoided so that grain size of Cr layer and the subsequent Co-alloy layer can be well controlled.

To study the angular distribution of the PO in the layers of recording media, an electron diffraction method has been developed [118–120]. Although this method needs TEM sample preparation, which is destructive to the disk, it has several advantages over the XRD rocking curves. First, the amorphous layer and the Al substrate can be completely removed. These contribute to either a diffuse background intensity (Glass, NiP) or overlapping peaks (e.g. Al (200) overlaps with Cr (110)). Secondly, the electron diffraction technique allows the study of the crystallographic textures layer by layer. This is useful when the XRD peaks of the different layers overlap with each other, for example, MgO (200) and Cr (200), NiAl (112) and Cr (112), etc.

Using this method, Tang et al. [121] analyzed the degree of texture of MgO (200), Ta (110), Co-alloy (11.0)/Cr (200), and Co-alloy (10.1)/Cr (110). The authors found that the texture axis distribution angle α, which is comparable to the full width at half maximum (FWHM) of XRD rocking curve, to be 12°, 10°, 6°, 6°, respectively for these films. Using the same method, Lu et al. [85] analyzed the different layers of media CoCrPtTa/Cr/NiAl. They determined α to be about 14° for the NiAl layer, and 10° for both the Cr and CoCrPtTa layers. Not only did they find that the texture degree of the Co-alloy (10.0) is weaker than that of Co-alloy (11.0) and Co-alloy (10.1), but also that there are randomly oriented Co-alloy grains, which are seldom found in (11.0) oriented media.

A similar media Co-alloy/CrV/NiAl has been investigated by J. Li et al. for the 10 Gbit/in2 areal density demonstration [4]. The results from a XRD pole figure also show a wide angular distribution (20°) of the (112) PO around the film normal. By using a CrV intermediate layer, the CrV (112) PO was more enhanced compared to the film with a Cr intermediate layer. Moreover, the magnetic layer was found to have fewer randomly oriented Co-alloy grains, which is due to the enhancement of the epitaxy through better lattice match.

Parker et al. [122] have found that CrV alloys have the effect of enhancing the (200) texture degree. The V content can be adjusted up to $Cr_{50}V_{50}$ because of the complete solubility of V in Cr. In their work 50 nm thick CrV films were sputter deposited onto NiP/Al-Mg substrates at 25 ± 1°C with 50 V bias. Electron diffraction patterns (EDPs) of the Cr film showed clearly that both (110) and (200) oriented grains exist in the underlayer, while the CrV EDP showed only a strong (200) texture. The XRD experiments in the same paper indicate that the Co-alloy layer has better epitaxy on a CrV underlayer than on a pure Cr underlayer. This brings up the second important consideration for obtaining a high degree of PO of the magnetic layer, namely the epitaxy between the underlayer and the magnetic layer.

2.3.3.2 Epitaxy

The lattice parameters of Co-alloys for magnetic recording are usually larger than those of pure Co because most of the alloying elements (e.g. Pt and Ta) have larger atomic radii than Co. Also, some of the atoms, such as B, are in interstitial sites, which also expand the lattice. It is critical to obtain lattice matching between the magnetic layer and the underlayer in order to obtain good epitaxial growth, which in turn increases the degree of texture, reduces the interfacial stress and stacking fault density. On the other hand, if the lattice match gets too poor for epitaxy, the underlayer will be nonfunctional. For example, when Ta concentration becomes too large in the CoCrPtTa/Cr media many CoCrPtTa grains tend to orient with the (00.2) planes parallel to the film plane [123].

Several CrX alloys have been tried to accomplish good lattice matching. Actually when one looks up the binary phase diagrams and lattice parameters, one finds that there are not many CrX alloys that are suitable for lattice matching with Co-alloys. CrMn, CrMo, CrOs, CrRu, CrSi, CrTc, CrTi, CrV, CrW are possible alloys. It appears that the CrV is the best in terms of solubility. The Cr-V phase diagram (Fig. 2.11) shows complete mutually solubility between V and Cr in the solid state. Since V has the same bcc structure as Cr, the lattice parameter of CrV underlayer can be adjusted continuously between Cr lattice parameter (0.2884 nm) and V lattice parameter (0.3027 nm).

In addition, alloys like CrMo (Fig. 2.12) and CrW whose equilibrium phase diagrams exhibit miscibility gaps may also be used because phase separation is unlikely to happen in deposited films. Moreover, since both W and Mo have large melting point, films of CrW and CrMo will form smaller grains. Most of the above mentioned alloys have been tried as underlayer materials. CrV [122], CrW [125], CrTi [126–128], CrRu [115], CrSi [127], CrMn [129] have all been investigated and have been found to improve the magnetic and recording properties of the media.

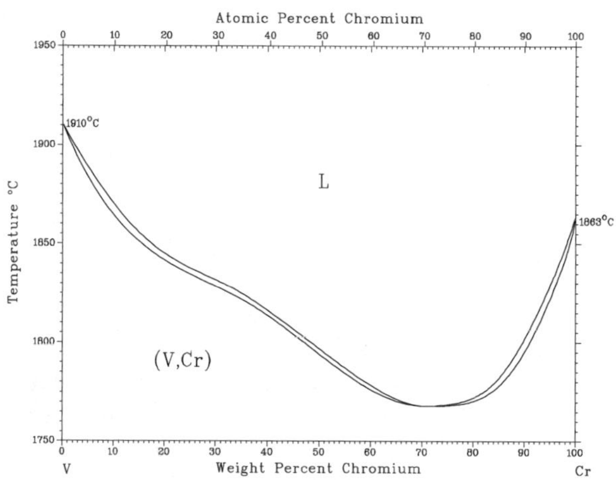

Fig. 2.11. Cr-V phase diagram [124]

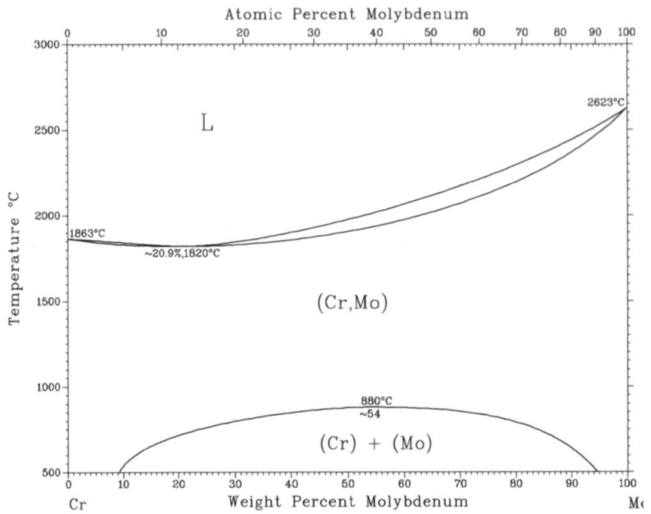

Fig. 2.12. Cr-Mo phase diagram [124]

Vegard's law [130,131] is commonly used in calculating the lattice parameters of CrX alloys. The law postulates a linear dependence of lattice spacings with composition, but does not often hold over large composition ranges for metallic solid solutions. Nevertheless, within small composition ranges a linear extrapolation may be acceptable. Wong and Ying [132] have shown that the lattice parameter of CrV and CrMo alloys can be predicted using Vegurd's law. Lu et al. have shown the lattice parameter of CoCrPtTa with different Pt contents varies linearly within the composition range [133].

Lattice mismatch problems can be further alleviated by introducing an intermediate layer, which can either be a bcc CrX alloy or an hcp Co-alloy [134–138]. The values of the lattice parameters of the intermediate layer are chosen to be between that of underlayer and magnetic layer. Thus the mismatch can be spread across two interfaces. Zhang et al. [139] also found that the media with an hcp non-magnetic intermediate layer, has better thermal stability as well as better recording properties.

In summary, Co-alloy magnetic films can have (10.1), (11.0), and (10.0) crystallographic textures when deposited onto CrX underlayers of (110), (200), (112) textures, respectively. The degree of the Co-alloy texture depends on the quality of the CrX texture and the extent of epitaxial growth. The underlayer texture depends on the CrX composition, the initial texture type, the interaction with the substrate, and processing parameters, while the quality of the epitaxy depends mainly on the chemical bond matching and the lattice matching between the magnetic layer and the underlayer.

2.4 Stacking Faults and fcc Phase

2.4.1 Introduction

The element Co has the hcp structure (space group $P6_3/mmc$) at room temperature and transforms to the fcc structure (space group $Fm\overline{3}m$) at 422°C [124]. This is unlike the elements Ti, Ru, Re, Zr, etc, which remain hcp until they melt. The hcp↔fcc transformation in Co is first order and therefore often occurs with superheating (hcp→fcc) or supercooling (fcc→hcp).

The unit cells and the stacking sequence of hcp and fcc structure are shown in Fig. 2.13. It can be seen that the hcp structure is formed by stacking the close packed planes in the sequence ABABAB . . . , while the fcc structure is formed by stacking the close packed planes in the sequence ABCABCABC . . . , where the letters represent the position of the centers of the atoms of each plane, as shown in the figure. Any error in the order of the stacking

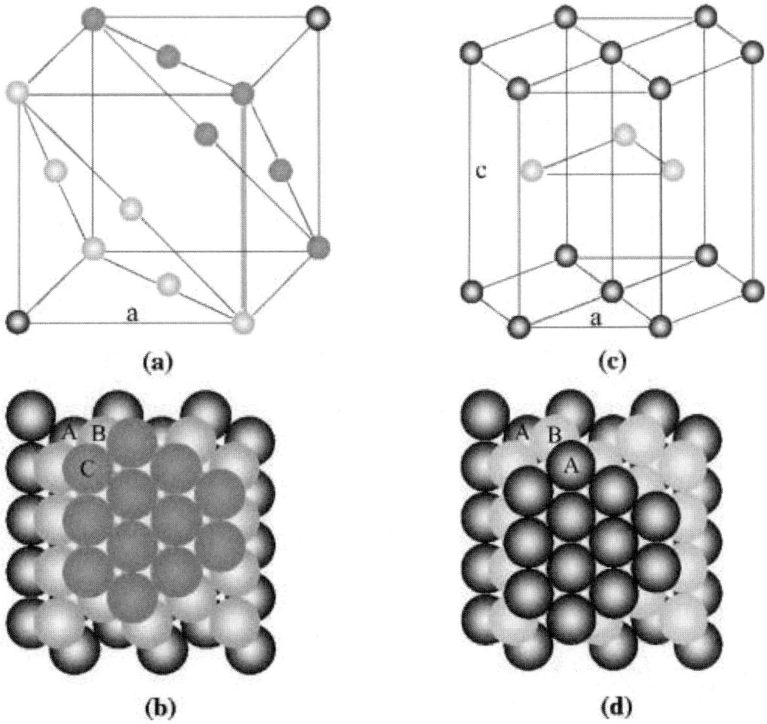

Fig. 2.13. **a** Unit cell of the fcc structure. **b** Stacking of (111) planes. **c** Unit cell of the hcp structure. **d** Stacking of (00.2) planes

of the close packed planes is referred to as a stacking fault. A stacking fault (SF) in an hcp phase produce a thin slab of an fcc phase, for example AB$\underline{\text{ABC}}$BCBCB ... , where the underlined planes represent the fcc slab. SFs exist in all close packed structures though the energy of formation differs from one alloy to another. The SF energy is an intrinsic property of a material. For hcp alloys that transform to fcc at higher temperatures, the SF energy is lower than in those hcp alloys which do not transform to fcc [140].

Stacking faults (SFs) can be classified as either growth faults or deformation faults. A growth fault occurs by incorrect stacking during the growth of the hcp or fcc phase. A deformation fault is caused by a shearing of the close packed plane as a result of resolved shear stress. Such a stress could be due to a lattice mismatch at the interface or by the difference in thermal expansion coefficient between the layers. The SF density is closely related to the process temperature: the closer to the transformation temperature, the smaller is the energy difference between hcp and fcc and the greater the number of faults [141,142].

In Co alloys the transformation temperature, $T_{hcp-fcc}$ and $T_{\mathrm{fcc-hcp}}$, varies with composition. Although it is difficult to study the phase transformation below 800°C due to slow diffusion in the bulk alloy, some results [124] suggest the transformation temperatures increase with the addition of Ir (up to 30%), Os, Re, Rh (up to 20%), Ru, Ca, Mo, Si, and decrease with the addition of Al, Mn, Ni, Pd. This implies that the amount of fcc regions and SF density may vary with Co-alloy composition. Since the process of sputter deposition is far from equilibrium, the normal substrate temperature (T_{sub}) of 250°C in producing of longitudinal media is well within the range of the fcc↔hcp phase transformation temperatures. Therefore a large amount of SFs and a large quantity of fcc grains are expected if the Co-alloy compositions and process parameters are not carefully selected.

Fcc grains are detrimental to magnetic recording because the Co fcc phase has a small magnetocrystalline anisotropy energy. Investigation of fcc grains in Co-alloy media can be dated back to 1982, when Coughlin et al. found that sputtering gas contamination might induce the formation of fcc grains in perpendicular CoCr thin films [143]. The existence of stacking faults in recording media was also discussed in the late 1980s [144]. SFs normally give rise to striation contrast in the TEM image of the Co-alloy grains. However, unlike the fcc phase, the effect of stacking faults on magnetic and recording properties has not been fully understood until recently. This is because of the complicated convolution of the effect of SF with other effects, such as, composition, grain size, chemical segregation, stress, etc. Next sections elucidate the effect of both fcc grains and SFs on the magnetic and recording properties of Co-alloy longitudinal media.

2.4.2 Fcc Grains

The magnetic anisotropy energy of Co fcc phase is about one order of magnitude lower than that of the Co hcp phase [145]. Consequently, the presence of fcc grains can significantly lower the H_c of magnetic recording media and cause thermal instability problems. These problems get more severe in the region of the recording transitions, where the grains experience strong demagnetizing fields. A micromagnetic simulation confirms that percolation within the written bits happens through the low coercivity grains. Figure 2.14 shows a map of such a transition. It can be seen that the low coercivity grains (circles) cause the transition to be rougher, generating more transition noise [146,147].

Hcp and fcc Co-alloys have many planes with similar d-spacings. There are however a few unique ones, such as, $(200)_{\text{fcc}}$, $(10.0)_{\text{hcp}}$, $(10.1)_{\text{hcp}}$ and $(10.2)_{\text{hcp}}$. Therefore the (200) reflection is used in x-ray diffraction (XRD) or electron diffraction (ED) to estimate the amount of fcc grains inside hcp Co-alloy thin films [148,149]. High-resolution transmission electron microscopy (HRTEM) can also distinguish a fcc grain from the hcp matrix by directly imaging the atomic stacking sequence of the close packed planes [150].

Many factors cause the existence of fcc grains in the magnetic layer. Processing temperature is one of the most important factors. When a CoCrTa film is deposited onto a substrate at temperatures greater than 250°C, the H_c of the magnetic layer decreases because more fcc grains are formed [151]. This indicates that at high temperatures the hcp phase transforms to fcc phase and

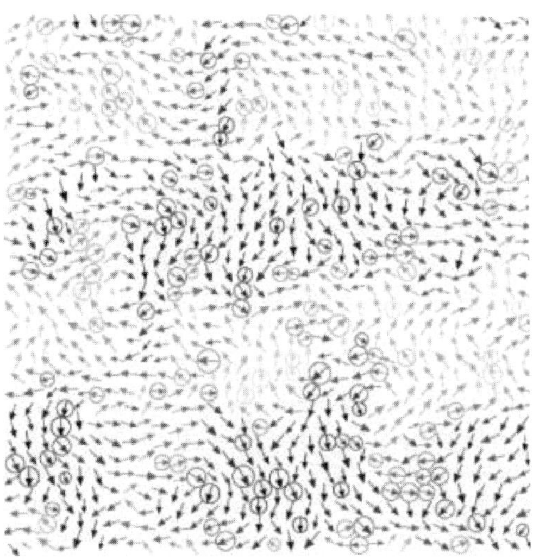

Fig. 2.14. A map of micromagnetic simulation at the transition showing the percolation happens where there are a large number of low coercivity grains [146] (© 1998 IEEE, courtesy of R. W. Chantrell)

that a substantial amount of fcc phase can be retained upon cooling to room temperature. The temperature effect gives rise to a temperature dilemma: the magnetic layer needs a higher temperature to obtain better Cr segregation to exchange de-couple the magnetic grains, but such high temperatures may increase the amount of fcc phase in the magnetic layer.

A second factor causing fcc grains to form is gas contamination. This has first been shown in studies of perpendicular media [143], where introducing nitrogen gas into the sputtering chamber dramatically increased the appearance of the fcc grains thereby lowering the H_c. An XRD scan showed that the (200) peak of fcc increased significantly while the (00.2) peak was reduced and was replaced by (10.0) and (10.1) peaks. This result shows that not only does the nitrogen induce the formation of the fcc phase but it also makes the grains orient more randomly.

Studies on the impurity gas effects in *longitudinal* media [152-154] also show that introduction of oxygen results in a drastic reduction of the coercivity of the films. The coercivity reduction due to nitrogen is moderate and the effect of air is intermediate between the effects of O_2 and N_2. A close investigation on the nitrogen gas effect reveals that magnetic and recording properties, such as, H_c, S^*, SNR, deteriorate significantly for the $Co_{72}Cr_{16}Pt_{12}$ alloy as the base pressure (the chamber was initially flushed with N_2) is increased from 10^{-8} to 10^{-4} Torr [149]. The authors confirmed by plan-view electron diffraction that a large amount of fcc grains were present when the base pressure was higher than 10^{-6} Torr.

The alloy composition is also a very important factor in controlling the amount of fcc phase in Co-alloy thin films. From Fig. 2.15 it can be seen that the fcc phase becomes stable at lower temperature as the Ni content increase. Ni is an example of an fcc stabilizing element, as is Pt.

Pt addition causes another dilemma in searching for the ideal composition for magnetic recording alloy. Pt has been found to increase H_c by enhancing the magnetocrystalline anisotropy of CoCr-alloy [155–157]. However, adding more Pt gives rise to a greater "fcc tendency" [158] of the alloy. Consequently, greater amounts of fcc grains, or fcc regions bounded by SFs, exist in high Pt CoCrPt-alloys. An H_c-Pt composition curve of CoCrPt alloy present a peak around 18% of Pt [156], though it is more appropriate to plot K_u versus Pt content.

The manifestation of the fcc tendency can also be found in the comparison of $Co_{84}Cr_{14}Ta_2$, $Co_{76}Cr_{12}Pt_{12}$, and $Co_{75}Ni_7Cr_6Pt_{12}$ thin films [159]. In [159] the authors reasoned that since in the $Co_{84}Cr_{14}Ta_2$ alloy the (00.2) lattice spacing matches almost exactly with the (110) spacing of Cr underlayer and that a 2% mismatch is found between $Co_{75}Ni_7Cr_6Pt_{12}$ and Cr, it is expected that more SFs and grains of fcc phase should be observed in the films of the latter composition. Consistently, $Co_{76}Cr_{12}Pt_{12}$, whose lattice match with Cr is in between the other two, shows an intermediate behavior. However, the phenomena can also be understood in terms of the greater amount of

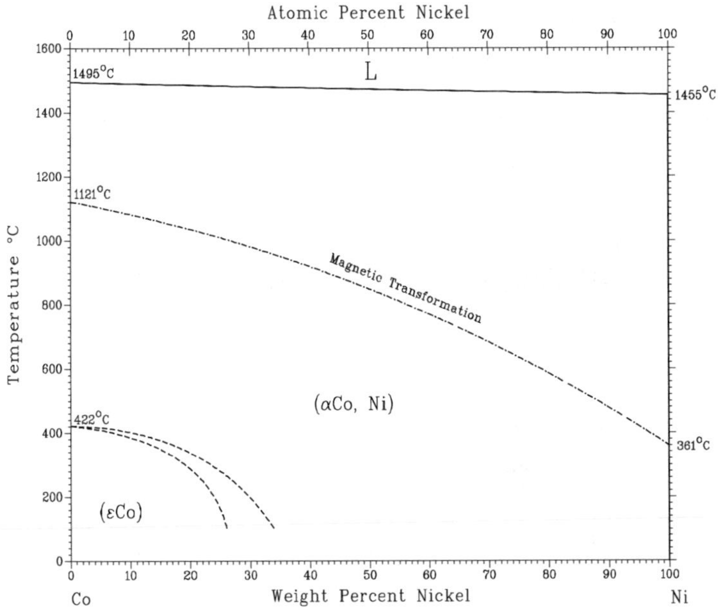

Fig. 2.15. Co-Ni phase diagram [124]

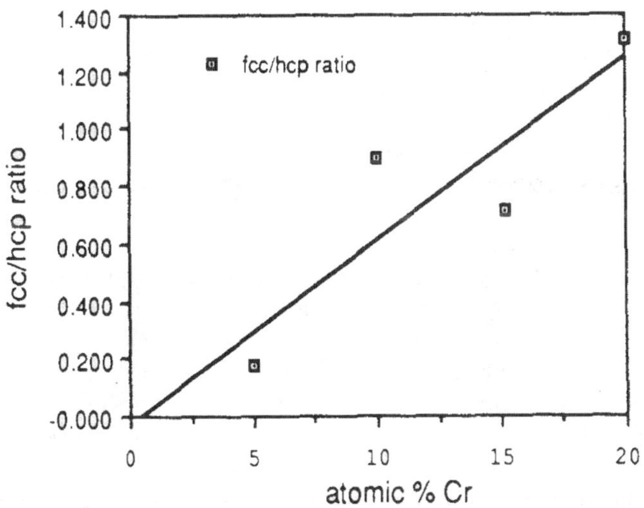

Fig. 2.16. Content of fcc phase in CoCr bulk alloys (reprinted from [160], © 1989, with permission from Elsevier Science)

fcc stabilizing elements (Ni and Pt) in the CoNiCrPt alloy. Since there is such a large difference in the composition of the three alloys, the effect of composition on H_k, M_s and K_u of the Co-alloy should not be over looked.

The phase equilibria in the Co rich region of the Co–Cr phase diagram below 700°C have not been firmly established. This is due to sluggish diffusion in the temperature range. However, the work of Chan et al. [160] carefully studied the *bulk* Co-Cr alloy from 1.44% to 20% Cr. They find that the amount of fcc phase increases systematically with an increase in the Cr content (Fig. 2.16). Moreover, the SF density in the bulk materials was never found to be zero, no matter how the samples were heat-treated.

2.4.3 Stacking Faults

Planar defects which appear as striations in TEM images have been observed for many years in plan-view TEM images of Co-alloy thin films [144,161-163]. These striations were determined to be stacking faults by electron nanodiffraction [164] and high-resolution TEM [20]. It was found that SFs exist all through the thickness of Co-alloy film [20], indicating that SFs occur even in the nucleation stage of the film growth.

Unlike the fcc phase, stacking faults show almost no trace in a regular $\theta - 2\theta$ XRD scans of Co-alloy media. According to Warren [165], SFs in a hcp system can cause streaking and hence peak broadening in all the diffraction spots along c-axis direction except the ones whose Miller indices satisfy $h - k = 3n$, $(n = 0, 1, 2, 3 \ldots)$. Accordingly, among the low index reflections, (10.0), (10.1), (10.2), (10.3) will have streaks, (00.2), (11.0) will not have streaks.

There are several ways to detect SFs. XRD is the most quantitative way. However, in longitudinal media, the c-axis directions are in the film plane. As a result, all the peak broadening from the SF lies in the film plane. These broadenings cannot be observed by normal $\theta - 2\theta$ XRD scans, but an in-plane $\theta - 2\theta$ scan using glazing incident beam does reveal them [148]. Synchrotron radiation has to be used to obtain greater signal to noise ratios in such studies. Quantitative information of growth fault and deformation fault probabilities can be obtained by curve fitting the XRD spectra. Figure 2.17 shows such XRD scans of two CoCrPt media grown at 10^{-6} and 10^{-5} Torr base pressures [149]. The fits to the experimental data are shown as solid lines. By curve fitting, the SF probabilities and total percentage of fcc-like region can be obtained. It was determined that the higher base pressure samples have more SFs and total fcc-like environment.

SF can also be detected by TEM imaging techniques, such as, diffraction contrast imaging [39,166] or high resolution imaging [166-168]. Since the atomic arrangement is clearly seen in HREM images, characteristics of the SFs can be easily determined [167]. By measuring the percentage of the faulted area of a high resolution image (Fig. 2.18), the SF density of that particular grain can be determined accurately. The value may be used to estimate the SF density of the whole sample.

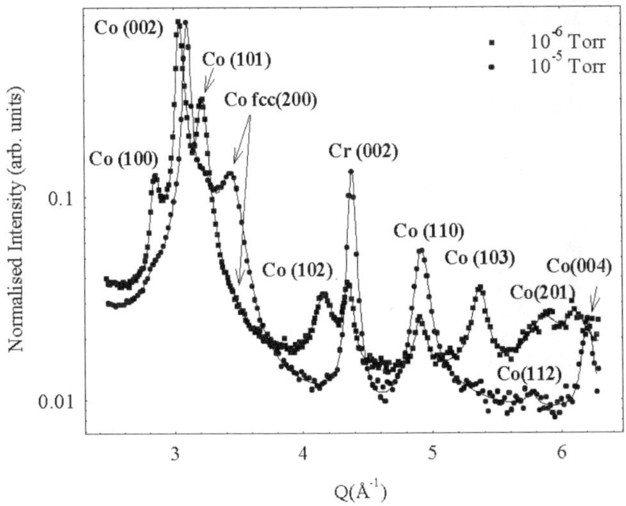

Fig. 2.17. In-plane $\theta - 2\theta$ scans of two CoCrPt media grown at 10^{-6} and 10^{-5} Torr. The fits to the experimental data are shown as solid lines [148] (courtesy of Helen Laidler)

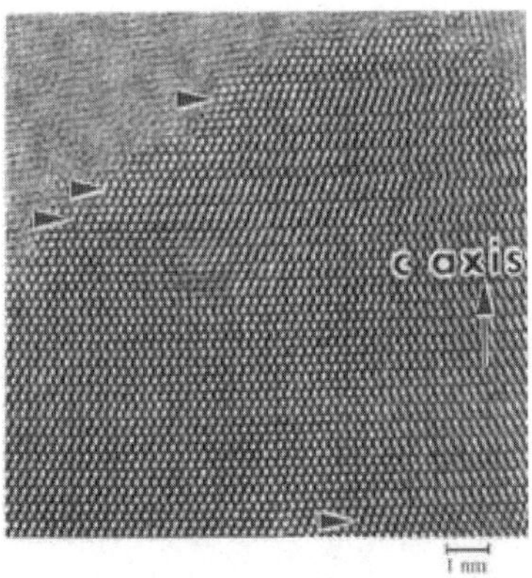

Fig. 2.18. High resolution TEM image showing SFs in a Co-alloy grain [168] (© 1996 IEEE)

Electron diffraction can also be used to study the SFs. Simply by taking plan-view electron diffraction patterns, one can determine qualitatively the SF densities [169]. Figure 2.19 shows an electron diffraction pattern (EDP) of a (11.0) oriented Co-alloy media. The streaks from the (10.1), (10.2), and (10.3) reflections can be clearly observed. However, it should be noted when the film texture is (10.0), only rings like (00.2), (11.0), with $h - k = 3n$ are present in a plan-view image or EDP. Hence, there will not be either any streaks in the EDP or SF contrast in bright field or high-resolution

Fig. 2.19. Electron diffraction pattern of a CoCrPtTa film in (11.0) texture. The streaks are due to the SFs [169] (© 2000 IEEE)

images. This problem can be solved by tilting the TEM specimen to a higher angle [169,170]. A similar problem exists in the XRD setup. Sample tilting or asymmetric XRD scans should be used to measure the SF density in (10.0) textured Co-alloy thin films.

There are several causes for SFs in a Co-alloy thin film. Since the atomic arrangement inside a SF region is same as that of fcc phase, all the above mentioned causes of the fcc grains are also applicable for SFs. For example, it has been found that high substrate temperature promotes the formation of SFs [171]. It has been found that the SF density increases with substrate temperature up to 260°C [167].

Fcc stabilizing elements, such as, Pt or Ni will also increase the SF density. The commonly sputtered Co films have a c/a ratio of less than the ideal close packing value of 1.633. The addition of the large Pt atoms causes the c/a ratio to approach that of the ideal value, which in turn makes it energetically easier for the crystallites to form with either an hcp stacking or fcc stacking.

Another important cause of SFs is stress. Stress may arise from the difference in the thermal expansion coefficients or lattice mismatch between the layers [159]. As determined by XRD [172] most of the SFs in the longitudinal media are deformation faults caused by stress. It has been demonstrated that $Co_{64}Cr_{22}Pt_{14}$ on Cr has a higher SF density than that of the media on $Cr_{82}Ta_{18}$. The latter film has better lattice match than the former. Moreover, the two samples have the same probability of producing growth faults; only the probability of deformation faults of the former is higher that the later. This strongly suggests that lattice mismatch between the magnetic and the underlayer plays an important role in controlling the stacking fault density. Stress as an cause for SFs is also supported by high resolution TEM observations [173]. By comparing CoCrTa/Cr and CoCrPt/Cr thin films, it was found by high resolution TEM that increasing the Pt or Ta content increases the SF density. Since Pt and Ta have much larger atomic size than Co or Cr, the increase of Pt or Ta causes an expansion of the Co-alloy lattices. These results indicate that internal stress at Co-alloy/Cr interface causes SFs.

Both composition and lattice match affects the SF density, which in turn affects the magnetic properties of the Co-alloy media. However, composition directly affects the M_s and $K_u$of the film, while lattice match strongly affects epitaxy and the spread of texture orientation of the magnetic layer, which are both closely related to the magnetic properties [57,122]. Therefore it is very difficult to discriminate the real effect of SFs on the magnetic and recording properties of Co-alloy media. Hence, it is not surprising that many research findings in the past contradict with each other.

For example, several groups [167,174,175] observed H_c increases with SF densities and improves media noise by promoting elemental segregation in the site of SFs due to Suzuki effect [176,177]. Later work did not find that SFs contribute to the improvement of magnetic performance [150,169]. The earlier findings may be due to multi-domain phenomena in the media, where Co-alloy grains could be as large as 50 nm. There is also a contradiction in the trend of SF density varying with substrate temperature [167,175]. This reflects the difficulty in estimating the SF density in Co-alloy thin films.

Recently the effect of SFs on magnetic properties and recording properties has been convincingly demonstrated [148,172,178]. The high SF density samples were shown to have more significant magnetic viscosity effects, which may cause significant thermal instability problems.

Moreover, research on the effect of the SF density in single crystal Co-alloy films has shown a strong correlation between SF density and magnetocrystalline anisotropy constant [170]. It was found that SF density negatively affects the anisotropy and the anisotropy temperature dependence.

2.5 Composition and Segregation

2.5.1 Introduction

As the areal density in longitudinal recording approaches 100 Gbits/in^2, there is a need for significant improvement of the intrinsic and extrinsic magnetic properties of the Co-alloys to achieve shorter transition lengths, low media noise, and higher thermally stability of thin film media. To make the recorded information thermally stable, a high value of the ratio of K_uV/k_BT is required [3,179,180]. Since V, the volume of magnetic grains, decreases with higher areal density, greater K_u materials are necessary for increasing the recording density.

Moreover, large H_c and low M_s values are essential for higher linear density and lower media noise [181]. It is also necessary to obtain intergranular magnetic separation to lower the exchange interaction [182]. The recent modeling of 100 Gbits/in^2 areal density [180] used media parameters of $K_u = 2.5\times10^6$ erg/cc, M_s = 300 emu/cc, while a recent laboratory demonstration of 30 Gbits/in^2 media used Co-alloy media with M_s of 320 emu/cc, K_u over 2.2×10^6 erg/cc and a remnant coercivity (H_{cr}) of 4900 Oe [183]. Obviously, more stringent control of the Co-alloy composition is necessary to achieve superior magnetic properties for higher density media. On the other hand, since element segregation is the best way of magnetically de-coupling the grains [184,185] finding the right composition to enhance chemical segregation at the grain boundary is another mission for alloy improvement. This section covers both the effect of composition on the intrinsic magnetic properties and on within Co-alloy media.

2.5.2 Alloy Composition

Pure Co has a $M_s = 1400$ emu/cc, $K_u = 4.5 \times 10^6$ erg/cc, and a Curie temperature (T_c) of 1404 K. Many elements have been alloyed with Co in order to decrease M_s while keeping K_u relatively high. Some of the alloy investigated for thin film media include: CoP [186], CoPt [187], CoNi [188], CoNiCr [189,190], CoCrTa [191], CoCrPr [192,193], CoCrPt [28,194,195], CoNiPt and CoNiPd [196]. High M_s values used to be needed to obtain high signal output for the reading was done by an inductive ring head. With the introduction of GMR heads (see Chap. 10) large M_s values are no longer necessary, which not only reduces magnetostatic coupling at the transitions by decreasing the $4\pi M_s/H_k$ value, but also inspires numerous studies of greater variety [197–200] of the Co-alloy with larger number of elements [201].

Throughout the 1990s, CoCrTa, CoCrPt and CoCrPtTa were the most popular alloys used by the hard disk industry. CoCrTa has very good noise properties, CoCrPt has high K_u, H_k and H_c, while CoCrPtTa is a compromise between the former two alloys. A large research effort has been devoted to understand the compositional dependence of the magnetic and recording properties as well as the microstructures of the thin films of Co–Cr–Pt–Ta systems.

The study of CoCrPtB alloys began in the early 1990s [202–205]. The coercivity of the CoCrPtB film was found to be very large and the noise level was low compared to other alloys of the same period of time. Therefore CoCrPtB has drawn increasing attention. Recently, CoCrPtB was also found to have a narrower grain size distribution and better intergranular de-coupling than the CoCrPtTa thin film [16,206]. Therefore, CoCrPtB is now being widely used at the beginning of the 21st century.

Research on the effect of Cr, Ta, and Pt on the magnetic properties of CoCrPtTa thin films has been extensively performed leading to a comprehensive understanding of the ternary/quaternary alloys. In 1993 Doerner et al. reported that increasing the Cr concentration in CoCrPt alloys results in an improved signal to noise ratio (SNR) [207]. The H_c of the CoCrPt film increased with higher Cr concentration. This is because of the Cr segregated toward grain boundaries causing better exchange de-coupling between the grains. A nearly linear decrease of M_s with increased Cr concentration has

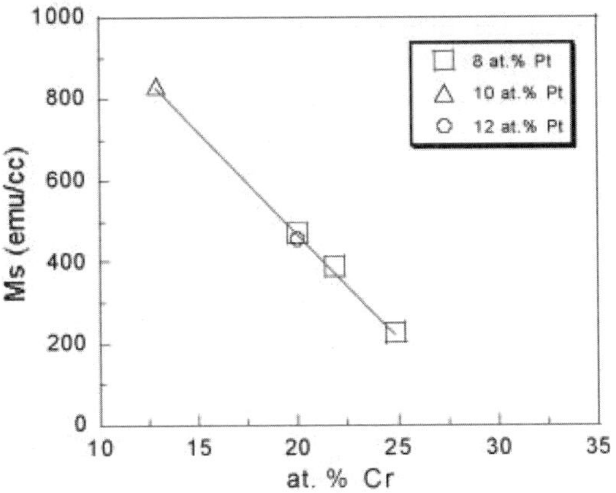

Fig. 2.20. M_s decrease in CoCrPt alloy due to Cr additions [207] (© 1993 IEEE)

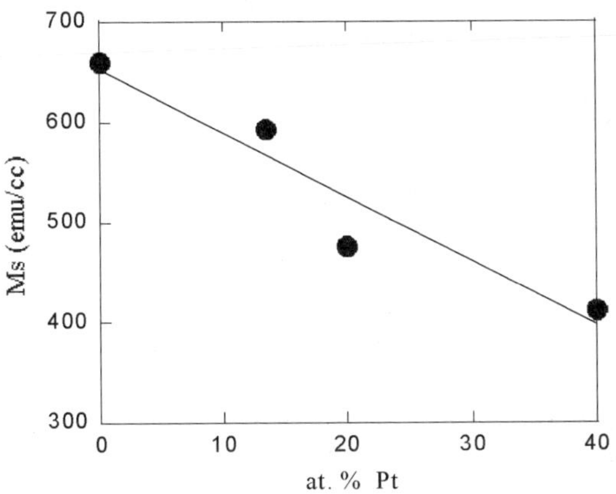

Fig. 2.21. M_s decrease in CoCrPt alloy due to Pt additions [156]

been demonstrated (Fig. 2.20). Doerner et al. also reported that an increase of Pt concentration has little effect on M_s, T_c and SNR; however, the H_c increased effectively with higher Pt content. This is believed due to the large K_u of the alloy.

The effect of Pt was further investigated in the low concentration range ($(Co_{86}Cr_{12}Ta_2)_{100-x}Pt_x$, x = 0–6%) [155] and high concentration range ($(Co_{83.6}Cr_{16.2})_{100-x}Pt_x$, $x = 0, 13, 20, 40\%$)) [156]. It was found that adding Pt causes a linear M_s decrease (Fig. 2.21). The M_s decrease caused by Pt addition is nearly 8 times less than that caused by the Cr addition. The maximum H_c exists for Pt $< 20\%$. The decrease of H_c with higher Pt additions is believed to be caused by a decrease in the epitaxy at the CoCrPt and Cr interface, and also by the presence of fcc phase.

There are two main difficulties in the measurement of the intrinsic magnetic properties of Co-alloys from polycrystalline thin films. First, there is significant elemental segregation both within the grains [208] and to the grain boundaries [209]. It is very difficult to relate the measured magnetic properties to the nominal composition. For example, in the case of a $Co_{79}Cr_{16}Ta_6$ thin film, the maximum Cr concentration at the grain boundary is determined to be 23%, while inside the grains, the Cr concentration fluctuates between 8 and 11% [210]. In fact, the intrinsic magnetic properties of the polycrystalline films are dominated by the core magnetic grains because the grain boundaries are thought to be non-magnetic due to high Cr concentration. To further complicate matters, the degree of chemical segregation varies with the deposition conditions. Thus, different research groups often report incongruent results on same CoCr-alloy compositions.

The second difficulty comes from the small granular structure and the 2-d random orientation of the polycrystalline film. It is difficult to measure the K_u of such films accurately, since the precise demagnetization energy of the film cannot be estimated. Therefore, it is much better to determine compositional dependence of basic magnetic properties of CoCr-alloy using single crystal thin films [55,58,157,201,211–214]. The effect of Cr and Pt composition on the magnetic properties of CoCrPt, CoCrTa, and CoCrPtTa single crystalline films has been studied by Inaba *et al.* [58,201]. The Co-alloy thin films were deposited onto single crystal MgO (110) substrates with Cr or CrTi as underlayers to obtain a good lattice match. As a result, single crystal films with (10.0) orientation were obtained. Three series of Co-alloys which have been studied are: $Co_{96-x}Cr_xTa_4$ (x = 8–19%), $Co_{88-y}Cr_yPt_{12}$ (y = 15, 19%) and $(Co_{73}Cr_{15}Pt_{12})_{100-z}Ta_z$ (z = 0–5%).

It has been found that the absolute magnetization (M_{s0}) and Curie temperature (T_c) values decrease linearly (Fig. 2.22) with an increase Cr content in the $Co_{96-x}Cr_xTa_4$ and $Co_{88-y}Cr_yPt_{12}$ systems and vanish at Cr concentrations of 33%. The magnetization at 285 K vanishes when the Cr concentration reaches 27%. CoCrPt has a higher T_c and K_u than the CoCrTa films. For example, K_u of the $Co_{88}Cr_8Ta_4$ at 290 K is 2.2×10^6 erg/cc, while K_u of $Co_{73}Cr_{15}Pt_{12}$ is about 3.7×10^6 erg/cc. Both M_s and K_u decrease with Ta additions, however as shown by Fig. 2.23, H_k remains constant up to 4% Ta. Beyond this concentration, K_u and H_k values decrease abruptly. This clearly demonstrates that excess Ta additions to the CoCrPt system degrades the magnetic properties. As discussed above, the magnetic properties of polycrystalline films depend strongly on the Co-rich core grains due to the Cr segregation at the grain boundaries. It was shown that the temperature dependence of M_s and K_u for the $Co_{81}Cr_{15}Ta_4$ polycrystalline thin film is more like that of $Co_{88}Cr_8Ta_4$ single crystal film than that of $Co_{81}Cr_{15}Ta_4$ single crystal film [201].

M_s and H_c of CoCrPtB thin film with B addition up to 15%, Pt up to 20%, Cr up to 20% have been studied [202,203]. The dependence of M_s on Cr

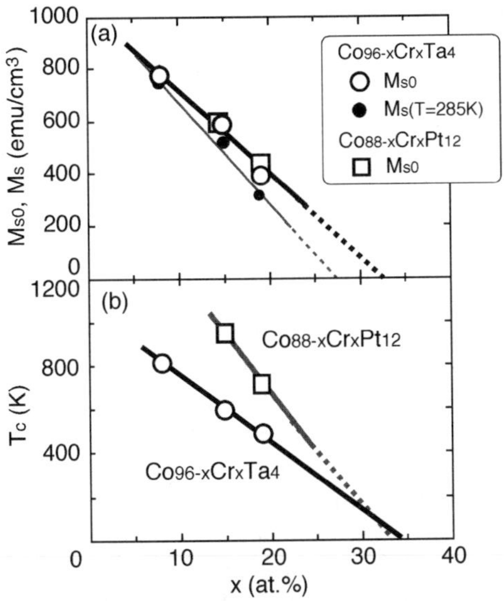

Fig. 2.22. M_s and T_c variation with Cr concentration [201] (© 2000 IEEE, courtesy of N. Inaba)

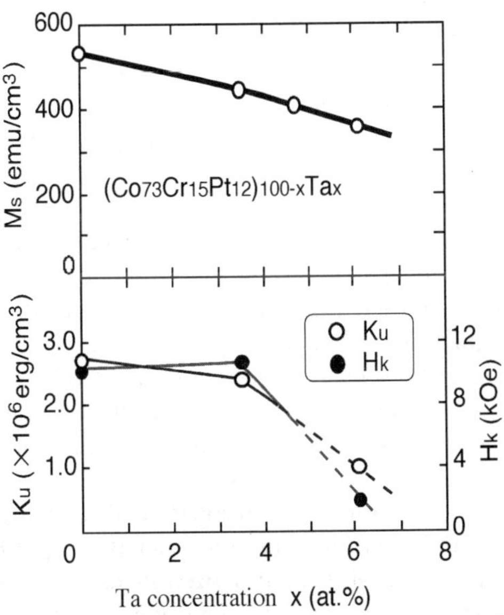

Fig. 2.23. M_s and K_u variation with Ta concentration [201] (© 2000 IEEE, courtesy of N. Inaba)

and Pt is similar to the case of CoCrPtTa media. M_s decreases linearly with B additions. A recent study [215] showed that the M_s decrease rate caused by adding B is similar to that caused by adding Pt. Moreover, it has also been showed that the H_k increase rate due to B addition is similar to that due to Pt addition. Magnetic force microscopy (MFM) studies [205] of the micro-magnetic structure of CoCrPtB/Cr thin films have shown that CoCrPtB/Cr films have a smaller magnetic volume and lower magnetic roughness than CoCrPt/Cr media.

2.5.3 Chemical Segregation

Experimental and modeling work have shown that exchange coupling between the grains of magnetic thin films is largely responsible for the media noise [216,217]. This intergranular coupling causes the formation of magnetic clusters, which are much larger than the individual crystalline grains. Inside the magnetic clusters, the grains are crystallographically randomly oriented but all have the same magnetization direction. Consequently, the transition region between two opposite recorded bits has a zigzag like profile instead of a sharp continuous line [71,216,218–220]. This zigzag transition makes the transition center fluctuate randomly from transition to transition resulting in a timing jitter-induced noise. A detailed review of the relationship between media noise and intergranular coupling can be found in Ref. [221]. The key to making low-noise media is to minimize strong exchange interaction between the magnetic grains. Since the exchange coupling acts over short distance and the interaction decreases exponentially with the separation, magnetic separations as little as 1 nm might be sufficient to de-couple the neighboring grains.

There are generally three ways to de-couple the magnetic grains. One is to physically separate the grains, which can be achieved by using low mobility conditions during deposition of the underlayer and (or) magnetic layer. According to Thornton's model [222] discussed in Sect. 2.2.3, voided films can be obtained under high Ar pressure and low substrate temperature with no bias [26–29,185]. Growing thicker columnar underlayers is also found to be effective in reducing media noise [30,31]. Physical separation can also be achieved by using extremely low melting point alloys for the underlayer materials [103]. The physical separation method is not utilized for higher density recording media, due to its undesirable crystallographic texture, poor crystallinity, poor corrosion resistance and large surface roughness.

A second method of decoupling grains is to have non-magnetic phases, such as oxides, in the grain boundaries. In the past O_2 has been shown to reduce media noise by forming oxidized grain boundaries [223]. CoO or SiO_2 have been deposited together with Co-alloys resulting in a complete separation of the magnetic grains by the nonmagnetic materials [66,195,224,225]. However, films made by this kind of deposition tend to have a poor degree of crystallographic texture, poor corrosion resistance and a larger amount of the fcc phase.

The third way to achieve intergranular decoupling is through chemical segregation at the grain boundary. Processing conditions required for composition segregation in thin magnetic films are opposite to those used for physical segregation. High substrate temperatures, low sputter pressures, and bias, are used in combination to drive the segregation of one or more elements to the grain boundaries. These deposition conditions are congruent with what is required for other microstructural features, such as, epitaxial growth, control of stacking fault density and crystallographic texture. Therefore this method is considered the best way to magnetically separate the Co-alloy grains.

The phenomena of chemical segregation in Co-alloy thin films was reported as early as the 1960s, when thick grain boundaries were observed in CoP and CoNiP thin films [186,226–229]. The thickness of the separation was about 4 nm. The grain boundaries were believed to be composed of phosphorous, which was later confirmed by X-ray energy dispersive analysis (EDAX) in a transmission electron microscope (TEM) investigation [230].

A large number of papers have been written on the studies of the chrysanthemum pattern (CP) phenomena [231]. Since this phenomena is mainly observed in very thick perpendicular films we will not discuss it in this review.

Precise determination of chemical segregation in CoCrPtTa thin films has been shown using Nano-probe EDAX and EFTEM since 1993 [184, 232–234]. These two methods have a spatial resolution of about 1–2 nm, which is about the same dimension as the width of the grain boundary [159]. The former can provide precise quantitative results while the latter also presents 2-D elemental mapping of the films.

Cr Segregation. It has been unambiguously found that Cr segregates to the grain boundaries in CoCrTa, CoCrPt and CoCrPtTa thin films. Figure 2.24 shows plan-view EFTEM images of $Co_{84}Cr_{12}Ta_4$/Cr media deposited at T_{sub}=250°C. It can be clearly seen from the Cr composition maps that Cr is depleted within the grains (dark contrast) and that Cr is enriched in the grain boundary regions (bright contrast). Similar results were found in a cross-sectional TEM specimen. Results [81] from Nano-Probe EDAX from 29 grain boundaries showed that the normal grain boundaries have a Cr content from 19 to 34% (mean 23% Cr), while bicrystal grain boundaries have a lower level of Cr (mean 17% Cr). The Cr composition in the Cr depleted regions was found to be less than 7%. The full width at half maximum (FWHM) of the composition profile across the grain boundaries is about 4 nm. More recently, Cr segregation has been demonstrated by Nano-Probe EDAX [210] that the thickness of the Cr-enriched shell is about 1nm in $Co_{81}Cr_{15}Ta_4$ alloy. This thickness is close to that observed by HRTEM [159], which shows $Co_{84}Cr_{14}Ta_2$ film has a 0.5 to 1 nm amorphous like grain boundaries.

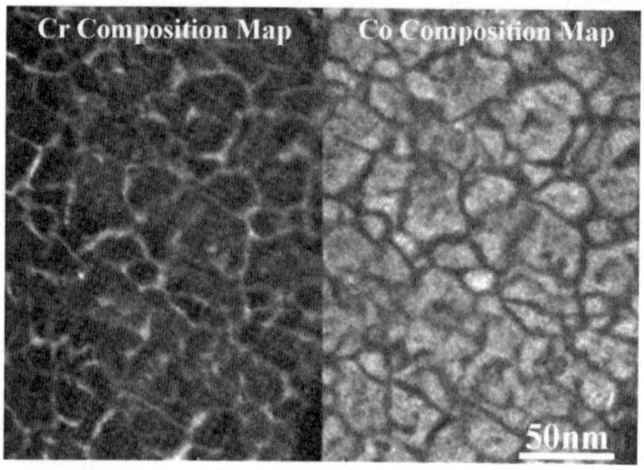

Fig. 2.24. Plan-view Cr and Co composition map of a CoCrTa thin film [81] (© 1998 IEEE, courtesy of J. E. Wittig)

It has also been discovered that more Cr segregates to the grain boundaries as substrate temperature (T_{sub}) increases [209]. The films will exhibit an increase in H_c with the increasing T_{sub} up to 250–300°C, partly because of the decrease of exchange coupling by the Cr segregation. Further increases in T_{sub} will cause H_c to drop due to the formation of fcc phase. It has also been found that S^* [235] and the ΔM peak [236], which are indicators of degree of exchange coupling, decreased with an increase of T_{sub}. In addition, an applied bias has a similar effect on the H_c, S^* and ΔM peak as T_{sub} [237].

Cr segregation in CoCrPt films has also been observed [232, 233], but the amount of Cr at the grain boundaries was found to be less than in the CoCrTa films. In the case of $Co_{74}Cr_{13}Pt_{13}$ (25 nm)/CrTi (50 nm) media deposited at $T_{sub} = 300$°C, the Cr concentration at grain boundaries was determined to be 16–18%, which is only 3–5% greater than the overall Cr composition. This may well explain why HRTEM could not observe thick amorphous like grain boundaries in the CoCrPt media.

Cr segregation is also observed within the grains [234]. The segregation sites within the grains are most likely to be defects, especially *small angle boundaries* between two CoCrTa crystallites that nucleate separately on one Cr grain. Such defects are discussed in [237].

Type of Grain Boundary. Wittig *et al.* [81] found that Cr segregated more to the high angle grain boundaries than to the coherent (semi-coherent) 90° bicrystal boundaries. The mean Cr concentration in high angle boundaries was found to be 23%, while the mean Cr concentration in bicrystal boundaries was 17%. Similar results have been reported by Inaba et al. [210], who found in $Co_{79}Cr_{15}Ta_6$ thin films at high angle boundaries the maximum Cr concentration was 23% with the Cr-rich region thickness of about 1 nm. At bicrystal boundaries the Cr composition was determined to be 18% and the width of the Cr enriched region was greater than 2 nm. These results suggest

Table 2.4. Grain boundary energy of high angle grain boundaries and coherent grain boundaries (reprinted from [140], © 1995, with permission from Elsevier Science)

Metal	high angle GBE (mJ/m^2)	coherent twin GBE (mJ/m^2)
Al	324	75
Cu	625	24
Ag	375	8
Au	378	15
Ni	866	43

that the driving force of Cr segregation is stronger toward the high angle boundaries than toward the bicrystal boundaries.

This phenomenon can be understood by the fact that high angle grain boundaries have much larger grain boundary energy (GBE) than the bicrystal grain boundaries. Physically a non-coherent grain boundary has a more open structure, which better accommodates solute atoms than would a coherent grain boundary. Table 2.4 list GBE of selected elements.

Inter-Layer Diffusion. Interdiffusion between the $Co_{84}Cr_{12}Ta_4$ layer and Cr layer (T_{sub} = 250°C) was not observed by EFTEM work [81,184]. On the other hand, magnetic measurements which are more sensitive than the structural analysis have found that M_s, S^* and the ΔM peak decrease and H_c increases with an increase of T_{sub} in Co/Cr bilayers [238]. The use of a pure Co layer eliminated the possibility of intergranular segregation leaving only interlayer diffusion as the mechanism for Cr segregation. However, the magnetic properties varied only slightly when the T_{sub} was less than at 250°C.

Interlayer diffusion can be enhanced by post-deposition annealing. Feng et al. [238] reported that after annealing at 550°C for 1 hr. the M_s and S^* of Co/Cr bilayer decreases by more than 30% and that the H_c increased from 300 to 800 Oe. In fact, interlayer diffusion is an effective way of transporting foreign elements into grain boundaries of the CoCr-alloy films, in order to improve and tailor the magnetic properties of the media without destroying the structure or intrinsic magnetic properties of the core grains. Elements, such as, Mn [239,240], Zn [241], Cu [242] and C [243] have also been used to enhance the magnetic properties by interlayer diffusion.

Effect of Ta and Pt. It is very difficult to map the elements Ta and Pt using the EFTEM technique [208]. It has been found by EDAX that Ta and Pt are randomly distributed inside grains of both $Co_{81}Cr_{15}Ta_4$ and $Co_{74}Cr_{13}Pt_{13}$ [184,233].

It has been widely accepted that Ta (not Pt) increases the segregation of Cr solute atoms to the grain boundaries. This is supported by magnetic

Table 2.5. Cr distribution in the CoCr-alloy films [210]

Alloy	Cr at GB	Cr inside grain	width
$Co_{81}Cr_{15}Ta_4$	25%	6–12%	1 nm
$Co_{79}Cr_{15}Ta_6$	23%	8–11%	1 nm
$Co_{70}Cr_{14}Pt_{12}Ta_4$	18%	10%	2.3 nm
$Co_{73}Cr_{15}Pt_{12}$	18%	14%	2.2 nm
$Co_{79}Cr_{19}Pt_{12}$	24%		

measurements demonstrating that CoCrPtTa media with more Ta content exhibit lower S^* and ΔM peaks and is also supported by elemental analysis indicating that Cr segregates more in CoCrTa films than in the CoCrPt films.

Inaba et al. [210] have used Nano-Probe EDAX to compare segregation in 5 CoCr-alloys (see Table 2.5). Thin films of the alloys were deposited under the same conditions as the CoCr-alloy (25 nm)/CrTi (50 nm)/Si at 300°C. Table 2.5 lists the measured Cr concentration at the grain boundaries and inside the grains as well as the average width of the Cr rich regions. The widths of the Cr rich region were determined as FWHM of the Cr distribution profile.

It can be seen from the table that increasing the Ta concentration from 4 to 6% does not further enhance Cr segregation. However, more Ta was found at the grain boundaries. Comparing the first and third rows of the table, we notice that adding 12% Pt into the CoCrTa alloy decreases the Cr segregation. The increase of width indicates a smaller Cr concentration gradient from core to boundary, which implies that the Cr-rich phase at the grain boundaries may not be paramagnetic. As a result, the magnetic grains would not be effectively exchange de-coupled. Comparing the last two rows of the table, it can be seen that adding 4% Cr increases the Cr concentration at grain boundaries, which is 5% more than the overall composition. Considering that before the addition, it is only 3% more than the overall composition, it is reasonable to conclude that the addition of Cr still preferentially goes to the grain boundaries.

Boron Diffusion. The Goldschmidt atomic radius of Boron atoms (0.097 nm) is much smaller than that of Co (0.125 nm), Cr (0.128 nm), Pt (0.138 nm) atoms. Therefore it is very likely that the B atoms can diffuse through interstitial sites as in the case of carbon diffusion [243], despite the fact that the close packed structure (hcp) does not have wide channels for interstitial diffusion compared with the body center cubic (bcc) systems. On the other hand the Cr atoms have to diffuse by vacancy exchange, which is much slower than interstitial diffusion [244]. Nuhfer et al. [245] have shown that B segregates to the grain boundaries in CoCrPtB alloys. TEM and magnetic studies [16] has shown that adding 3–8% of boron causes a wider grain boundary phase and smaller S* and ΔM peak. These results indicate that better exchange de-coupled magnetic grains can be achieved by adding boron in the CoCrPt alloy.

Post-Deposition Annealing. annealingMuch effort has been devoted to post-deposition annealing of Co-alloy media [26,115]. It has been demonstrated that post-deposition annealing can effectively magnetically de-couple the grains, providing excellent magnetic properties of the media. This is because annealing can take advantage of using high Pt compositions, which enhances the K_u of the CoCr-alloy media. At the same time, enough Cr can be driven to the grain boundary to exchange de-couple the grains. Without annealing, one has to add Ta and more Cr into CoCrPt to get better exchange de-coupling media, which in turn decreases K_u and H_k of the film.

For example, after rapid thermal annealing in vacuum, CoCrPt media can have an H_c as high as 6000 Oe and an $M_r t = 0.3$ memu/cc [247]. The H_c/H_k ratio is another way of monitoring the degree of coupling between the magnetic grains. This ratio has a maximum value of 0.5 for 2-D random media [246]. Mukai et al. [248] has shown that by depositing at lower temperatures and annealing at higher temperatures, the H_c of CoCrPt can be increased from 2000 to 4571 Oe and the H_c/H_k value increase from 0.2 to 0.46. Granular CoCrPt thin films were prepared by alternatively depositing $Co_{78}Cr_{10}Pt_{12}$ (2.5 nm) and SiO_2 (0.1–0.6 nm) layers onto a Cr underlayer on glass substrate at room temperature. The films were vacuum annealed up to 580°C for 10 min. A large H_c increase from 2555 Oe to 5611 Oe and M_s drop from 701 emu/cc to 242 emu/cc was observed [249].

However, it should be noted that post-deposition annealing is not a cost-effective way for manufacturing of large quantities of disks. Seeking novel Co-alloy compositions is still the number one task to make high-density media with superior magnetic and recording properties.

2.6 Closure

We have shown that microstructural features such as grain size and distribution, preferred orientation, stacking faults, chemical segregation and composition of the media are all important in determining the magnetic properties and recording performance of longitudinal thin film media. Optimum media should have small grains with a narrow distribution, strong crystallographic texture with c-axis in plane, low stacking fault density and a high degree of chemical segregation to the grain boundaries.

Further research is important for the continuous improvement of these microstructural features. However, as the grain size gets smaller, the superparamagnetic limit will necessitate the use of high K_u materials [179] such as L1o-alloys or new media designs such as AFC media [250,251].

Acknowledgements. We wish to thank our colleagues at Carnegie Mellon University and Seagate Technology for many helpful discussions. We especially thank Prof. David Lambeth for long term collaboration on thin film research. BL wishes to thank Dr. Dieter Weller for the constant support of the writing. He also wants to thank Ms. Marge Cain for organizing most of the references. DEL wishes to acknowledge the many students and postdoctorals that have worked with him and also the ongoing financial support of the Data Storage Systems Center at CMU under a grant ECD-8907068.

References

1. Relation of Properties to Microstructure, American Society for Metals, Cleveland Ohio, (1954)
2. Magnetic Properties of Metals and Alloys, American Society for Metals, Cleveland Ohio, (1959)
3. S. H. Charap, P. -L. Lu, Y. He, IEEE Trans. Magn. 33, 978 (1997)
4. J. Li, M. Mirzamaani, X. Bian, M. Doerner, S. Duan, K. Tang, M. Toney, T. Arnoldussen, M. Madison, J. Appl. Phys. **85**, 4286 (1999)
5. D. Weller, and A. Moser, IEEE Trans. Magn. **35**, 4423, (1999)
6. E. T. Yen, S. Z. Wu, T. Thomson, R. Ristau, R. Ranjan, G. Rauch, IEEE Trans. Magn. **34**, 2730 (1999).
7. T. P. Nolan, M. Hara, K Yoshida, M. Futamoto, J. Appl. Phys. **81**, 3922 (1997)
8. H. Sato, J. Nakai, A. Kikuchi, H. Mitsuya, IEEE, Trans. Magn. **32**, 3596 (1996)
9. A. KiKuchi, S. Kawakita, J. Nakai, T. Shimatsu and Mj. Takahashi, J. Appl. Phys. **79**, 5339 (1996)
10. M. Lu, T. Min, Q. Chen, J. Judy, IEEE Trans. Magn. **28**, 3255 (1992)
11. C. Gao, Z. S. Shan, R. Maimhall, Y. Liu, H. J. Richter, A. Barneu, G. C. Rauch, D. J. Sellmyer, J. Appl. Phys. **81**, 15 (1997)
12. M. Takahashi, A. Kikuchi, H. Hara, H. Shoji, IEEE Trans. Magn, **34**, 1573 (1998)
13. G. C. Rauch, C. Byun, E. R. C. Johns, and C. Messinger, J. Greggi, Jr., IEEE Trnas. Magn. **28**, 3105 (1992)
14. Y. Takahashi, Y. Yajima, S. Kojima, Y. Hosoe, J. Appl. Phys. **87**, 5699 (2000)
15. J. Soderlund, L.B. Kiss, G. A. Niklasson , C. G. Granqvist, Phys. Rev. Lett. **80,** 2386 (1998)
16. Y. Kubota, L. Folks and E. E. Marinero, J. Appl. Phys. **84**, 6202 (1998)
17. M. Doerner, K. Tang, T. Arnoldussen, H. Zeng, M. G. Toney, D. Weller, IEEE Trans. Magn. **36**, 43 (2000)
18. J. P. Hirth and K, L. Moazed, *Physics of Thin films*, edited by G. Hass and R. Thum, Academic Press Inc., pp.97, (1967)
19. J. J. K. Chang, Q. Chen, G. L. Chen, R. Sinclair, J. Appl. Phys. **81**, 3943 (1997)
20. C. A. Ross, F. M. Ross, G. Bertero, K. Tang, IEEE Trans. Magn. **34**, 282 (1998)
21. K. L. Chopra, *Thin Film Phenomena*, McGraw-Hill Book Company, pp142, (1969)

22. R. W. Vook, Int. Metals Rev. **27**, 209 (1982)
23. J. A. Thornton, Ann. Rev. Mater, Sci. **7**, 239 (1977)
24. R. Messier, R. C. Ross, J. Appl. Phys. **53,** 6220 (1982)
25. C. R. M. Grovenor, H. T. G. Hentzell, D.A. Smith, Acta Metall. **32**, 773 (1984)
26. G. Choe, IEEE. Trans. Magn. **31**, 2809 (1995)
27. R. Ranjan, J. A. Christner, D. P. Ravipati, IEEE Trans. Magn. **26**, 322 (1990)
28. T. Yogi, C. Tsang, T. A. Nguyen, K. Ju, G. L. Gorman, G. Castillo, IEEE Trans. Magn. **26**, 2271 (1990)
29. T. Yogi, T. A. Nguyen, S. E. Lambert, G. L. Gorman, G. Castillo, IEEE Trans. Magn. **26**, 1578 (1990)
30. T. Yogi, G. L. Gorman, C. H. Hwang, M. A. Kakalec, S. E. Lambert, IEEE Trans. Magn. **24**, 2727 (1988)
31. T. Yamashita, L. H. Chan, T. Fujiwara, T. Chen, IEEE. Trans. Magn. **27**, 4727 (1991)
32. A. Mazor, D. J. Srolovitz, P. S. Hagan B. G. Bukiet, Phys. Rev. Lett. **60**, 424 (1988)
33. C. V. Thompson, Mat. Res. Sco. Symp Proc. **343**, 3 (1994)
34. S. Yoshimura, D.D.Djayaprawira, T. K. Kong, Y. Masuda, H. Shoji, M. Takahashi, IEEE Trans. Magn. **36**, 2363 (2000)
35. Y. Matsuda, Y. Yahisa, K. Sakamoto, Y. Takahashi, A. Katou, Y. Hosoe, IEEE Trans. Magn. **35**, 2640 (1999)
36. W. Xiong, H. L. Hoo, IEEE Trans. Magn. **34**, 1570 (1998)
37. S. I. Pang, J. P. Wang, E. W. Soo, T. C. Chong, IEEE Trans. Magn. **36**, 2366 (2000)
38. L.L. Lee, D. E. Laughlin, D. N. Lambeth, IEEE Trans. Magn. **30**, 3951 (1994)
39. D. E. Laughlin, B. Lu, Y. N. Hsu, J. Zou, D. N. Lambeth, IEEE Trans. Magn. **36**, 48 (2000)
40. M. Takahashi, A. Kikuchi S. Kawakita IEEE Trans. Magn. **33**, 2938 (1997)
41. J. Nakai, A. Kikuchi, M. Kuwabara, T. Sakurai, T. Shimatsu, M. Takahashi, IEEE Trans. Magn. **31**, 2833 (1995)
42. S. Yoshimura, D. D. Djayaprawira, T. K. Kong, Y. Masuda, H. Shoji, M. Takahashi, J. Appl. Phys. **87**, 6860 (2000)
43. Y. Hosoe, T. Kanbe, K. Tanahashi, I. Tamai, S. Matsunuma and Y. Takahashi, IEEE Trans. Magn. **34**, 1528 (1998)
44. T. Kanbe, I. Tamai, Y. Takahashi, K. Tanahashi, and A. Ishikawa, Y. Hosoe J. Appl. Phys. **85**, 4717 (1999)
45. T. Kanbe, Y. Takahashi, K. Tanahashi, A. Ishikawa, Y. Hosoe, IEEE Trans. Magn. **35**, 2667 (1999)
46. S. Sun, C. B. Murray, D. Weller, L. Folks, A. Moser, Science **287**, 1989 (2000)
47. S. Sun, C. B. Murry, H. Doyle, Mat. Res. Soc. Symp. Proc. **577**, 385 (1999)
48. N. Mahvan, A. M. Zeltser, D. N. Lambeth, D.E. Laughlin and M. H. Kryder, IEEE Trans. Magn. 26, 2277 (1990)
49. J. P. Lazzari, I. Melnich and D. Randet, IEEE Trans. Magn. **3**, 205 (1967)
50. J. P. Lazzari, I. Melnich and D. Randet, IEEE Trans. Magn. **5,** 955 (1969)
51. D. E. Laughlin, B. Cheong, Y. C. Feng, D. N. Lambeth, L. L. Lee, and B. Wong, Scripta Metall. Mater. **33,** 1525 (1995)
52. H. Gong, W. Yang, M. Rao, D. E. Laughlin, D. N. Lambeth, IEEE Trans. Magn. **35**, 2676 (1999)
53. W. Yang, D. N. Lambeth, L. Tang, D. E. Laughlin, J. Appl. Phys. **81**, 4370 (1997)

54. Y. P. Deng, D. N. lambeth, D. E. Laughlin IEEE Trans. Magn. **28,** 3096 (1992)
55. W. Yang, D. N. Lambeth, D. E. Laughlin, J. Appl. Phys. **85**, 4723 (1999)
56. M. Futomoto, M. Suzuki, N. Inaba, A. Nakamura, Y. Honda, IEEE Trans. Magn. **30**, 3975 (1994)
57. N. Inaba, A. Nakamura, T. Yamamoto, Y. Hosoe, and M. Futamoto, J. Appl. Phys. **79**, 5354 (1996)
58. N. Inaba, M. Futamoto A. Nakamura, IEEE Trans. Magn. **34**, 1558 (1998)
59. B. Y. Wong, D. E. Laughlin, D. N. Lambeth, IEEE Trans. Magn. **27**, 4733 (1991)
60. J. Ding, J. Zhu, IEEE Trans. Magn. **30**, 3978 (1994)
61. T. Min, J. Zhu, J. Appl. Phys. **75**, 6129 (1994); M. Mirzamaani, C. V. Jahnes, M. A. Russak, J. Appl. Phys. **69**, 5169 (1991); O. Kitakami, N. Kikuchi, S. Okamoto, Y. Shimada, K. Oikama, Y. Otani, K. Fukamichi, J. Magn. Magn. Mater. **202**, 305 (1999). y
62. D. N. Lambeth, W. Yang, H. Gong, D. E. Laughlin, B. Lu, L. L. Lee, J. Zou, P. S. Harlee, Mat. Res. Soc. Symp. Proc. **517**, 181 (1998)
63. M. Futomoto, N. Inaba, A. Nakamura, Y. Honda, Acta Mater. **46**, 3777 (1998)
64. D. E. Laughlin, B. Y. Wong, IEEE Trans. Magn. **27**, 4713 (1991)
65. K. Hono, B. Wang, and D. E. Laughlin, J. Appl. Phys. **68**, 4734 (1990)
66. A. Murayama, S. Kondoh, and Mj. Miyamura, J. Appl. Phys. **75,** 6147 (1994)
67. H. J. Lee, J. Appl. Phys. **63**, 3269 (1988)
68. J. K. Howard, R. Ahlert and G. Lim, J. Appl. Phys. **61**, 3834 (1987)
69. S. L. Duan, J. O. Artman, B. Wong, and D.E. Laughlin, IEEE, Trans. Magn. **26**, 1587 (1990)
70. M. Mirzamaani, C.V. Jahnes, and M. A. Russak, J. Appl. Phys. **69**, 5169 (1991)
71. J. D. Daval and D. Randet, IEEE Trans. Magn. **6**, 768 (1970)
72. K. E. Johnson, M. Mirzamaani, and M. F. Doerner, IEEE Trans. Magn. **31**, 2721 (1995)
73. T. Hirose, H. Teranishi, M. Ohsawa, A. Ueda, O. Ishiwata, T. Ataka, K. Ozawa, S. Komiya, A. Iida, IEEE Trans. Magn. **33**, 2971 (1997)
74. M. Mirzamaani, K. Johnson, D. Edmonson, P. Ivett, and M. Russak, J. Appl. Phys. **67**, 4695 (1990)
75. A. Kawamoto, F. Hikami, J. Appl. Phys. **69**, 5151 (1991)
76. T. P. Nolan, and R. Sinclair, R. Ranjan T. Yamashita, J. Appl. Phys. **73**, 5117 (1993)
77. Y. Hosoe, Y. Yahisa, R. Tsuchiyama, A. Ishikawa, K. Yoshida, Y. Shiroishi, IEEE Trans. Magn. **31**, 2824, (1995)
78. J. J. K. Chang, Q Peng, Q. Chen, Z. H. Lin, E. Yen, G. L. Chen, H. N. Bertram, R. Sinclair, IEEE Trans. Magn. **33**, 885 (1997)
79. Q. Chen, J. J. K. Chang, G. L. Chen, R. Sinclair, IEEE Trans. Magn. **32**, 3599 (1996)
80. Z. H. Lin, E. T. Yen, J. J. K. Chang, Q. Chen, M. C. Tang, G. L. Chen, J. Appl. Phys. **81,** 3946 (1997)
81. J. E. Wittig, T. P. Nolan, C. A. Ross, M. E. Schabes, K. Tang, R. Sinclair, J. Bently, IEEE Trans. Magn. **34**, 1564 (1998)
82. X. G. Ye, J. G. Zhu, IEEE Trans. Magn. **28**, 3087 (1992)
83. Q. Peng, H. N. Bertram, N. Fussing, M. Doerner, M. Mirzamaani, D. Margulies, R. Sinclair, S. Lambert, IEEE Trans. Magn. **31**, 2821 (1995)

84. L.-L. Lee, D. E. Laughlin, L. Fang and D. N. Lambeth, IEEE Trans. Magn. **31**, 2728 (1995)
85. B. Lu, D. E. Laughlin, D. N. Lambeth, S. Z. Wu, R. Ranjan and G. C. Rauch, J. Appl. Phys. **85**, 4295 (1999)
86. C. L. Fu and M. H. Yoo, Acta Metall. Mater., **40**, 703 (1992),
87. Z. W. Lu, S. H. Wei and A. Zunger, Acta Metall. Mater. **40**, 2155 (1992)
88. B. Lu, D. N. Lambeth and D. E. Laughlin, IEEE Tran. Magn. 36, 2357 (2000)
89. E. Bauer, *Single-Crystal Films*, Edited by M. H. Francombe and H. Sato, Pergamon Press, pp.64, (1964)
90. L.L. Lee, B. K. Cheong, D.E. Laughlin, D.N. Lambeth, Appl. Phys. Lett. **67**, 3638 (1995)
91. H. J. Lee, J. Appl. Phys. **57**, 4037 (1985)
92. S. L. Duan, J. O. Artman, B. Wong, D. E. Laughlin, J. Appl. Phys. **67**, 4913 (1990)
93. D. P. Ravipati, W.G. Haines, J. L. Dockendorf, J.Vac. Sci. Technol. **A5**, 1968 (1987)
94. J. Pressesky, S. Y. Lee, S. L. Duan, D. Williams, J. Appl. Phys. **69**, 5163 (1991)
95. Y. C. Feng, D. E. Laughlin, D. N. Lambeth, J. Appl. Phys. **76**, 7311 (1994)
96. L. L. Lee, DE. Laughlin, D. N. Lambeth, US patent **5693426**, (1997)
97. L. L. Lee, D. E. Laughlin, D. N. Lambeth, J. Appl. Phys. **81**, 4366 (1997)
98. L. L. Lee, PhD Thesis, "Improving cobalt alloy thin film media through novel engineering of underlayers" Carnegie Mellon University, (1997)
99. R. Ranjan, J. Appl. Phys. **67**, 4698 (1990)
100. T. Kogure and S. Katayama, N. Ishii, J. Appl. Phys. **67**, 4701 (1990)
101. H. Suzuki, N. Tsumita, M. Hayashi, Y. Shiroishi, Y. Matsuda, IEEE Trans. Magn. **26**, 2280 (1990)
102. J. K. Howard, J. Appl. Phys. **63**, 3263 (1987)
103. M. Mirzamaani, C. V. Jahnes, M. A. Russak, IEEE Trans. Magn. **28**, 3090 (1992)
104. R. Ranjan, J. Chang, T. Yamashita, and T. Chen, J. Appl. Phys, **73**, 5542 (1993)
105. X. Tang, B. Reed, R. Zubeck, IEEE Trans. Magn. **30**, 3963 (1994)
106. S. Duan, B. Zhang, C. Gao, G. C. Rauch, J. L. Pressesky, A. Schwartz, IEEE Trans. Magn. **30**, 3966 (1994)
107. Y. Okumuram, M. Yashi, T. Akita, M. Maeda, X. B. Yang, H. Fujimori, IEEE Tran. Magn. **33**, 2974 (1997)
108. H. Kataoka, T. Kanbe and H. Kashiwase, E. Fjita, Y. Yahisa K. Furusawa, IEEE, Tans. Magn. **31**, 2734 (1995)
109. Y. Onishi, H. Matsumura, T. Hase, H. Hayashi, M. Sato K. Katsumoto, IEEE, Trans. Magn. **25**, 3887 (1989)
110. M. Kuwabara, M. Sato Y. Onishi, M. R. Visokay, H. Hayashi H. Inoue, K. Muramatsu, J. Appl. Phys. **73**, 6686 (1992)
111. T. D. Lee, IEEE Trans. Magn. **35**, 2628 (1999)
112. S. Y. Hong, T. D. Lee, K. H. Shin, J. Appl. Phys. **85**, 4298 (1999)
113. K. Sato, Y. Yoshida, M. Yamagishi, H. Ueno, H. Akimoto, E. N. Abarra, H. Kanai, Y. Uehara, I. Okamoto, Y. Yematsu, IEEE Trans. Magn. **35**, 2655 (1999)
114. T. Kawanabe, J.G. Park and M. Naoe, Mat. Res. Soc. Symp. Proc. **232**, 21 (1991)

115. Z. S. Shan, C. P. Luo, M. Azarisooreh. K. Honardoost, M. Russak, Y. Liu, J. P. Liu, D. J. Sellmyer, IEEE Trans. Magn. **35**, 2643 (1999)
116. M. Mirzamaani, X. Bian, M.F. Doerner, J. Li, M. Parker, IEEE Trans. Magn. **34**, 1588 (1998)
117. L. L. Lee, D. E. Laughlin, D. N. Lambeth, J. Appl. Phys. **79**, 4902 (1996)
118. L. Tang and D. E. Laughlin, J. Appl. Crystal, **29**, 411 (1996)
119. B. Lu, L. Tang, D. N. Lambeth, and D. E. Laughlin, "Grain Growth in Polycrystalline Materials", Proceedings of 3rd International Conference on Grain Growth, 529 (1998)
120. L. Tang, S. Duan, D. E. Laughlin, Mat. Res. Soc. Symp. **562**, 105 (1999)
121. L. Tang, Y. C. Feng, L. L. Lee, D. E. Laughlin, J. Appl. Crystal, **29**, 411 (1996)
122. M. A. Parker, J. K. Howard, R. Ahlert and K. R. Coffey, J. Appl. Phys. **73**, 5560 (1993)
123. K. Sin, J. M. Sivertsen, J. H. Judy, IEEE Trans. Magn. **30**, 4008 (1994)
124. T. B. Massalski, Binary Alloy Phase Diagrams, 2nd edition, ASM International.
125. S. Malhorta, D. Strafford, B. Lal, C. Gao, M. Russak, J. Appl. Phys. **85**, 6157 (1999)
126. Y. Shiroishi, Y. Hosoe, A. Ishikawa, Y. Sugita, H. Suzuki, T. Ohno, M. Ohura, J. Appl. Phys. **73**, 5569 (1993)
127. G. Choe, J. Appl. Phys. **79**, 4923 (1996)
128. Y. Matsuda, Y. Yashi, J. Inagaki, E. Fujita, A. Ishikawa, J. Appl. Phys. **79**, 5351 (1996)
129. L. L. Lee, D. E. Laughlin, D.N. Lambeth, IEEE Trans. Magn. **34**, 1561 (1998)
130. L. Vegard, Z. Physik, **5**, 17 (1921)
131. L. Vegard, Z. Krist. **67**, 239 (1928)
132. B. Y. Wong, J. F. Ying, IEEE Trans. Magn. **35**, 2646 (1999)
133. B. Lu, D. N. Lambeth, and D. E. Laughlin, S. D.Harkness, W. A. Lewis, Proc. Mater. Res. Soc., **562**, 117 (1999)
134. M. Fotomoto, F. Kugiya, M. Suzuki, H. Takano. Y. Matsuda IEEE Trans. Magn. **27**, 5280 (1991)
135. L. Fang, D. N. Lambeth, Appl. Phys. Lett. **65**, 3137 (1994)
136. B. Zhang, W. R. Bennett, C. Gao. G. C. Rauch, IEEE Trans. Magn. **32**, 3590 (1996)
137. J. Zou, D. E. Laughlin, D. N. Lambeth, IEEE Trans. Magn. **34**, 1582 (1998)
138. K. Tanahashi, I. Tamai, S. Matsunuma, T. Kanbe, A. Ishikawa, J. Appl. Phys. **87**, 6857 (2000)
139. L. Zhang, B. B. Lal, M. A. Russak, IEEE Trans. Magn. **35**, 2649 (1999)
140. F. J. Humphreys and M. Hatherly, Recrystallization and Related Annealing Phenomena, p.69 Pergamon 1995
141. T. Ericsson, Acta Metall. **14**, 853 (1966)
142. T. C. Tisone, Acta Metall. **21**, 229 (1973)
143. T. M. Coughlin, E. R. Wuori, J. H. Judy, J. Vac. Sci. Tech. **20**, 171 (1982)
144. J. W. Lee, B. G. Demczyk, K. R. Mountfield and D. E. Laughlin, J. Appl. Phys. **61**, 3813 (1987)
145. W. Sucksmith and J. E. Thompson, Proc. R. Soc. **A225**, 362 (1954)
146. K. O Grady, N. S. Walmsley, C. F. Wood, R. W. Chantrell, IEEE Trans. Magn. **34**, 1579 (1998)

147. N. S. Walmsley and R. W. Chantrell, J. Appl. Phys. **85**, 6154 (1999)
148. L. Holloway, H. Laidler, J. Appl. Phys. **87**, 5690 (2000)
149. C. Gao, S. Wu, J-P Chen, R. Malmhall, C. Habermeier, R. Sinclair, H. Laidler, K. O'Grady, IEEE Trans. Magn. **34**, 1576 (1998)
150. S. E. McKinlay, N. Fussing, R. Sinclair, Mary Doerner, IEEE Trans. Magn. **32**, 3587 (1996)
151. Y. Shen, B.Y. Wong and D.E. Laughlin, J. Appl. Phys. **76**, 8167 (1994)
152. D. J. Mapps, N. Mahvan, M. A. Akhter, IEEE Trans. Magn. **23**, 2473 (1987)
153. J. A. Thompson, D. A. Stevenson, IEEE Trans. Magn. **21**, 1441 (1985)
154. S. Iwasaki, K. Ouchi, K. Saiki, M. Kimura, IEEE Trans. Magn. **22**, 1158 (1986)
155. K. Utsumi, T. Inase, A. Kondo, J. Appl. Phys. **73**, 6680 (1993)
156. P. Glijer, J. M. Sivertsen, J. Judy, J. Appl. Phys. **73**, 5563 (1993)
157. O. Kitakami, N. Kikuchi, S. Okamoto, Y. Shimada, K. Oikama, Y. Otani, K. Fukamichi, J. Magn. Magn. Mater. **202**, 305 (1999).
158. H. Yamaguchi M. Yanagisawa, IEEE Trans. Magn. **22**, 576 (1986)
159. T. P. Nolan, R. Sinclair, R. Ranjan, T. Yamashita, J. Appl. Phys. **73**, 5566 (1993)
160. L. H. Chan, G. Thomas, J. S. Gau, J. Magn. Magn. Mater. **79**, 95 (1989).
161. J. W. Lee, B. G. Demczyk, K. R. Mountfield, and D. E. Laughlin, J. Appl. Phys. **63**, 2905 (1988)
162. B. G. Demczyk, and D. E. Laughlin, Mater. Res. Soc. Symp. Proc. **119**, 158 (1988)
163. J. W. Lee, K. R. Mountfield, and D. E. Laughlin, J. Appl. Phys. **63**, 3266 (1988)
164. K. Hono. B. G. Demczyk, D. E. Laughlin, Appl. Phys. Lett. **55**, 229 (1989)
165. B. E. Warren, *X-ray diffraction*, Dover Publications INC, pp. 298, (1969)
166. T. P. Nolan, R. Sinclair, R. Ranjan, T. Yamashita, IEEE Trans. Magn. **29**, 292 (1993)
167. B. Y. Wong, Y. Shen, D. E. Laughlin, J. Appl. Phys. **73**, 418 (1993)
168. A. Ishikawa, and R. Sinclair IEEE Trans. Magn. **32**, 3605 (1996)
169. B. Lu, J. Zou, D. N. Lambeth, D. E. Laughlin, Intemag 2000, #BP-04
170. B. Bian, W. Yang, D. N. Lambeth, D. E. Laughlin, The 8th Joint MMM-Intermag Conference, #HA-03, 2001
171. Y. Shen, B. Y. Wong, and D. E. Laughlin, J. Appl. Phys. **76**, 8174 (1994)
172. P. Dova, H. Laidler, K. O'Grady, M. F. Toney, M. F. Doerner, J. Appl. Phys. **85**, 2775 (1999)
173. R. Sinclair T. P. Nolan, G. A. Bertero, M. R. Visokay, Mat. Res. Soc. Symp. Proc. **313**, 705 (1990)
174. L. Tang G. Thomas, M. R. Khan, S. L. Duan, N. Heiman, J. Appl. Phys. **69**, 5166 (1991)
175. R. Ranjan, W. R. Bennett, G. J. Tarnopolsky, T. Yamashita, T. Nolan, R. Sinclair, J. Appl. Phys. **75**, 6144 (1994)
176. R. Herschitz and D. N. Seidman, Acta Metall. **33**, 1547 (1985)
177. H. Suzuki, J. Phys. Soc. Jpn. **17**, 322 (1962)
178. G. J. Parker, B. Lu, T. J. Klemmer and R. W. Chantrell, The 8th Joint MMM-Intermag Conference, #HF-02, 2001
179. D. Weller, A. Moser, L. Folks, M. E. Best, W. Lee, M. F. Toney, M. Schwickert, J. U. Thiele, M. F. Doerner, IEEE Trans. Magn. **36**, 10 (2000)

180. H. N. Bertram, H. Zhou, R. Gustafson, IEEE Trans. Magn. **34**, 1845 (1998)
181. K. E. Johnson, K J. Schulz, J. M. Severson, IEEE Trans. Magn. **29**, 3670 (1993)
182. J. G. Zhu, H. N. Bertram, J. Appl. Phys. **63**, 3248 (1988)
183. M. F. Doerner, X. Bian, K. Tang, M. F. Tony, K. Rubin, D. Weller, A. Moser, M. Mirzamaani, A. Polcyn, T. Minvielle,K. Takano, R. White, #GA-01, Intermag 2000
184. M. Futamoto, N. Inaba, Y. Hirayama, K. Ito, Y. Honda, Mat. Res. Soc. Symp. Pros. **517**, 243 (1998)
185. K. E. Johnson, J. B. Mahlke, K. J. Schulz, A. C. Wall, IEEE Trans. Magn. **29**, 215 (1993)
186. T. Chen, D. A. Rogowski, R. M. White, J. Appl. Phys. **49**, 1816 (1978)
187. M. Yanagisawa, N. Shiota, H. Yamaguchi, Y. Suganuma, IEEE Trans. Mag. **19**, 1638, (1983)
188. H. Maeda, J. Appl. Phys. **53,** 3735, (1982)
189. T. Yamada, N. Tani, M. Ishikawa, Y. Ota. K. Nakamura, A. Itoh, IEEE Trans. Magn. **21**, 1429 (1985)
190. M. Ishikawa, N. Tani, T. Yamada, Y. Ota, K. Nakamura, A. Itoh, IEEE Trans. Magn. **22**, 573, (1986)
191. R. D. Fisher, J. C. Allan, J. L. Pressesky, IEEE Trans. Magn. **22**, 352 (1986),
192. T. Suzuki, N. Yoshida, U. Hwang, Y. Uchiyama, K. Ishibashi, J. Appl. Phys. **63**, 2929 (1988)
193. M. R. Khan, R. D. Fisher, Neil Heiman, IEEE Trans. Magn. **26**, 118 (1990)
194. I. L. Sanders, T. Yogi, J. K. Howard, S. E. Lambert, G. L. Gorman, C. Hwang, IEEE Trans. Magn. **25**, 3869 (1990)
195. T. Shimizu, Y. Ikeda, S. Takayama, IEEE Trans. Magn. **28**, 3102 (1992)
196. T. Inase, A. Kondo, J. Appl. Phys. **69**, 5160 (1991)
197. H. Akimoto, I. Okamoto, M. Shinohara, IEEE Trans. Magn. **34**, 1597 (1998)
198. T. Miyamoto, J. Nakai, H. Matsumura, K. Yoshikawa, M. Kuwabara, M. Visokay, H. Saffari, IEEE Trans. Magn. **31**, 2839 (1995)
199. S. Yoshimura, D. D. Djayaprawira, H. Shoji, M. Takahashi, J. Appl. Phys. **87**, 6866 (2000)
200. M. Takahashi, H. Shoji, S. Kadowaki, D.D. Djayaprawira, Y. Komori H. Domon, Appl. Phys. Lett. **76**, 3457 (2000)
201. N. Inaba, Y. Uesaka, M. Futamoto, IEEE Trans. Magn. **36**, 54 (2000)
202. N. Tani, T. Takahashi, M. Hashimoto, M. Ishikawa, Y. Ota, K. Nakamura, IEEE Trans. Magn. **27**, 4736 (1991)
203. C. R. Paik, I. Suzuki, N. Tani, M. Ishikawa, Y. Ota, K. Nakamura, IEEE Trans. Magn. **28**, 3084 (1992)
204. Y. Murakami, T. Yamamoto, A. Okabe, T. Yanada, K. Hayashi, K. Aso, J. Appl. Phys. **73**, 6674 (1993)
205. P. Glijer, J. M. Sivertsen, J. H. Judy. IEEE Trans. Magn. **31**, 2842 (1995)
206. G. R. Jones, K. O'Grady, X. Bian, M. Mirzamaani M. F. Doerner, #BP06, Intermag 2000
207. M. F. Doerner, T. Yogi, D. S. Parker, S. Lambert, B. Hermsmeier, O. C. Allegranza, IEEE Trans. Magn. **29**, 3667 (1993)
208. J. Bentley, J. E. Wittig, T. P. Nolan, Mat. Res. Soc. Symp. Pros, **517**, 205 (1998)
209. Y. Yahisa, K. Kimoto, K. Usami, Y. Matsuda, J. Inagaki, K. Furusawa, S. Narishige, IEEE Trans. Magn. **31**, 2836 (1995)

210. N. Inaba, M. Futamoto, J. Appl. Phys. **87**, 6863 (2000)
211. A. Nakamura, M. Futamoto, Jpn. J. Appl. Phys. **32**, L1410 (1993)
212. A. Nakamura, M. Kouguchi, M. Futamoto, Jpn. J. Appl. Phys. **34**, 2307 (1993)
213. N. Inaba, Y. Uesaka, A. Nakamura, M. Futamoto, Y. Sugita, S. Narishige, IEEE Trans. Magn. **33**, 2989 (1997)
214. W. Yang, D. N. Lambeth, D. E. Laughlin, J. Appl. Phys. **87**, 6884 (2000)
215. R. Ranjan, H. J. Richter, J. Chen, S. D. Harkness, S. Z. Wu, R. Ristau, Er. Girt, C. Chang, R. M. Brochie, G. C. Rauch, S. Gangopadhyay, K. Subramanian, #HA-01, the 8th Joint MMM-Intermag Conference, 2001.
216. T. Chen, IEEE Trans. Magn. **17**, 1181 (1981)
217. J. G. Zhu, H. N. Bertram, IEEE Trans. Magn. **24**, 2706 (1988)
218. J. N. Chapman, J. Phys. D **17**, 623 (1984)
219. T. Chen, G. B. Charlan, J. Appl. Phys. **50**, 4285 (1979)
220. D. D. Dressler, J. J. Judy, IEEE Trans. Magn. **Mag-10**, 674 (1974)
221. J. H. Judy, IEEE Trans. Magn. **29**, 209 (1993)
222. J. A. Thornton, J. Vac, Sci. Tech. **A4**, 3059 (1986)
223. J. K. Howard, R. G. Simmons T. Yogi, US Patent No. **5066552** (1991)
224. A. Murayama, M. Miyamura, S. Kondoh, J. Appl. Phys. **76**, 5361 (1994)
225. R. Ranjan, M. Lu, T. Yamashita, T. Chen, IEEE. Trans. Magn. **30**, 3942 (1994)
226. J. S. Judge, J. R. Morrison, D. E. Speliotis, J. Electronchem. Soc. **113**, 547 (1966)
227. M. G. Miksic, R. Travieso, A. Arcus, R. H. Wright, J. Electrochem. Soc. **113**, 360 (1966)
228. J. S. Judge, J. R. Morrison, D. E. Speliotis, J. Appl. Phys. **36**, 948 (1965)
229. M. Aspland, G. A. Jones, B. K. Middleton, IEEE Trans. Magn. **Mag-5**, 314 (1969)
230. K. Hono, D. E. Laughlin, J. Magn. Magn. Mater. **80**, L137 (1989)
231. Y. Maeda, D. J. Rogers, O. Song, K. Takei, T. Ohkubo. S. Hirono, J. Suzuki, Y. Morii, IEEE Trans. Magn. **33**, 879 (1997)
232. M. R. Kim, S. Guruswamy, K. E. Johnson, IEEE Trans. Magn. **29**, 3673 (1993),
233. N. Inaba, T. Yamamoto, Y. Hosoe, M. Futamoto J. Magn. Magn. Mats. **168**, 222 (1997)
234. J. E. Wittig, T. P. Nolan, R. Sinclair, J. Bentley, Mat. Res. Soc. Symp. Pros, **517**, 211 (1998)
235. B. R. Natarajan, E. S. Murdock, IEEE Trans. Magn. **24**, 2724 (1988)
236. P. I. Mayo, K. O'Grady, P. E. Kelly, J. Cambridge, I. L. Sanders, T. Yogi, R. W. Chantrell, J. Appl. Phys. **69**, 4733 (1991)
237. B. Y. Wong, Y. Shen, D. E. Laughlin, J. Appl. Phys. **73**, 418 (1993)
238. Y. C. Feng, D. E. Laughlin, D. N. Lambeth, IEEE Trans. Magn. **30**, 3948 (1994)
239. J. Zou, B. Lu, T. Leonhardt, D. E. Laughlin, D. N. Lambeth, J. Appl. Phys. **87**, 6869 (2000)
240. J. Zou, B. Lu, D. E. Laughlin, D. N. Lambeth, IEEE. Trans. Magn. **35**, 2661 (1999)
241. J. Zou, PhD Thesis "Improvement of Co-Alloy Magnetic Media via Grain Boundary Diffusion and Intermediate Layers" Carnegie Mellon University, 2000

242. W. Peng, Z. Qian, C. Yang, J. M. Sivertson, J. H. Judy, J. Appl. Phys. **85**, 4702 (1999)
243. J. J. K. Chang, K. E. Johnson, H. Kawayoshi, P. Ling, M. Strathman, IEEE Trans. Magn. **34**, 1567 (1998)
244. D. A. Porter, K. E. Esterling, Phase Transformations in Metals and Alloys. Van Nostrand Reinhold Company, 1980
245. T. Nuhfer, T. F. Fliervoet, Y. N. Hsu, M. Doerner, D. E. Laughlin, unpublished research
246. M. Sato, Y. Onishi, A. Nakaue, IEEE. Trans. Magn. **29**, 3685 (1993)
247. J. P. Wang, L. P. Tan, M. L. Yan, T. C. Chong, J Appl. Phys. **87**, 6352 (2000)
248. R. Mukai, K. Yamanaka, M. Oshiki. J. Appl. Phys. **81**, 3931, (1997)
249. Y. Xu, J. P. Wang, Y. Su, J. Appl. Phys. **87**, 6971 (2000)
250. E. N. Abarra, I. Okamoto, Y. Mizoshita, #AA-06, Intermag 2000
251. M. Madison, T. Arnoldussen, M. Pinarbasi, M. Parker, M. Doerner, X. Bian, K. Tang. M. Mirzamaani, S. Yuan, J. Li, L. Lauchlan, L. Ingall, J. Raniseski, P. Fang, D. Margulies, E. Marinero, K. Rubin, K. Takano, E. Fullerton, D. Weller, A. Moser, M. E. Best, #HT-01, Intermag 2000

3 Magnetization Dynamics and Thermal Fluctuations in Fine Grains and Films

Vladimir L. Safonov and H. Neal Bertram

3.1 Introduction

Advanced magnetic recording systems are designed for extremely high areal densities and data rate. These two aspects require both magnetization reversal at very short times (<1 ns) and long term (~5–10 years) stability against thermal fluctuations. There are two basic physics problems associated with these requirements. The first is an understanding of the physics of the relaxation mechanisms for large magnetization motions. The second is a characterization of thermally agitated magnetization reversal over very wide time range. Both these subjects will be considered in this chapter.

Hard disk media consist of individual magnetic nanograins. Some degree of interaction occurs, but the magnetostatic and exchange energies are much smaller than the uniaxial anisotropy energy that dominates the intrinsic coercivity. Therefore the problems of dynamic and thermal magnetization reversal of magnetized medium may reduced to the corresponding problems for a single-domain nanograin.

The goal of this chapter is to present an in depth overview of dynamic and thermal reversal mechanisms in magnetic grains. In Sect. 2 we give a brief overview of the conventional dynamic magnetization equations. Our studies of fundamental relaxation mechanisms have involved a two step approach. The results of computer simulations of magnetization reversal without phenomenological damping will be discussed in Sect. 3. In this case coherent rotation of the magnetization in a grain excites nonlinear spin waves during which an excess of Zeeman energy is transformed to anisotropy and exchange energies. However, for fine grains whose size is sufficiently small so that the grain magnetization is virtually uniform, non-linear spin waves cannot assist reversal. We also briefly discuss results of dynamic magnetization reversal without phenomenological damping in thin film.

In Sect. 4 we propose a new analytic model of reversal that couples coherent rotation to impurity ions by an anisotropic exchange mechanism. These impurity ions are assumed to relax at a very high rate to the lattice through spin-orbital coupling.

Thermal dynamic reversal requires solution of the Landau–Lifshitz–Gilbert equation with fluctuations. We discuss the conventional approach in Sect. 5. In Sect. 6 we solve the problem of thermal magnetization reversal in terms of the "random walk" dynamics of a nonlinear oscillator. The expressions for the switching field versus pulse time are analytic and show good agreement with experiment.

Finally, in Sect. 7 we discuss a possibility to enhance stability of magnetic memory by delocalization of thermal fluctuations. Such collective processes are expected to be important in nearly identical systems of grains, such as proposed perpendicular recording.

3.2 Magnetization Dynamics with Phenomenological Damping

The magnetization dynamic equation

$$\frac{d\mathbf{M}(t)}{dt} = -\gamma\mathbf{M}(t) \times \mathbf{H}_{\text{eff}}(t) \tag{3.1}$$

is fundamentally gyromagnetic where macroscopic descriptions follow directly from the dynamics of the quantum spin Hamiltonian. In (3.1) $\mathbf{H}_{\text{eff}} = -\partial\mathcal{U}/\partial\mathbf{M}$ is the net effective field providing torque to the magnetization $\mathbf{M}$, $\mathcal{U}$ is the total system energy density and $\gamma = 1.76 \times 10^7$ (Oe sec)$^{-1}$ is the gyromagnetic ratio. Multiplying both sides of (3.1) by $\mathbf{M}$ causes the right hand side to vanish, thus conserving the magnetization magnitude:

$$|\mathbf{M}(t)| = M_{\text{s}}. \tag{3.2}$$

Here M_{s} is the material bulk saturation magnetization. In the spirit of micromagnetics M_{s} as well as other bulk properties are taken as the equilibrium values at a given system temperature.

All magnetic motion is "damped" by relaxation to the lattice (magnons, phonons, itinerant electrons, impurities, etc.). Due to the difficulty of fundamental analysis of relaxation, phenomenological expressions are commonly used in conjuction with the basic gyromagnetic motion. Physically, damping is a result of 'viscosity' and corresponding 'memory' in the system. Mathematically, the effect of memory means a retardation. The simplest modification of the gyromagnetic equation (3.1) with memory can be written, for example, as

$$\frac{d\mathbf{M}(t)}{dt} = -\gamma\mathbf{M}(t) \times \mathbf{H}_{\text{eff}}(t - \delta t), \tag{3.3}$$

where δt is the memory time. Expanding

$$\mathbf{H}_{\text{eff}}(t - \delta t) = \mathbf{H}_{\text{eff}}(t) - \delta t \, \frac{d\mathbf{H}_{\text{eff}}}{d\mathbf{M}} \cdot \frac{d\mathbf{M}(t)}{dt} + \dots \tag{3.4}$$

for short memory and utilizing (3.3) in the first order approximation, we obtain

$$\frac{d\mathbf{M}}{dt} = -\gamma\mathbf{M} \times \mathbf{H}_{\text{eff}} + \frac{\mathbf{M}}{M_{\text{s}}} \times \left(\hat{\alpha}\,\frac{d\mathbf{M}}{dt}\right), \tag{3.5}$$

where $\hat{\alpha} = \delta t\, \gamma M_\mathrm{s} \left(d\mathbf{H}_\mathrm{eff}/d\mathbf{M} \right)$ represents a dimensionless phenomenological damping tensor. Usually, for simplicity, this damping tensor is considered to be a diagonal $\hat{\alpha} = \alpha \delta_{ij}$, where α is a scalar constant. Thus we obtain the magnetization dynamic equation proposed by Gilbert [1]:

$$\frac{d\mathbf{M}}{dt} = -\gamma \mathbf{M} \times \mathbf{H}_\mathrm{eff} + \alpha \frac{\mathbf{M}}{M_\mathrm{s}} \times \frac{d\mathbf{M}}{dt}. \tag{3.6}$$

The gyromagnetic equation with Gilbert damping (3.6) is equivalent to the gyromagnetic equation with Landau–Lifshitz damping [2]:

$$\frac{d\mathbf{M}}{dt} = -\tilde{\gamma}\, \mathbf{M} \times \mathbf{H}_\mathrm{eff} - \alpha \frac{\tilde{\gamma}}{M_\mathrm{s}} \mathbf{M} \times \left(\mathbf{M} \times \mathbf{H}_\mathrm{eff} \right), \tag{3.7}$$

where

$$\tilde{\gamma} = \gamma/(1+\alpha^2). \tag{3.8}$$

For a given magnetization (or discretization unit in a simulation of non-uniform reversal) the length of $\mathbf{M}$ in (3.7) is also conserved (see, (3.2)).

Multiplying both sides of (3.7) by the effective field, one obtains:

$$\frac{d\mathcal{U}}{dt} = -\frac{\alpha}{1+\alpha^2} \frac{\gamma}{M_\mathrm{s}} \left(\mathbf{M} \times \mathbf{H}_\mathrm{eff} \right)^2. \tag{3.9}$$

Thus the rate of energy decrease varies as the square of the instantaneous torque. This equation demonstrates that the relaxation torque is always orthogonal to the magnetization motion, characteristic of damping in a gyromagnetic system. So far as $\alpha \ll 1$ for almost all magnetic materials, the α^2 in (3.8) and (3.9) may be omitted.

There are more complicated phenomenological damping terms to the gyromagnetic equation (3.1) [3]. The microscopic derivation of the relaxation terms has been performed, for example, in [4–8].

3.2.1 Damping Mechanisms for Small Magnetization Motions

The dimensionless damping parameter α in (3.7) for small magnetization motions can be obtained from ferromagnetic resonance (FMR) experiments (see, e.g., [9]):

$$\alpha = \Delta\omega/2\omega_\mathrm{r} = \Delta H/2H_\mathrm{r}, \tag{3.10}$$

where $\omega_\mathrm{r} = \gamma H_\mathrm{r}$ is the resonance frequency and $\Delta\omega = \gamma \Delta H$ is the linewidth.

The main problem in measuring α in magnetic recording media is the dispersion of local FMR fields due to inhomogeneities. Recent experimental estimates are $\alpha \simeq 0.01 - 0.04$ [10] and $\alpha \simeq 0.03$ [11] The problem of FMR in polycrystalline structures has been considered in [9,12,13].

Theoretically the energy transfer (relaxation) from coherent magnetization rotation to the thermal bath can occur by interactions with itinerant electrons, spin waves (magnons), lattice vibrations (phonons), and crystal imperfections (impurities, defects) [9,14,15]. Theories of magnon damping in itinerant ferromagnets [16,17] give relaxation rates $\propto k^2$, where k is the wave vector. Therefore this channel of energy loss for coherent magnetization rotation ($k = 0$) is expected to be very small. Suhl considered a mechanism for direct coupling to the lattice via magnetoelastic interaction [6] and relaxation by valence fluctuations [18].

It is well known that impurity magnetic ions of transition and rare-earth elements with strong spin-orbital coupling give rise to ferromagnetic resonance line broadening [9,19]. Two different (but not contradictory) mechanisms of relaxation have been proposed. In the, so-called, 'slow-relaxing ion' mechanism [20–25] small magnetization oscillations modulate the impurity levels in the vicinity of thermal equilibrium. This energy modulation is absorbed by the lattice at an extremely fast rate. The 'fast-relaxing ion' mechanism [22,26,27] is effective when the transverse relaxation rate of the impurity is so fast that the magnetization oscillations can excite direct transitions between impurity levels.

According to recent FMR experiments in ultra thin films [28,29], spin-wave scattering on the surface [30] is a possible mechanism of dephasing of coherent rotation of the magnetization. To prove this interpretation it is necessary to carry out additional experiments to demonstrate a temperature independence of the FMR linewidth.

3.3 Pure Dynamics in Non-Uniform Systems: Non-linear Damping by Spin Waves

In this section we study large magnetization motions simulating magnetization reversal in a system of coupled gyromagnetic equations. According to Suhl's speculation [6], the switching process should be rapid, governed by nonlinear spin-wave excitations. By neglecting phenomenological damping, the total system energy is conserved: the goal is to explore magnetization reversal in a finite field as a reduction of the Zeeman energy by the excitation of the non-uniform magnetization modes. Certainly, in real physical systems, loss of energy from the entire spin system will eventually occur. However, in a variety of magnetic systems, energy from the uniform mode initially is absorbed by spin waves leading to an effective damping.

3.3.1 Reversal in Almost Single Domain Fine Particle

A micromagnetic grain was considered as a system of n, $4 \times 4 \times 4 = 64$ or $10 \times 10 \times 10 = 1000$ sub-cubes coupled by exchange and dipole-dipole interactions (Fig. 3.1). Each subgrain is characterized by the magnetization

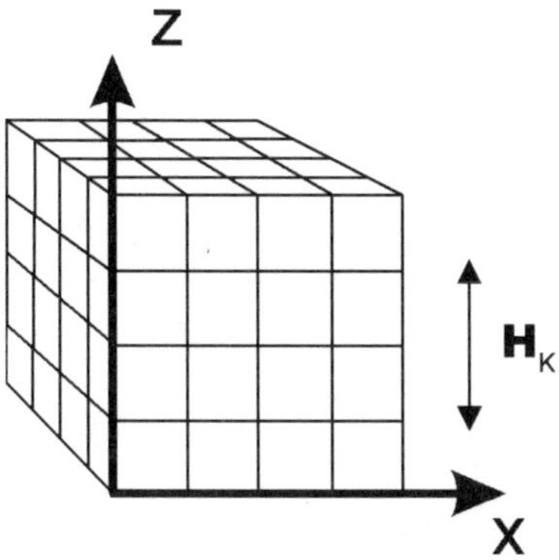

Fig. 3.1. Model of micromagnetic grain

vector $\mathbf{M}_j$. The grain is "quasi" single domain as far as the cube size δ_s is in the order of the domain wall width $\delta_w \simeq 4\sqrt{A/K_u}$, where K_u the uniaxial anisotropy constant and $A = 10^{-6}$ erg/cm is the exchange constant. The exchange interaction can also be characterized by the so-called, "exchange length" $\delta_{ex} = A^{1/2}/M_s$. We study the problem of time evolution of the net magnetization $\mathbf{m} = (1/n)\sum_j \mathbf{M}_j/M_s$. The system of coupled gyromagnetic equations was numerically solved without any phenomenological damping terms. Initially, the magnetizations of all sub-cubes are oriented "upward" along the z axis, which corresponds to the anisotropy axis. Before the field is applied each cell magnetization is given a slight deviation from perfect orientation. The total energy of these deviations is nk_BT, where T is the temperature of the system. To get a mixed state corresponding to thermal equilibrium, the above initial configuration is allowed to evolve dynamically for a scaled time $\gamma H_K t \sim 200$. To start the reversal process, the direction of the magnetic field is oriented slightly away from the anisotropy axis (z). In order to obtain high-accuracy solutions to ordinary differential equations with minimal computational effort, the Bulirsch-Stoer method with adaptive stepsize control [31] was used. Steps of integration were about four orders less than the characteristic time $1/\gamma H_K$.

We have found that the simulations of time evolution of the magnetization for a grain consisting of either 4^3 or 10^3 subgrains in virtually all cases give qualitatively and quantitatively similar results. This means that the level of discretization of the system is fine enough to obtain all physical details of the intrinsic mechanism of magnetization reversal.

In Fig. 3.2 the time dependences $m_z(t)$ and $|\mathbf{m}(t)|$ are shown. When the external magnetic field equals the "dynamic nucleation field", the process of a rapid magnetization reversal accompanied by a decrease in the average magnetization is started (see Fig. 3.2a). The system exhibits a nonlinear excitation of non-uniform magnetic oscillations (spin waves) driven by the uniform mode. The dynamic nucleation field or, the onset of spin-wave excitations

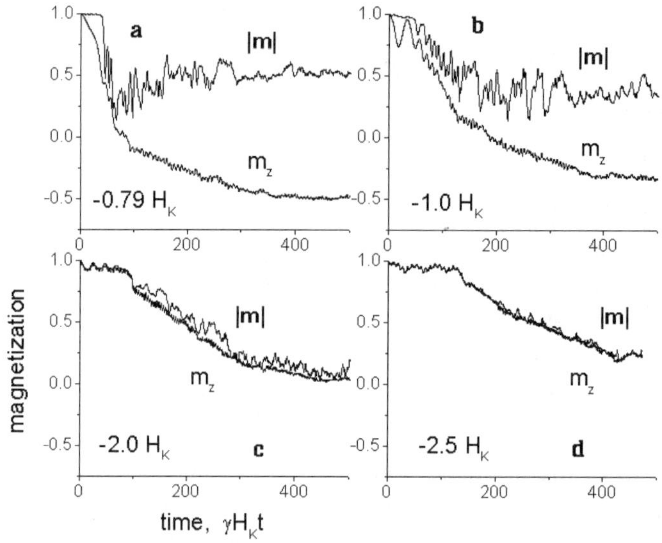

Fig. 3.2. Time evolution of magnetization for various reversal fields at $H_x/H_z = 0.05$ and $\delta_{\mathrm{w}}/\delta_{\mathrm{ex}} = 1.08$

(in Fig. 3.2a $H_z/H_{\mathrm{K}} = -0.79$ for this particular deviation angle) was always slightly less than the nucleation field for uniform rotation. Nonlinear spin-wave dynamics mixes all the energies and forms an overheated state in the spin-wave system. In this example, as seen in Fig. 3.2a, the relative magnetization reverses to about -0.6. The field component of magnetization vanishes ($m_z(t_0) = 0$) at about $\gamma H_{\mathrm{K}} t_0 \simeq 125$. For example, for $H_K = 5000$ Oe, this corresponds to a reversal time of $t_0 \simeq 1.4$ nsec. The characteristic time t_0, which corresponds to the point $m_z(t_0) = 0$, increases with increasing $|\mathbf{H}|$.

The total energy is conserved ($\alpha = 0$) and reversal occurs by transferring Zeeman energy to nonuniform spin waves that include anisotropy and exchange energies (Fig. 3.3a,b). The dipole-dipole energy is relatively small ($M_{\mathrm{s}}/H_{\mathrm{K}} \ll 1$) and therefore cannot absorb sufficient Zeeman energy. Nevertheless the dipole-dipole interaction was found to give an additional channel for the energy transfer and to increase the rate of nonlinear relaxation [32–34].

In Fig. 3.2b the time dependence $m_z(t)$ is shown for the magnetic reversal field slightly higher than the nucleation field. We have compared this case with a simple model of coherent reversal in the framework of the Landau–Lifshitz–Gilbert equation (3.7) with phenomenological damping. It was found that the intrinsic mechanism of nonlinear damping gives faster dynamics for longitudinal magnetization in the beginning of the reversal and then is followed by slow saturation. The equivalent LLG damping parameter defined for $m_z(t_0) = 0$ corresponds to $\alpha \simeq 0.02$.

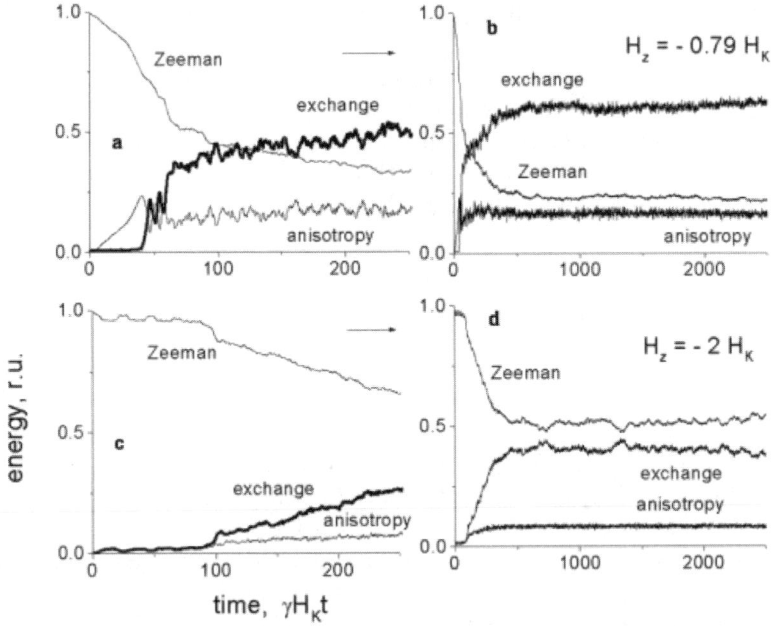

Fig. 3.3. Transformations of Zeeman, anisotropy and exchange energies for short (a,c) and long (b,d) time intervals. Figures a,b and c,d correspond, respectively to fields $H_z = -0.79 H_K$ and $-2H_K$ at $H_x/H_z = 0.05$ and $\delta_w/\delta_{ex} = 1.08$

Figures 3.2c,d and 3.3c,d show suppression of the reversal process for high reverse fields ($H_z/H_K = -2$ and -2.5). For $H_z/H_K < -3$ we did not observe any reversal dynamics at all. There are similar limitations on magnetization reversal for small grain sizes (less than the wall size). These limitations have a general origin typical for Hamiltonian systems of nonlinear oscillators [35]. The energy transfer in such systems (nonlinear diffusion) is defined by thresholds of nonlinear resonances, which depend on a system nonlinearity. This nonlinear diffusion is forbidden for either a strong reversal field and/or a small grain size.

3.3.2 Simulation of Damping

In order to estimate the damping parameter α for small perturbations we have performed numerical simulation of uniform magnetization precession in the vicinity of stable (or metastable) states. The α can be obtained by the following procedure: 1) apply a small static transverse field and allow the magnetization to precess; 2) remove the transverse field and observe damped precession of the uniform magnetization; 3) plot magnitude of transverse component versus time and fit to $m_\perp \exp(-\alpha\omega_0 t)$, where $\omega_0 \simeq \gamma(H_K + H)$ is the FMR frequency. The results of our simulation are $\alpha \sim (1-2) \times 10^{-4}$ for $-0.5 \leq H/H_K \leq 1$.

As mentioned above, the damping parameter for recording media was measured utilizing FMR. It was found that $\alpha \sim 0.01 - 0.04$ [10,11]. This experimental estimation of α is much higher than that of obtained by our simulation. It is likely that FMR linewidth measurements as well as reversal in fine grained materials are dominated by medium non-uniformities and impurities (e.g., on the surface).

3.3.3 Reversal in Soft Thin Film

The dynamic magnetization reversal in thin film platelets without the addition of the phenomenological damping was studied in [36–38]. The material was discretized into cubic cells (side length equal to the exchange length) with the plane of the film. Within each cell, the magnetization was uniform. The crystalline anisotropy field was taken to be zero, and therefore, there is only a shape anisotropy. Parameters approximating Permalloy were chosen: $M_s = 800$ emu/cm^3 and $\delta_{ex} = 12.5$ nm. The system of coupled gyromagnetic equations was solved. Again, the total magnetic energy is conserved.

The effect of platelet size is shown in Fig. 3.4. The average magnetization component along the long axis of the platelet is plotted versus time for various platelet sizes, while maintaining the same aspect ratio. The unit cell size was equal to the exchange length. As the platelet becomes smaller, the amount of average reversal decreases. This result is qualitatively in agreement with the previous results on fine grains: smaller sizes restrict the excitation of nonlinear spin waves. In each case the applied field is approximately 5% greater than the intrinsic field.

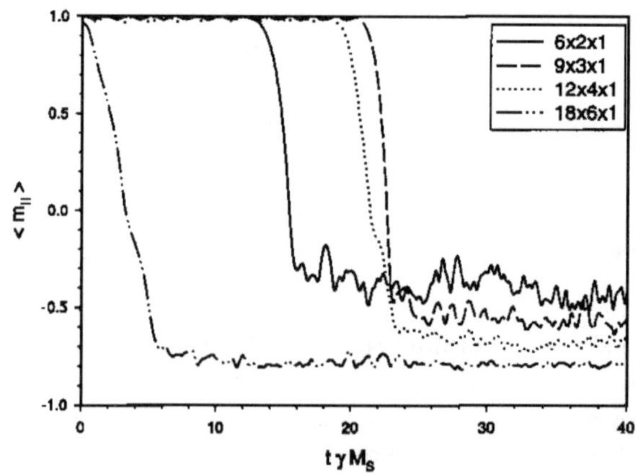

Fig. 3.4. Magnetization versus time for several platelet sizes [38]

Again, following the previous analysis on fine grains, as the applied field is increased, it becomes more difficult to excite spin waves and therefore less reversal is obtained. With increasing reverse field there is initially a dramatic reduction in switching time, but eventually the switching time starts to increase slightly as it becomes more difficult to excite spin waves modes.

3.4 Impurity Relaxation Model: Application to a Fine Grain

In this section we focus on the impurity relaxation process. We believe that in fine grains the role of imperfections is expected to be much larger than in bulk material: in addition to paramagnetic ions (such as Fe^{2+}, Fe^{4+}, Mn^{3+}, Cr^{2+}, Cr^{4+} etc.), the surface magnetic atoms are expected to behave as impurities. Here we present a theory of magnetization relaxation of impurities for large magnetization oscillations, including magnetization reversal, which extends the previous work on small angle (FMR) relaxation. Our approach is rigorous in that no phenomenological damping parameter is assumed. The fundamental magnetization dynamics of the uniformly magnetized "host" grain is described by gyromagnetic precession only:

$$\frac{d\mathbf{m}}{dt} = -\gamma \mathbf{m} \times \mathbf{H}_{\text{eff}}, \quad \mathbf{m} \equiv \mathbf{M}/M_{\text{s}}. \tag{3.11}$$

We also include the kinetics of impurity spin-lattice relaxation for the typical case of a low energy two-level splitting. Anisotropic exchange couples the magnetic host atoms with the impurity spins and thereby provides an effective field for the gyromagnetic precession of the ferromagnetic grain.

We approximate the impurities as independent two-level systems (Fig. 3.5). The use of the word "impurity" applies strictly to the lowest energy splittings of ions whose angular momentum is not completely quenched by the crystal field. However, here, the term also applies to the surface ions of the bulk material where the crystal field symmetry is substantially decreased and normally quenched ions will regain some angular momentum.

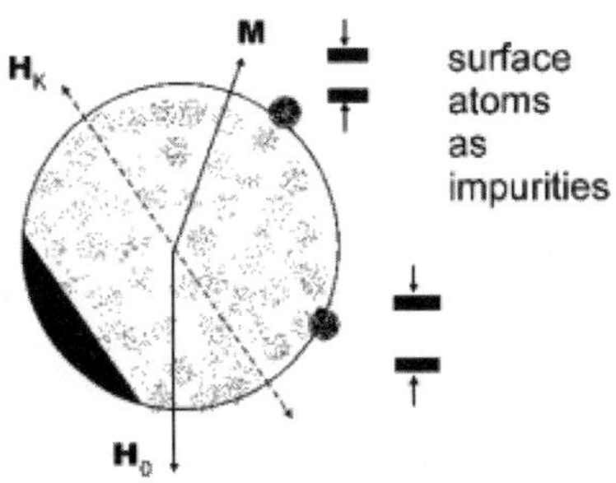

Fig. 3.5. Illustration of single domain grain and impurities on the surface

Let us write the energy density in the form $\mathcal{U} = \mathcal{U}_{\mathrm{m}} + \mathcal{U}_{\mathrm{imp}}$, where $\mathcal{U}_{\mathrm{m}}$ is the magnetic energy density of the host grain (including only anisotropy and Zeeman energies). $\mathcal{U}_{\mathrm{imp}}$ is the energy density arising from the impurities and their coupling to the host magnetization. We shall derive $\mathcal{U}_{\mathrm{imp}}$ from the microscopic model.

The two-state impurity ion can be described in terms of an effective spin $s = 1/2$. In this case the Hamiltonian of the exchange interaction between the impurity and the neighboring host atoms may be written in a general form:

$$\mathcal{H}_{\mathrm{ex}} = \sum_{j,\nu} \sum_{a_j,a} B_{a_j,a}(\mathbf{R}_j, \mathbf{r}_\nu) s_{a_j}(\mathbf{R}_j) S_a(\mathbf{R}_j + \mathbf{r}_\nu). \tag{3.12}$$

Here $B_{a_j,a}(\mathbf{R}_j, \mathbf{r}_\nu)$ are the parameters of anisotropic exchange, $a_j = x_j, y_j, z_j$ are the local principal axes for the j-th impurity "spin" and $a = x, y, z$ are the principal axes of magnetic matrix. $\mathbf{s}(\mathbf{R}_j) = \{s_{x_j}(\mathbf{R}_j), s_{y_j}(\mathbf{R}_j), s_{z_j}(\mathbf{R}_j)\}$ is the effective impurity spin located at $\mathbf{R}_j$ and $\mathbf{S}(\mathbf{R}_j + \mathbf{r}_\nu) = \{S_x(\mathbf{R}_j + \mathbf{r}_\nu), S_y(\mathbf{R}_j + \mathbf{r}_\nu), S_z(\mathbf{R}_j + \mathbf{r}_\nu)\}$ is the spin of the host matrix located in the vicinity of j-th impurity. We can express the j-th impurity spin components in terms of creation c_j^+ and annihilation c_j operators (these operators describe transitions between the two levels):

$$s_{z_j}(\mathbf{R}_j) = c_j^+ c_j - 1/2, \quad s_+(\mathbf{R}_j) = c_j^+, \quad s_-(\mathbf{R}_j) = c_j, \tag{3.13}$$

$$s_\pm(\mathbf{R}_j) = s_{x_j}(\mathbf{R}_j) \pm i s_{y_j}(\mathbf{R}_j). \tag{3.14}$$

The following commutation relations are valid:

$$(c_j)^2 = \left(c_j^+\right)^2 = 0, \quad c_j c_j^+ + c_j^+ c_j = 1,$$

$$[c_j, c_k] = 0, \quad [c_j, c_k^+] = (1 - 2c_j^+ c_j)\delta_{jk}.$$

In the case of only coherent motion of the host spins $\mathbf{S}(\mathbf{R}_j + \mathbf{r}_\nu) = (V_0 M_{\mathrm{s}}/\hbar\gamma)\mathbf{m}$, where V_0 is the volume of one host spin. Substituting (3.14) into (3.12) and writing $\mathcal{U}_{\mathrm{ex}} = \langle \mathcal{H}_{\mathrm{ex}} \rangle / V$, where V is the grain volume, we obtain the impurity energy density:

$$\mathcal{U}_{\mathrm{imp}} = (\hbar/V) \sum_j \Omega_{0,j} n_j + \mathcal{U}_{\mathrm{ex}}, \tag{3.15}$$

$$\mathcal{U}_{\mathrm{ex}} = \frac{\hbar}{V} \sum_j [\mathbf{w}_j \cdot \mathbf{m}(n_j - 1/2) + \mathbf{e}_j \cdot \mathbf{m} \left\langle c_j^+ \right\rangle + \mathbf{e}_j^* \cdot \mathbf{m} \left\langle c_j \right\rangle].$$

Here j indicates a sum over all impurities, $\Omega_{0,j}$ is the splitting (due to crystal fields, exchange and spin-orbit interactions) of the j-th impurity unperturbed by coupling to the host magnetization. $n_j = \left\langle c_j^+ c_j \right\rangle$ is the upper level impurity population, and $\langle ... \rangle$ denotes quantum mechanical expectation values. $\mathbf{w}_j$, $\mathbf{e}_j$ (and its complex conjugate $\mathbf{e}_j^*$) characterize the anisotropic exchange interaction between each impurity (as an effective spin 1/2) and host nearest neighbor spins:

$$w_{j,a} = (V_0/\hbar^2\gamma M_s)\sum_{\nu} B_{z_j,a}(\mathbf{R}_j, \mathbf{r}_\nu), \quad a = x, y, z, \tag{3.16}$$

$$e_{j,a} = (V_0/2\hbar^2\gamma M_s)\sum_{\nu} \left[B_{x_j,a}(\mathbf{R}_j, \mathbf{r}_\nu) - iB_{y_j,a}(\mathbf{R}_j, \mathbf{r}_\nu)\right]. \tag{3.17}$$

The effective field which drives (3.11) is thus given by

$$\mathbf{H}_{\text{eff}} = -\frac{\partial \mathcal{U}_{\text{m}}}{\partial \mathbf{M}} - \frac{\hbar}{M_{\text{s}}V}\sum_{j}\left[\mathbf{w}_j(n_j - 1/2) + \mathbf{e}_j\left\langle c_j^+\right\rangle + \mathbf{e}_j^*\left\langle c_j\right\rangle\right], \tag{3.18}$$

where the second term is the contribution of the impurity interactions.

3.4.1 Equations of Motion

Equations (3.11) and (3.18) describe the dynamic magnetization including coupling to the impurities. The characteristic frequencies of magnetization rotation ($\sim$ 10 GHz) are usually much smaller than Ω_j (infrared region) [9]. In this case the interaction of j-th impurity with the thermal bath can be described by longitudinal $\Gamma_{\parallel,j}$ and transverse $\Gamma_{\perp,j}$ relaxation rates. Both $\Gamma_{\parallel,j}$ and $\Gamma_{\perp,j}$ are determined by elementary spin-lattice relaxation processes [9,24] and can depend on the impurity splitting Ω_j, temperature T and external static magnetic field $\mathbf{H}_0$.

With the above assumptions, the kinetics of the j-th impurity population is defined by the following equation:

$$\frac{d}{dt}n_j = -\Gamma_{\parallel,j}(\Omega_j)[n_j - n_T(\Omega_j)]. \tag{3.19}$$

At each instant of time the population n_j relaxes to the equilibrium value $n_T(\Omega_j) = [\exp(\hbar\Omega_j/k_{\text{B}}T)+1]^{-1}$ corresponding to the dynamic splitting $\Omega_j = \Omega_{0,j} + \delta\Omega_j(t)$, where $\delta\Omega_j(t) \equiv \mathbf{w}_j \cdot \mathbf{m} = w_{j,x}m_x(t) + w_{j,y}m_y(t) + w_{j,z}m_z(t)$, as follows from (3.15). Note that the rate $\Gamma_{\parallel,j}(\Omega_j)$ also depends on this dynamic splitting. The mechanism of relaxation described by (3.19), which involves only the population of the impurity states, is called "slow" relaxation [9].

Another relaxation mechanism, called "fast" relaxation involves only the transverse components of the impurity effective spin. Similar to (3.19), the kinetics of this process is given by:

$$\left[\frac{d}{dt} + \Gamma_{\perp,j}(\Omega_j)\right]\langle c_j\rangle = -i\Omega_j\langle c_j\rangle - i\mathbf{e}_j \cdot \mathbf{m}[(1 - 2n_j)]. \tag{3.20}$$

In right hand side of this equation the first term represents the natural precession of the impurity spin. The second term explicitly reflects the coupling to the host magnetization.

General Solution. The general solution of (3.19) is:

$$\delta n_j(t) = \int_{-\infty}^{t} \exp[\Phi_{\|,j}(t_1) - \Phi_{\|,j}(t)] \times \Gamma_{\|,j}[\Omega_j(t_1)]\{n_T[\Omega_j(t_1)] - n_T(\Omega_{0,j})\}dt_1, \quad (3.21)$$

where

$$\Phi_{\|,j}(t) = \int_{-\infty}^{t} \Gamma_{\|,j}[\Omega_j(t')]dt', \quad \delta n_j(t) \equiv n_j(t) - n_T(\Omega_{0,j}). \quad (3.22)$$

Analogously, the solution of the (3.20) can be written as:

$$\langle c_j(t)\rangle = -i \int_{-\infty}^{t} \exp[\Phi_{\perp,j}(t_1) - \Phi_{\perp,j}(t)] \times \mathbf{e}_j \cdot \mathbf{m}(t_1)[1 - 2n_T(\Omega_{0,j}) - 2\delta n_j(t_1)]dt_1 \quad (3.23)$$

with

$$\Phi_{\perp,j}(t) = \int_{-\infty}^{t} \{\Gamma_{\perp,j}[\Omega_j(t')] + i\Omega_j(t')\}dt'. \quad (3.24)$$

By substitution of (3.21), (3.23) into the effective field (3.18) and the latter into the gyromagnetic equation (3.11), the system of integro-differential equations for the magnetization $\mathbf{m}$ is obtained.

Slow Relaxation. In the case of small modulation $\hbar|\delta\Omega_j(t)|/k_\mathrm{B}T \ll 1$ we can write the following expansions:

$$n_T[\Omega_{0,j} + \delta\Omega_j(t)] = n_T(\Omega_{0,j}) + [\partial n_T(\Omega_{0,j})/\partial\Omega_{0,j}]\delta\Omega_j(t) + \dots ,$$
$$\Gamma_{\|,j}[\Omega_{0,j} + \delta\Omega_j(t)] = \Gamma_{\|,j}(\Omega_{0,j}) + [\partial\Gamma_{\|,j}(\Omega_{0,j})/\partial\Omega_{0,j}]\delta\Omega_j(t) + \dots .$$

Utilizing only the first expansion terms, (3.21) becomes

$$\delta n_j^{(1)}(t) = \Gamma_{\|,j}(\Omega_{0,j})\frac{\partial n_T(\Omega_{0,j})}{\partial\Omega_{0,j}} \times \int_{-\infty}^{t} \exp[-\Gamma_{\|,j}(\Omega_{0,j})(t - t_1)]\mathbf{w}_j \cdot \mathbf{m}(t_1)dt_1. \quad (3.25)$$

For the case of short memory (longitudinal relaxation $\Gamma_{\|,j}(\Omega_{0,j})$ is much faster than the characteristic frequency of magnetization rotation) we can introduce a variable $\tau = t - t_1$ and make a moment expansion in powers of τ. The result has the form

$$\delta n_j^{(1)}(t) = \frac{\partial n_T(\Omega_{0,j})}{\partial \Omega_{0,j}} \mathbf{w}_j \cdot \left[\mathbf{m}(t) - \frac{1}{\Gamma_{\|,j}(\Omega_{0,j})} \frac{d\mathbf{m}(t)}{dt} + \dots \right] . \tag{3.26}$$

This first order expansion to the kinetic equation (3.19) for the impurity relaxation yields an additional term to the effective magnetic field (3.18):

$$\delta \mathbf{H}_{\mathrm{eff}}^{(\mathrm{slow})} \simeq -\frac{\hbar}{M_{\mathrm{s}} V} \sum_j \mathbf{w}_j \left\{ \left[n_T(\Omega_{0,j}) - \frac{1}{2} \right] + \frac{\partial n_T(\Omega_{0,j})}{\partial \Omega_{0,j}} \mathbf{w}_j \cdot \mathbf{m}(t) \right.$$
$$\left. - \frac{\partial n_T(\Omega_{0,j})}{\partial \Omega_{0,j}} \frac{1}{\Gamma_{\|,j}(\Omega_{0,j})} \mathbf{w}_j \cdot \frac{d\mathbf{m}(t)}{dt} \right\} . \tag{3.27}$$

The first and the second terms in (3.27) can be included into the effective anisotropy fields. The term containing $d\mathbf{m}(t)/dt$ in (3.27) gives the magnetization relaxation.

Fast Relaxation. The kinetics of the transverse impurity components (3.23) in the simplest approximation (neglecting level modulation, $\Omega_j = \Omega_{0,j}$) can be written as:

$$\langle c_j(t) \rangle \simeq -i[1 - 2n_T(\Omega_{0,j})] \int_{-\infty}^{t} \exp[(\Gamma_{\perp,j} + i\Omega_{0,j})(t' - t)] \mathbf{e}_j \cdot \mathbf{m}(t_1) dt_1 . \tag{3.28}$$

For a short memory the moment expansion gives:

$$\langle c_j(t) \rangle \simeq -i \frac{1 - 2n_T(\Omega_{0,j})}{\Gamma_{\perp,j} + i\Omega_{0,j}} \mathbf{e}_j \cdot \left[\mathbf{m}(t) - \frac{1}{\Gamma_{\perp,j} + i\Omega_{0,j}} \frac{d\mathbf{m}(t)}{dt} \right] . \tag{3.29}$$

This solution (and its complex conjugate $\langle c_j^+(t) \rangle = \langle c_j(t) \rangle^*$) may directly be added to the effective field (3.18).

3.4.2 Example

Consider for simplicity $\Omega_{0,j} = \Omega_0$ and the case of slow relaxation where $\mathbf{w}_j = (w, 0, 0)$ and $\mathbf{e}_j = 0$. Let us write the magnetic energy in the form $\mathcal{U}_{\mathrm{m}} = K_{\mathrm{u}}(1 - {m_z}^2) - H_0 M_{\mathrm{s}} m_z$, which assumes an applied field H_0 parallel to a uniaxial anisotropy axis. In this case the combination of (3.11), (3.18) and (3.27) may be written as:

$$\begin{aligned}
\frac{dm_x}{d\tau} &= -(m_z + h_0) m_y , \\
\frac{dm_y}{d\tau} &= (m_z + h_0) m_x + \alpha_1 (dm_x/d\tau) m_z , \\
\frac{dm_z}{dt} &= -\alpha_1 (dm_x/d\tau) m_y ,
\end{aligned} \tag{3.30}$$

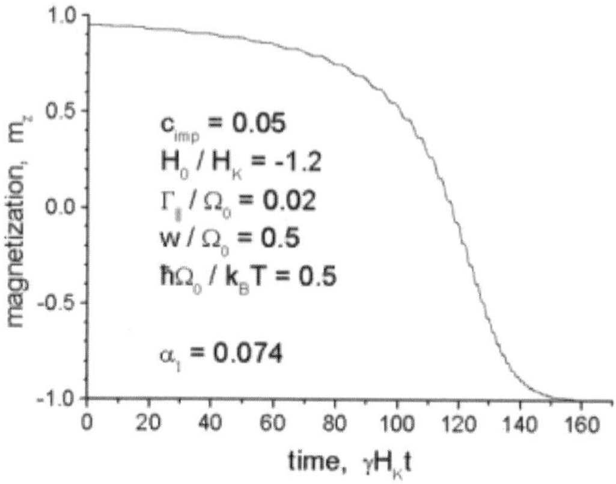

Fig. 3.6. Field component of the magnetization versus scaled time

where $\tau = \gamma H_{\mathrm{K}} t$, $h_0 = H_0/H_{\mathrm{K}}$, $H_{\mathrm{K}} = 2K_{\mathrm{u}}/M_{\mathrm{s}}$ and

$$\alpha_1 = c_{\mathrm{imp}} \left(\frac{w}{\Gamma_{\|}}\right)\left(\frac{\hbar w}{k_{\mathrm{B}}T}\right)\frac{\exp\left(\hbar\Omega_0/k_{\mathrm{B}}T\right)}{\left[\exp\left(\hbar\Omega_0/k_{\mathrm{B}}T\right)+1\right]^2}, \tag{3.31}$$

$$c_{\mathrm{imp}} = \hbar\gamma N_{\mathrm{imp}} z_{\mathrm{imp}}/M_{\mathrm{s}}V. \tag{3.32}$$

N_{imp} is the number of impurities in the grain and z_{imp} is the average number of magnetic neighbors for one impurity.

From the literature (see, e.g., [9]) it is known that typical spin-orbital splittings $\Omega_{0,j}$ are $\sim 10^2$ cm^{-1} (or, $\sim$ 144 K in temperature units). Spin-lattice relaxation times $\Gamma_{\|}^{-1}$ of the impurity at room temperature are $\sim 10^{-12}$ s and anisotropic exchange constants are of the order of (or, smaller than) the exchange interaction in bulk material. In order to estimate (3.31), let us take T = 300 K, $\hbar\Omega_0/k_{\mathrm{B}}$ = 150 K, $\hbar w/k_{\mathrm{B}}$ = 75 K and $\Gamma_{\|} = 8\times 10^{11}$ s^{-1}. For a spherical particle of radius R with all surface atoms as impurities: $N_{\mathrm{imp}} = 4\pi R^2 \Delta R/V_0$, where $V_0 = \Delta R^3$ is the volume of one host spin S. Substituting $V = 4\pi R^3/3$ and $M_{\mathrm{s}} = \hbar\gamma S/V_0$ into (3.32), one obtains $c_{\mathrm{imp}} = (3z_{\mathrm{imp}}/S)(\Delta R/R)$. For $z_{\mathrm{imp}} = 1$, $S = 1$, $\Delta R = 2$ Å and $R = 10$ nm (i.e. typical magnetic disk drive grain) we have $\alpha_1 = 0.074$. Figure 3.6 shows a typical reversal of magnetization described by the (3.30). It should be noted that the Landau–Lifshitz–Gilbert equation with phenomenological damping $\mathcal{R} = \alpha\,\mathbf{m}\times(d\mathbf{m}/d\tau)$ gives virtually the identical evolution (within 1% of accuracy) of $m_z(\tau)$ if $\alpha = \alpha_1/2$. This correspondence is a particular case of the rotational invariance of this analysis, which appears after averaging the damping terms for a distribution of impurities over a period of magnetization rotation.

Thus we have derived effective fields (see, (3.18), (3.27) and (3.29)) for magnetization dynamics (3.11) in the presence of two-level impurities. These fields exhibit both the supplementary anisotropy and the damping. The relaxing ions can be impurities in the bulk material or simply the bulk ions on the sample surface. A realistic estimate of the dimensionless damping parameter α for a magnetic nanograin is obtained. In certain cases it is shown that the magnetization reversal dynamics can be well approximation by the utilization of the Landau–Lifshitz–Gilbert equation with phenomenological damping. Although this analysis was for a single domain grain, we expect this formalism to be directly applicable to numerical simulations of large magnetic objects. In this case the impurity coupling mechanism derived here can be applied to each discretization cell.

3.5 Thermal Fluctuations of the Magnetization

Thermal fluctuations agitate the magnetic moment of a nanograin and cause a 'Brownian motion' of the magnetization. Néel [39] first proposed that the magnetization reversal time τ from one energy minimum to another can be estimated by the Arrhenius exponent:

$$1/\tau = f_0 \exp\left(-\mathcal{E}_b/k_B T\right). \tag{3.33}$$

Here $\mathcal{E}_b$ is the energy barrier, T is the temperature, k_B is the Boltzmann constant and f_0 is the attempt frequency. Experimental estimates of the attempt frequency vary from $f_0 \simeq 2.5 \times 10^8$ s^{-1} [40] to 2.5×10^{11} s^{-1} [41] for Co nanoparticles and $f_0 \sim 10^{10}$ s^{-1} [42] for α–Fe nanoparticles. Recent data on thermal decay are published in [43,44].

Here we consider the dynamics of a single-domain magnetic grain as a coherent rotation of the magnetization in the effective magnetic field $\mathbf{H}_{\mathrm{eff}} = -(\partial\mathcal{E}/V)/\partial\mathbf{M}$, where

$$\mathcal{E} = K_{\mathrm{u}} V \sin^2\theta - M_{\mathrm{s}} V H_0 \cos\theta, \tag{3.34}$$

or,

$$\mathcal{E} = K_{\mathrm{u}} V[1 - (M^z/M_{\mathrm{s}})^2] - V H_0 M^z \tag{3.35}$$

is the energy, V is the grain volume, K_{u} is the uniaxial anisotropy constant, θ is the angle between the magnetization $\mathbf{M}$ ($|\mathbf{M}| = M_{\mathrm{s}}$) and the axis of anisotropy (z), H_0 is the external magnetic field oriented along the axis of anisotropy.

From (3.34) the energy barrier for reversal from antiparallel ($H_0 < 0$) to along the applied field is:

$$\mathcal{E}_b = K_{\mathrm{u}} V (1 + H_0/H_{\mathrm{K}})^2. \tag{3.36}$$

This expression together with (3.33) can be rewritten as [45]:

$$-H_0 = H_{\mathrm{K}} \left\{ 1 - \left[\frac{k_B T}{K_{\mathrm{u}} V} \ln(f_0 \tau) \right]^{1/2} \right\}. \tag{3.37}$$

From this formula we have the dependence for remanent coercivity $H_{cr} = -H_0$ versus pulse time τ.

Brown [46] was the first who analyzed the damped motion of **M** supplemented by random Gaussian fluctuations to solve the problem of coherent reversal of magnetization under thermal agitation. He constructed a "Langevin" equation by adding a random magnetic field $\mathbf{h}(t)$ to the effective field $\mathbf{H}_{\text{eff}}$ in the Landau–Lifshitz–Gilbert equation (3.7). For high energy barriers $K_{\text{u}}V \gg k_{\text{B}}T$ he analyzed the corresponding Fokker–Planck equation and obtained an explicit expression for f_0 in (3.33). Since that time the stochastic Landau–Lifshitz–Gilbert equation has been solved in many papers (see, e.g., [47–53]). A phenomenological modification of the LLG equation was also suggested in [54].

It should be emphasized that the conventional approach utilizing the stochastic Landau–Lifshitz–Gilbert equation exhibits two physical problems. First is a problem of inconsistency of the conservation of the magnetization length $|\mathbf{M}| = M_{\text{s}}$ and the correct initial and final conditions at a finite temperature. Second, as shown in [55], the Landau–Lifshitz phenomenological damping is consistent with thermodynamics only for small deviations of the magnetization from the equilibrium state. Because of these difficulties, the conventional approach utilizing the stochastic Landau–Lifshitz–Gilbert equation can give only an approximate solution of the thermal decay problem. The higher the energy barrier the better the approximation since in this case the magnetization vector spends most of the time in the vicinity of the bottom of the potential well.

In the following section this stochastic problem is solved analytically using a formalism that completely avoids the abovementioned difficulties.

3.6 Magnetization Reversal Dynamics with Thermal Fluctuations in Fine Grain: New Efficient Technique

Recently, a new theoretical approach has been proposed [56] and developed [55,57] in which the coherent rotation of the magnetization is analyzed in terms of the motion of a damped nonlinear oscillator subject to random Gaussian forces introduced by thermal agitation. Below, we present an exact analytical calculation of the reversal time using this new technique.

3.6.1 Model

To describe the coherent motion of the magnetization we utilize the classical spin as $\mathbf{S} = \mathbf{M}V/\hbar\gamma$. We rewrite the vector spin S in terms of complex variables a_j^* and a_j (classical analog of creation and annihilation operators)

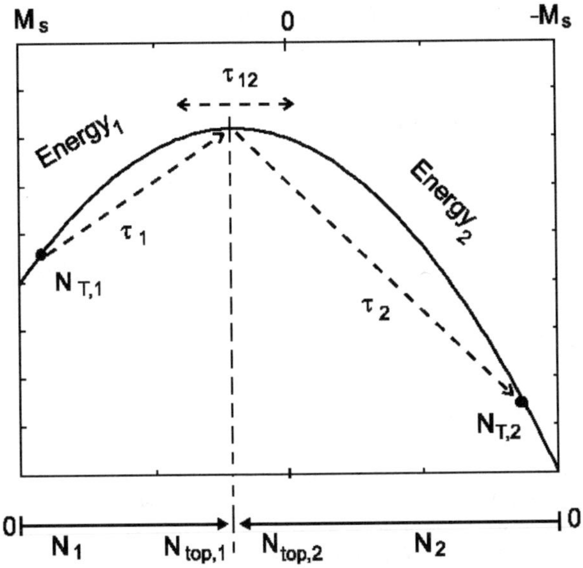

Fig. 3.7. The energy barrier (in relative units) and coordinates of nonlinear oscillators. $H_0/H_K = -0.2$

in terms of the Holstein–Primakoff transformation [58]:

$$
\begin{aligned}
S^+ &= a_1 \sqrt{2S - N_1}, \quad S^z = S - N_1, \\
S^- &= a_1^* \sqrt{2S - N_1}, \quad N_1 = a_1^* a_1.
\end{aligned}
\tag{3.38}
$$

This transformation gives a convenient representation of the spin as an oscillator in the vicinity of equilibrium state $\mathbf{S} = (0, 0, S)$. It is used within the first well when $0 \leq N_1 < N_{\text{top},1}$, $N_{\text{top},1} = S(1 + H_0/H_K)$. In this regime the energy (3.34) has the form:

$$
\mathcal{E}_1(N_1) = -\hbar\gamma S H_0 + \hbar\omega_1(1 - N_1/2N_{\text{top},1})N_1, \quad \omega_1 = \gamma(H_K + H_0).
\tag{3.39}
$$

Here ω_1 is the frequency of ferromagnetic resonance in this well.

At the top of the energy barrier $N_{\text{top},1}$ (see, Fig. 3.7) the magnetization changes its direction of rotation. In the second well it is convenient to use another form of Holstein-Primakoff transformation:

$$
\begin{aligned}
S^+ &= a_2^* \sqrt{2S - N_2}, \quad S^z = -S + N_2, \\
S^- &= a_2 \sqrt{2S - N_2}, \quad N_2 = a_2^* a_2,
\end{aligned}
\tag{3.40}
$$

which describe an oscillator in the vicinity of the equilibrium state $\mathbf{S} = (0, 0, -S)$ for $0 \leq N_2 < N_{\text{top},2}$, where $N_{\text{top},2} = S(1 - H_0/H_K)$. The energy (3.34) in this case is

$$
\mathcal{E}_2(N_2) = \hbar\gamma S H_0 + \hbar\omega_2(1 - N_2/2N_{\text{top},2})N_2, \quad \omega_2 = \gamma(H_K - H_0), \tag{3.41}
$$

where ω_2 is the ferromagnetic resonance frequency in this well. It should be mentioned that $\mathcal{E}_1(N_{\text{top},1}) = \mathcal{E}_2(N_{\text{top},2})$ and $N_{\text{top},1} + N_{\text{top},2} = 2S$.

3.6.2 Stochastic Differential Equations

The dynamic equations for complex variables a_j and a_j^* with Langevin random forces have the form [55]:

$$\left[\frac{d}{dt} + \eta_j(N_j)\right] a = -i\,\tilde{\omega}_j\, a_j + f_j(t), \tag{3.42}$$

$$\left[\frac{d}{dt} + \eta_j(N_j)\right] a_j^* = i\,\tilde{\omega}_j\, a_j^* + f_j^*(t). \tag{3.43}$$

Here $\eta_j(N_j)$ are the phenomenological relaxation rates,

$$\tilde{\omega}_1 = \omega_1 \left[1 - \frac{N_1}{S(1 + H_0/H_{\text{K}})}\right], \quad \tilde{\omega}_2 = \omega_2 \left[1 - \frac{N_2}{S(1 - H_0/H_{\text{K}})}\right], \tag{3.44}$$

$\tilde{\omega}_j$ are the nonlinear frequencies of magnetization oscillations and ω_j are corresponding FMR frequencies in the first and second wells, respectively. The following expressions

$$\begin{aligned}\langle f_j(t)\rangle &= \langle f_j(t_1) f_j(t_2)\rangle = 0,\\ \langle f_j(t_1) f_j^*(t_2)\rangle &= 2D_j\,\delta(t_1 - t_2)\end{aligned} \tag{3.45}$$

are valid for the Gaussian forces $f_j(t)$ and $f_j^*(t)$, where $\langle \ldots \rangle$ denotes the mean value and D_j are the diffusion parameters. It should be noted that (3.42), (3.43) with $\eta_j = f_j = 0$ exactly describe gyromagnetic rotation of the magnetization (3.1). Phenomenological damping terms and random forces in (3.42), (3.43) play the role of a thermal bath.

Employing Ito's calculus (see, e.g. [59]), from (3.42), (3.43) we can obtain the stochastic differential equations for N_j [55]:

$$dN_j = 2(-\eta_j N_j + D_j)dt + 2\sqrt{N_j D_j} dW_j, \tag{3.46}$$

where W_j describe independent Wiener (random) processes. It should be noted that these equations do not depend on phases, therefore the random magnetization motion is reduced to the one dimensional case. The corresponding Fokker–Planck equations are:

$$\frac{\partial P_j}{\partial t} = -2\,\frac{\partial}{\partial N_j}\left[(-\eta_j N_j + D_j)P_j\right] + 2D_j\,\frac{\partial^2}{\partial N_j^2}\,N_j P_j. \tag{3.47}$$

According to statistical mechanics, the stationary solutions of these equations should have the Boltzmann forms:

$$P_j = const_j \cdot \exp\left(-\frac{\mathcal{E}_j}{k_{\text{B}}T}\right). \tag{3.48}$$

This is possible if

$$D_j\,\hbar/k_{\mathrm{B}}T = \eta_j(N_j)/\tilde{\omega}_j(N_j). \tag{3.49}$$

This condition establishes the relation between dissipation and thermal fluctuations for the system. The left hand side of this equation is a constant; therefore the ratio of nonlinear relaxation to nonlinear frequency does not depend on N_j. Thus

$$\eta_j(N_j) = \alpha_j\,\tilde{\omega}_j(N_j) \tag{3.50}$$

and

$$D_j = \alpha_j k_{\mathrm{B}}T/\hbar. \tag{3.51}$$

The dimensionless damping parameters α_1 and α_2 can be obtained by FMR experiments and, in general, they can differ from each other.

Substituting the complex variables a_j and a_j^* into the Landau–Lifshitz–Gilbert equation (3.7), we obtain the following relations [55]:

$$\eta(N_j) = \alpha\,\tilde{\omega}_j(N_j)\cdot(1 - N_j/2S). \tag{3.52}$$

Thus, by comparison with (3.50), we can see that Landau–Lifshitz–Gilbert relaxation is consistent with thermodynamics only for small $N_j \ll 2S$.

3.6.3 Mean First Passage Time

For the one-dimensional stochastic differential equation

$$d\xi(t) = \mathcal{A}[\xi(t),t]dt + \sqrt{\mathcal{B}[\xi(t),t]}dW(t) \tag{3.53}$$

the mean first passage time can be written as [59]:

$$\tau(x) = 2\int_x^b \frac{dy}{\psi(y)}\int_a^y \frac{\psi(z)}{\mathcal{B}(z)}dz. \tag{3.54}$$

Here $\xi(t)$ is a stochastic variable in the interval (a,b), a is reflecting and b absorbing boundaries, $x \equiv \xi(0)$ is the starting point, and

$$\psi(y) \equiv \exp\left\{\int_a^y dy\prime\left[\frac{2\mathcal{A}(y\prime)}{\mathcal{B}(y\prime)}\right]\right\}. \tag{3.55}$$

Let us divide the interval (a,b) into two sub-intervals: 1) (a,c), $a < x < c$, and 2) (c,b). In this case we can rewrite (3.54) in the form

$$\tau(x) = \tau_1(x) + \tau_2(c) + \tau_{12}, \tag{3.56}$$

where

$$\tau_1(x) = 2\int_x^c \frac{dy}{\psi_1(y)} \int_a^y \frac{\psi_1(z)}{\mathcal{B}(z)} dz, \tag{3.57}$$

$$\psi_1(y) \equiv \exp\left\{\int_a^y dy\prime \left[\frac{2\mathcal{A}(y\prime)}{\mathcal{B}(y\prime)}\right]\right\}, \quad a < y < c$$

is the mean first passage time with a, the reflecting boundary, c, the absorbing boundary, and x, the starting point.

$$\tau_2(c) = 2\int_c^b \frac{dy}{\psi_2(y)} \int_c^y \frac{\psi_2(z)}{\mathcal{B}(z)} dz, \tag{3.58}$$

$$\psi_2(y) \equiv \exp\left\{\int_c^y dy\prime \left[\frac{2\mathcal{A}(y\prime)}{\mathcal{B}(y\prime)}\right]\right\}, \quad c < y < b$$

is the mean first passage time with c, the reflecting boundary, b, the absorbing boundary, and c, the starting point.

$$\tau_{12} = 2\int_c^b \frac{dy}{\psi(y)} \int_a^c \frac{\psi_1(z)}{\mathcal{B}(z)} dz \tag{3.59}$$

describes an additional delay due to diffusion between (a, c) and (c, b) subintervals.

3.6.4 Thermal Reversal

Let us consider the case $0 \leq -H_0 < H_\mathrm{K}$ when the energy barrier separates two wells. It is natural to choose the starting point of random motion as $N_{T,1}$ and the finishing point as $N_{T,2}$, where

$$N_{T,j} = \frac{\int_0^{N_{\mathrm{top},j}} dN N \exp\left[-\mathcal{E}_j(N)/k_\mathrm{B}T\right]}{\int_0^{N_{\mathrm{top},j}} dN \exp\left[-\mathcal{E}_j(N)/k_\mathrm{B}T\right]} \tag{3.60}$$

are the thermal occupation numbers of these nonlinear oscillators. To calculate the mean first passage time from the thermodynamically equilibrium state of the metastable well to the thermodynamically equilibrium stable reversed state we shall use our theorem (3.56). In this case $x = N_{T,1}$, $a \to 0$, $c = N_{T,2}$ and b corresponds to the top of the barrier. Taking into account formulas (3.57), (3.58) and (3.39), (3.41), one can obtain:

$$\psi_1(y) = (y/a) \exp\{-[\mathcal{E}_1(y) - \mathcal{E}_1(a)]/k_\mathrm{B}T\}, \tag{3.61}$$

$$\psi_2(y) = (y/c) \exp\{-[\mathcal{E}_2(y) - \mathcal{E}_2(c)]/k_\mathrm{B}T\} \tag{3.62}$$

and

$$\tau_1 = \frac{1}{2\alpha_1}\frac{\hbar}{k_\mathrm{B}T}\int_{N_{T,1}}^{N_{\mathrm{top},1}} \frac{dN}{N}\int_0^N \exp\left(\frac{\mathcal{E}_1(N) - \mathcal{E}_1(N')}{k_\mathrm{B}T}\right) dN', \tag{3.63}$$

$$\tau_2 = \frac{1}{2\alpha_2}\frac{\hbar}{k_B T}\int_{N_{T,2}}^{N_{\mathrm{top},2}}\frac{dN}{N}\int_{N}^{N_{\mathrm{top},2}}\exp\left(\frac{\mathcal{E}_2(N)-\mathcal{E}_2(N')}{k_B T}\right)dN'. \qquad (3.64)$$

To calculate τ_{12} (3.59), we take $\psi(y)$ in the form

$$\psi(y) = (y/a)\exp\{-[\mathcal{E}_2(y)-\mathcal{E}_1(a)]/k_B T\} \qquad (3.65)$$

and obtain

$$\tau_{12} = \frac{1}{2\alpha_1}\frac{\hbar}{k_B T}\int_{N_{T,2}}^{N_{\mathrm{top},2}}\frac{dN}{N}\int_{0}^{N_{\mathrm{top},1}}\exp\left(\frac{\mathcal{E}_2(N)-\mathcal{E}_1(N')}{k_B T}\right)dN'. \qquad (3.66)$$

In the absence of the external magnetic field $H_0 = 0$ both wells are symmetrical and the following relation is valid: $\tau_1 + \tau_2 = \tau_{12}$.

After cumbersome but simple calculations one can write τ_1, τ_2 and τ_{12} in the form convenient for numerical calculations:

$$\begin{aligned}\alpha_1\gamma H_K\tau_1 = {}& \frac{\sqrt{\pi\xi}}{2}\int_0^1\frac{du}{\sqrt{1-u^2}}\exp\left[\xi(1+h)^2\frac{1-u}{1+u}\right]\\ &\times\{\mathrm{erf}\left[\sqrt{\xi}(1+h)u\sqrt{\frac{1-u}{1+u}}\right]\\ &+\mathrm{erf}\left[\sqrt{\xi}(1+h)\sqrt{\frac{1-u}{1+u}}[1-\overline{n}_1(1+u)]\right]\},\end{aligned} \qquad (3.67)$$

$$\begin{aligned}\alpha_2\gamma H_K\tau_2 = {}& \frac{\sqrt{\pi\xi}}{2i}(1-\overline{n}_2)\int_0^1\frac{dv}{\zeta_2(v)}\\ &\times\exp\{-\xi[(1-\overline{n}_2)(1-h)(1-v)]^2\}\\ &\times\mathrm{erf}\left[i\sqrt{\xi}(1-\overline{n}_2)(1-h)(1-v)\right],\end{aligned} \qquad (3.68)$$

$$\begin{aligned}\alpha_1\gamma H_K\tau_{12} = {}& \frac{\sqrt{\pi\xi}}{2i}\exp\left[-\xi(1-h)^2\right]\mathrm{erf}\left[i\sqrt{\xi}(1+h)\right]\\ &\times\int_0^1\frac{dv}{\zeta_2(v)}\exp\{-\xi(1-h)^2\zeta_2(v)[2-\zeta_2(v)]\}.\end{aligned} \qquad (3.69)$$

Here $h \equiv H_0/H_K$ ($h < 0$), $H_K = 2K_u/M_s$, $\xi \equiv K_u V/k_B T$, and $\zeta_j(v) = (1-\overline{n}_j)v+\overline{n}_j$. Even with the complex notation in (3.68) and (3.69), numerical evaluation of the single integrals is straightforward since the complex terms are actually real: $\mathrm{erf}[ix]/i = 2\int_0^x dt\exp[t^2]/\sqrt{\pi}$.

The above expressions utilize the initial and final thermal equilibrium magnetization values. The thermal occupation numbers $\overline{n}_j \equiv N_{T,j}/N_{\mathrm{top},j}$ can be calculated by the following formulas:

$$\overline{n}_1 = \frac{\int_0^1 dx\, x\exp\left[-\xi(1+h)^2 x(2-x)\right]}{\int_0^1 dx\,\exp\left[-\xi(1+h)^2 x(2-x)\right]}, \qquad (3.70)$$

$$\overline{n}_2 = \frac{\int_0^1 dx\, x \exp\left[-\xi(1-h)^2 x(2-x)\right]}{\int_0^1 dx\, \exp\left[-\xi(1-h)^2 x(2-x)\right]}. \tag{3.71}$$

3.6.5 Dynamic Reversal

For the case $-H_0 \geq H_K$ there is only one well and the mean first passage time from the initial saturated state to the thermodynamically equilibrium reversed state is defined by τ_2 only ($\tau_1 = \tau_{12} = 0$):

$$\tau_2 = \frac{1}{2\alpha_2}\frac{\hbar}{k_B T}\int_{N_{T,2}}^{2S}\frac{dN}{N}\int_N^{2S}\exp\left(\frac{\mathcal{E}_2(N)-\mathcal{E}_2(N')}{k_B T}\right)dN', \tag{3.72}$$

or

$$\begin{aligned}\alpha_2\gamma H_K\tau_2 = {} & \frac{\sqrt{\pi\xi}}{2i}(1-\overline{n}_d)\int_0^1\frac{dv}{\zeta_d(v)}\{\mathrm{erf}\left[i\sqrt{\xi}\,(1+h)\right]\\ & +\mathrm{erf}\left[i\sqrt{\xi}(2(1-\overline{n}_d)(1-v)-1-h)\right]\}\\ & \times\exp\{-\xi[2(1-\overline{n}_d)(1-v)-1-h]^2\}.\end{aligned} \tag{3.73}$$

Here $\zeta_d(v) = (1-\overline{n}_d)v + \overline{n}_d$ and $\overline{n}_d \equiv N_{T,2}/2S$, or

$$\overline{n}_d = \frac{\int_0^1 dx\, x \exp[-4\xi(1-h-x)x]}{\int_0^1 dx\, \exp[-4\xi(1-h-x)x]}. \tag{3.74}$$

Dynamic Time. If we neglect diffusion due to thermal fluctuations in (3.54) (i.e. formally consider $D_2 = 0$), then we get a simple dynamic equation for N_2:

$$\frac{d}{dt}N_2 + 2\eta_2(N_2)N_2 = 0, \tag{3.75}$$

where $\eta_2(N_2) = \alpha_2\omega_2[1 - N_2/S(1-h)]$ is the thermodynamically consistent damping. From Eq.(3.75) one gets

$$\tau_d = -\frac{1}{2}\int_{\text{start}}^{\text{finish}}\frac{dN}{\eta_2(N_2)N}. \tag{3.76}$$

Taking into account the starting ($2S$) and finishing ($N_{T,2}$) points, we can obtain

$$\alpha_2\gamma H_K\tau_d = \frac{1}{2(1-h)}\ln\left[\frac{h-1+2\overline{n}_d}{(h+1)\overline{n}_d}\right].$$

In the limit of $-h \gg 1$ both the dynamic time τ_d and τ_2 (3.73) are equal to each other

$$\tau_d \approx \tau_2 \approx \frac{1}{2(-h)\alpha_2 \gamma H_K} \ln\left(\frac{1}{\overline{n}_d}\right). \tag{3.77}$$

Simple scaling relations apply for this simple geometry where the field is applied coincident with the anisotropy axes and $\alpha_1 = \alpha_2 = \alpha$. The reversal field, which is determined by specifying initial and final magnetization directions, is a function only of H_K, $\alpha\gamma H_K$ and $K_u V/k_B T$ [60].

3.6.6 Coercivity versus Pulse Time for a Single Grain

Recent measurements of reversal field versus pulse time have been performed on tapes of CrO_2 [61]. The particle parameters are: *Diameter* = 26 nm, *Length* = 248 nm, $M_s = 490$ emu/cc and $V = 133 \times 10^{-18}$ cm^3. The particle diameter and length are reduced from that given in [61] by an assumed 3 nm non-magnetic surface coating.

In Fig. 3.8 experimental measurements of reversal field versus pulse time is plotted (circles). The solid curve is a fit of (3.67), (3.68), (3.69) and (3.73) to the data. A simple $(1-p)$ factor has been applied to the theory ($H_{cr} = -H_0(1-p)$) to approximately account for interactions [62]. We use $p = 0.25$ which includes the nonmagnetic particle surface coating. The agreement between theory and experiment is extremely good and yields the (single particle) fitting parameters $H_K^0 = 1129$ Oe, $K_u^0 V^0/k_B T = 192$, and $\alpha = 0.12$. $L^0 = 4V^0/\pi(Diameter)^2$ is an "effective" thermal reversal length. The sensitivity of the fitting is such that $K_u^0 V^0/k_B T$ and H_K^0 are determined at long times; the damping parameter α dominates the short time behavior where there is a rapid increase in the reversal field as the pulse time is reduced.

In Fig. 3.8 the simple Néel-Arrhenius model [45] is also plotted (dashed). Utilizing $f_o \approx 2\alpha\gamma H_K^0$ results in good agreement with the data at long times.

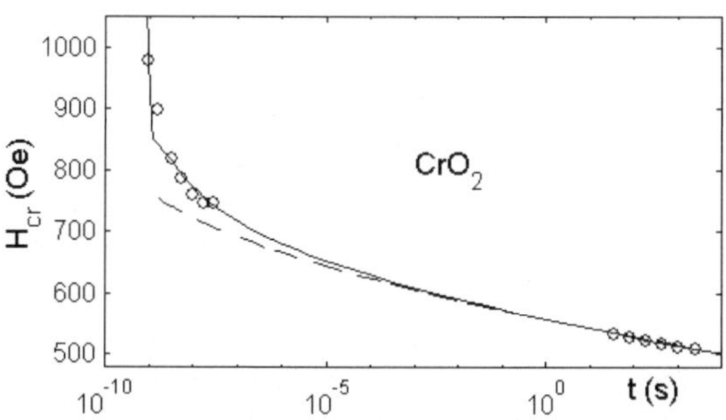

Fig. 3.8. Relative reversal field versus pulse time. Circles are experimental data [61], solid curve is theoretical fit, dashed curve is Néel-Arrhenius model

It must also be emphasized that the Néel-Arrhenius relation is only applicable, in this case, for times greater than about 1 ms.

The good fit in Fig. 3.8 is remarkable because the model is only for a single particle that reverses by uniform rotation where the field is applied parallel to the anisotropy axis. In contrast high performance tape is a collection of reasonably well oriented, elongated particles. These particles are expected to reverse by non-uniform processes [62]. In spite of these seeming differences, our model should apply to systems with a single dominant reversal mode. Micromagnetic analysis of nucleation field for an elongated particle (rectangular or cylindrical) of aspect ratio about 5 : 1 has been performed in [62] For these particles with negligible crystalline anisotropy the reversal field scaled by the particle, magnetization M_s decreases monotonically with particle radius scaled to the exchange length $l_{ex} = \sqrt{A}/M_s$. At very small radii the reversal is by uniform rotation, which for very elongated particles (such as CrO_2), yields $H_{cr} = H_K^{(shape)} = 2\pi M_s$. Utilizing $H_K^0 = 1129$ Oe, *radius* = 13 nm, and $M_s = 490$ emu/cc to deduce $l_{ex} \simeq 12$ nm [63]. This yields a value of $A \simeq 0.33 \times 10^{-6}$ erg/cm, which is consistent with the low Curie temperature of CrO_2 ($T_C \simeq 120°C$).

The model applies strictly to perfectly oriented particles; however for $r/l_{ex} \geq 1$, numerical simulations [62] show that the coercivity is insensitive to field angle with the particle axis until about 45°. Thus, the distribution of orientations in this well oriented medium ($S \simeq 0.9$) should not effect the application of this model. Numerical micromagnetics of the reversal process for $r/l_{ex} \simeq 1$ indicate a mode that initiates at the end of the particle. An effective length of this end mode L^0 has been estimated from the effective volume $V^0 = 48.8 \times 10^{-18}$ cm^3 determined from the values of $K_u^0 V^0/k_B T$ and H_K^0. A value of $L^0 \simeq 72$ nm is consistent with this mode.

Thus the model of a single oriented grain accurately describes measurements on reversal field versus pulse time in magnetic tape over a very wide range of pulse times. The effective parameters H_K^0 and V^0 are in agreement with non-uniform reversal in these particles. The fitted damping constant is reasonable. A theory of elongated particles that includes nonuniform excitations should give direct physical parameters with no fitting constants.

3.6.7 Field versus Pulse Time for 2D Random Ensemble

The above approach is generalized to arbitrary field directions [64,65]. Applicability is over the entire time range from purely dynamic reversal to low barrier thermal reversal to long times where Néel-Arrhenius applies. Compared to the uniaxial case, the stochastic differential equations or corresponding Fokker-Planck equation are much more complicated, due to the lack of uniaxial symmetry. In general, the magnetization motion is represented by a trajectory moving on a two dimensional unsymmetric energy surface. A stochastic averaging method [66] is used to transform the inherent 2D system into a 1D problem based upon a small damping parameter $\alpha \sim 0.01$–0.1

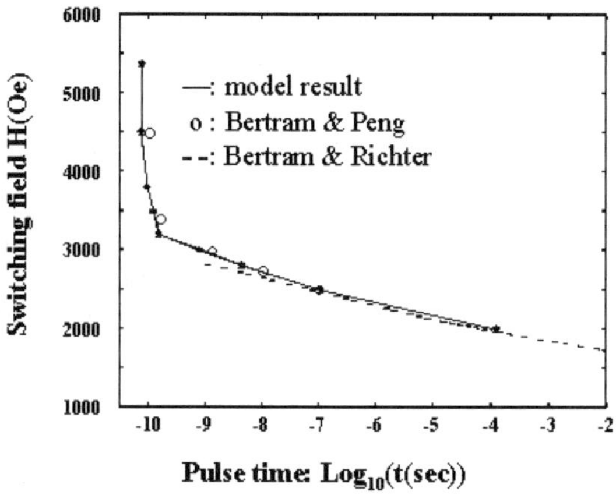

Fig. 3.9. Remanent coercivity versus pulse time for 2D random media

typical for magnetic disc drive media. The technique is to integrate the equation around the energy trajectory on the two dimensional energy surface to obtain a one dimensional stochastic differential equation. This approximation is justified by the fact that the trajectory moves many cycles around the energy surface before moving some distance across the energy level.

The magnetization as a function of time for 2D random media is the ensemble average of the magnetization for each individual grain. In the initial study, we assume uniform planar anisotropy and identical grain size. If the magnetic field is below the switching field, for a certain grain oriented at a particular angle, the mean exit time is calculated and the magnetization as a function of time is determined by using Kramer's approximation [59]. If the magnetic field exceeds the switching field for a certain grain oriented at a particular angle, the magnetization as a function of time can, to good approximation, be directly calculated from the purely dynamic equations. A detailed theory with guidance for evaluation will be published elsewhere [65].

In Fig. 3.9 this model is used to plot remanent coercivity versus pulse time for 2D random planar media (solid dots with connecting line). The grains are non-interacting and a finite rise time on the field pulse is not included. The parameters are M_s = 360 emu/cc, H_K = 6700 Oe, α = 0.05, grain volume = (12.5 nm)^3 and T = 300 K. The dashed curve corresponds to modification of Néel-Arrhenius formula [67] and shows good agreement at long times. The open circles are a numerical integration of the Langevin equation, which, due to the requirement of extremely small time steps, extend only to 100 nsec [68] (see also, [69] for numerical integration). For these short times the model also agrees well, both in the purely dynamic regime and in the low energy barrier non Néel-Arrhenius thermal regime. This evaluation of the model for a range of times as shown only requires about 15 min on a PC. It should be noted that this model is based on a theory that only needs to be evaluated for a given set of parameters. In contrast the stochastic Langevin equation must be evaluated many timers (100–500) to obtain reasonable statistical averaging. Note, that at times below about 100 nsec the Néel-Arrhenius model is no longer applicable.

In the plot of the model results there is a definite slope change at $H_{\rm cr} = H_{\rm K}/2$, where the energy barrier first vanishes (45° grain). This phase change was also seen in the application of the model to a single oriented particle (at $H_{\rm cr} = (1-p)H_{\rm K}^0$, see, Fig. 3.8). As the field is increased to the point where the energy barrier vanishes, the mean first passage time contains the dynamic reversal time only; this yields the slope change. This slope change is not seen in the plot of the numerical solution of the Langevin equation, because grain interactions are included. These interactions, even when small, do play an important role exactly when the energy barrier is extremely small.

3.7 Concept of Collective Modes and Delocalization of Thermal Fluctuations

Coupled magnetic grains can not be considered as independent if there are collective modes in the combined system. The collective modes (or, spin waves) prevent a localization of thermal fluctuations. In this case the use of mean fields to compute energy barriers, which may be numerically efficient, is not correct. The difficulty is that the fluctuating thermal energy must be applied to the collective system. Even in high density recording media, the weak magnetostatic and exchange interactions can cause nominally independent grains to behave collectively, through resonance interaction [56,70]. A continuous antiferromagnetic substrate (either synthetic or natural) may also be used as a source of additional resonance interaction between grains.

Recently, evidence has been presented of collective effects in 2D random media [71]: a type of spin waves has been shown to occur. Computer simulations with a Langevin approach in interacting particle chains have demonstrated the occurrence of correlated magnetization fluctuations, which can be interpreted in terms of spin waves [72].

Collective motions of grains result in delocalization of thermal fluctuations. The effective energy transfer between coupled grains prevents an accumulation of reverse energy at a given grain and therefore may enhance thermal stability. The thermal fluctuations are spread spatially, or "delocalized", over a very large region. The criterion of resonance interaction is that the interactions must only be large compared to the difference in ferromagnetic resonance frequencies of the grains. Simple computer simulations demonstrate a dynamic cluster structure in both well-oriented longitudinal and perpendicular recording media. These clusters help reduce thermal degradation of written magnetic data.

References

1. T. L. Gilbert, Phys. Rev. **100**, 1243 (1955); T. L. Gilbert and J. M. Kelly, Anomalous Rotational Damping in Ferromagnetic Sheets. Report. Armour Research Foundation of Illinois Institute of Technology, Chicago, Illinois (1955, unpublished)
2. L. Landau and E. Lifshitz, Phys. Z. Sowjet. **8**, 153 (1935); in Landau L. D. *Collected Papers.* ed. by D. ter Haar (Gordon and Breach, New York, 1967) p. 101
3. V. G. Bar'yakhtar, Sov. Phys. JETP **60**, 863 (1984); Physica B **159**, 20 (1989)
4. V. Kambersky and C. E. Patton, Phys. Rev. B **11**, 2668 (1975)
5. C. Heide and P. E. Zilberman, Phys. Rev. B **60**, 14756 (1999); J. Magn. Magn. Mater. **213**, L7 (2000)
6. H. Suhl, IEEE Trans. Mag. **34**, 1834 (1998)
7. D. . Fredkin and A. Ron, Phys. Rev. B **61**, 8654 (2000)
8. V. L. Safonov and H. N. Bertram, ePrint `http://xxx.lanl.gov`, cond-mat/9912014 1 Dec 1999; Phys. Rev. B **61**, R14893 (2000)
9. A. G. Gurevich and G. A. Melkov, *Magnetization Oscillations and Waves* (CRC Press, Boca Raton, FL, 1996)
10. N. Inaba, Y. Uesaka, A. Nakamura, M. Futamoto, and Y. Sugita, IEEE Trans. Mag. **33**, 2989 (1997)
11. M. Igarashi, T. Kambe, K. Yoshida, Y. Hosoe, and Y. Sugita, J. Appl. Phys. **85**, 4720 (1999)
12. E. Schlömann, J. Phys. Chem. Solids **6**, 242 (1958)
13. C. E. Patton, IEEE Trans. Mag., 1972 Intermag Conf., 433 (1972)
14. C. W. Haas and H. B. Callen, in: *Magnetism, Vol. 3.* ed. by G. T. Rado, H. Suhl (Academic Press, New York 1963) p. 449
15. M. Sparks, *Ferromagnetic-Relaxation Theory* (McGraw-Hill, New York 1964)
16. V. S. Lutovinov and M .Yu. Reizer, Sov. Phys. JETP **50**, 355 (1979)
17. A. Z. Solontsov and A. N. Vasil'ev, Phys. Lett. A **177**, 362 (1993); J. Magn. Magn. Mater. **140–144**, 213 (1995)
18. H. Suhl, The 8th Joint MMM-Intermag, J. Appl. Phys. **89**, 7478 (2001)
19. A. J. Heeger, T. G. Blocker III, and S. K. Ghosh, Phys. Rev. **134**, A399 (1964)
20. A. M. Clogston, Bell System Tech. J. **34**, 739 (1955)
21. J. H. Van Vleck and R. Orbach, Phys. Rev. Lett. **11**, 65 (1963)
22. J. H. Van Vleck, J. Appl. Phys. **35**, 882 (1964)
23. F. Hartmann-Boutron, J. Appl. Phys. **35**, 889 (1964)
24. A. S. Mikhailov and R. M. Farzetdinova, Sov. Phys. JETP **53**, 782 (1981); ibid. **57**, 109 (1983)
25. V. L. Safonov and R. M. Farzetdinova, Phys. Stat. Solidi (b) **163**, 259 (1991)
26. P. G. de Gennes, C. Kittel, and A. M. Portis, Phys. Rev. **116**, 323 (1959)
27. R. L. White, Phys. Rev. Lett. **2**, 465 (1959)
28. A. Azevedo, A. B. Oliveira, F. M. de Aguiar, and S .M. Rezende, Phys. Rev. B **62**, 5331 (2000)
29. J. R. Fermin, A. Azevedo, F. M. de Aguiar, B. Li, and S. M. Rezende, J. Appl. Phys. **85**, 7316 (1999)
30. R. Arias and D.L. Mills, Phys. Rev. B **60**, 7395 (1999)
31. W. H. Press, B. P. Flannery, S. A. Teukolsky, and W. T. Vetterling, *Numerical Recipes* (Cambridge University Press, Cambridge 1986)

32. V. L. Safonov and H. N. Bertram, J. Appl. Phys. **85**, 5072 (1999)
33. V. L. Safonov and H. N. Bertram, J. Appl. Phys. **87**, 5508 (2000)
34. V. L. Safonov and H. N. Bertram, ePrint http://xxx.lanl.gov, cond-mat/0009136 8 Sep 2000; Phys. Rev. B **63**, 094419 (2001)
35. B. V. Chirikov, Phys. Reports **52**, 265 (1979)
36. C.Y. Mao, J.-G. Zhu, and R. M. White, J. Appl. Phys. **85**, 5870 (1999)
37. E. D. Boerner, H. N. Bertram, and H. Suhl, J. Appl. Phys. **87**, 5389 (2000)
38. E. D. Boerner, Simulation of Thermal Decay and Dynamic Relaxation in Ferromagnetic Materials. PhD Thesis, University of California, San Diego (2000)
39. L. Néel, Ann. Geophys. **5**, 99 (1949); Compt. Rend., Acad. Sci. Paris **228**, 664 (1949)
40. W. Wernsdorfer, E.vB. Orozco, K. Hasselbach, A. Benoit, B.vBarbara, N.vDemoncy, A.vLoiseau, H.vPascard, and D.vMailly, Phys. Rev. Lett. **78**, 1791 (1997)
41. M. Respaud, M. Goiran, J. M. Broto, F. Lionti, L. Thomas, B. Barbara, T. Ould Ely, C. Amiens, and B. Chaudret, Europhys. Lett. **47**, 122 (1999)
42. F. Bødker, S. Mørup, M.S. Pedersen, P. Svedlindh, G. T. Jonsson, J. L. Garcia-Palacios, and F. J. Lazaro, J. Magn. Magn. Mater. **177–181**, 925 (1998)
43. H.J. Richter, J. Phys. D Appl. Phys. **32**, R147 (1999)
44. D. Weller and A. Moser, IEEE Trans. Mag. **35**, 4423 (1999)
45. M. P. Sharrock, J. Appl. Phys. **76**, 6413 (1994)
46. W. F. Brown, Jr., Phys. Rev. **130**, 1677 (1963)
47. A. Aharoni, Phys. Rev. **135**, A477 (1964)
48. D. A. Garanin, V. V. Ishchenko and L. V. Panina, Theor. Math. Phys. (USA) **82**, 169 (1990)
49. I. Klik and L. Gunther, J. Appl. Phys. **67**, 4505 (1990)
50. W. T. Coffey, D. S. F. Crowthers, Yu. P. Kalmykov, E. S. Massawe, and J. T. Waldron, J. Magn. Magn. Mater. **127**, L254 (1993)
51. W. T. Coffey, Yu. P. Kalmykov and J. T. Waldron, *The Langevin Equation* (World Scientific, Singapore, 1996)
52. S. I. Denisov and A. N. Yunda, Physica B **245**, 282 (1998)
53. W. T. Coffey, D. A. Garanin, H. Kachkachi, and D. J. McCarthy, J. Magn. Magn. Mater. **221**, 110 (2000)
54. K. Zhang and D. R. Fredkin, J. Appl. Phys. **87**, 4795 (2000)
55. V. L. Safonov, J. Magn. Magn. Mater. **195**, 526 (1999)
56. V. L. Safonov and T. Suzuki, IEEE Trans. Mag. **34**, 1860 (1998)
57. V. L. Safonov, J. Appl. Phys. **85**, 4370 (1999)
58. T. Holstein and H. Primakoff, Phys. Rev. **58**, 1098 (1940)
59. C. W. Gardiner, *Handbook of Stochastic Methods* (Springer, Berlin 1985)
60. V. L. Safonov and H. N. Bertram, J. Appl. Phys. **87**, 5681 (2000)
61. S. M. Stinnett, J. W. Harrell, A. F. Khapikov, and W. D. Doyle, IEEE Trans. Mag. **36**, 148 (2000)
62. H. N. Bertram and C. Seberino, J. Magn. Magn. Mater. **193**, 388 (1999)
63. H. N. Bertram and V. L. Safonov, IEEE Trans. Magn. **36**, 2447 (2000)
64. H. N. Bertram, X. Wang, and V. L. Safonov, The 8th Joint MMM-Intermag, IEE Trans. Magn., in press (2001)
65. X. Wang, H. N. Bertram, and V. L. Safonov, submitted to Phys. Rev. B (2001)
66. M. I. Freidlin and A. D. Wentzell, *Random Perturbations of Dynamical Systems* (Springer, New York 1998)

67. H. N. Bertram and H. J. Richter, J. Appl. Phys. **85**, 4991 (1999)
68. H. N. Bertram and Q. Peng, IEEE Trans. Magn. **34**, 1543 (1998)
69. R. W. Chantrell and J. Hannay, J. Mag. Soc. Japan **21**, suppl. S2, 283, (1997)
70. V. L. Safonov and H. N. Bertram, The 8th Joint MMM-Intermag, IEEE Trans. Magn., in press (2001)
71. A. Kikitsu, A. Murayama, K. Hyomi, and C. M. Falco, J. Appl. Phys. **87**, 6944 (2000)
72. M. A. Wongsam, J. D. Hannay, G. W. Roberts, and R. W. Chantrell, J. Magn. Magn. Mater. **200**, 649 (1999)

4 Measurement of Dynamic Properties in Thin-Films

Thomas J. Silva

4.1 Regimes of Magnetodynamic Response

When confronted with the task of characterizing the dynamic properties of a particular magnetic thin film, one must first consider the basic physics of the switching (or reversal) process. While the fundamental magnetic component of any ferromagnet is the electron spin, the many-body interactions between the spins, as mediated by exchange and dipolar interactions, result in a rich variety of non-linear phenomena that confound any simple classification scheme. Nevertheless, one typically finds that the dynamic properties can be generally categorized according to the applicability of various phenomenological descriptions of magnetization switching: 1) magnetic viscosity and thermal effects, where the reversal process seems to obey Boltzmann statistics at a fundamental level; 2) domain wall motion, where reversal occurs via two distinct processes of nucleation and domain growth, generally described by a viscous form of precessional dynamics; 3) incoherent rotation, a process typified by an extremely inhomogeneous spatial distribution of initial conditions and anisotropy strength, thereby requiring a statistical form of inhomogeneous dynamics; 4) coherent rotation, whereby a large fraction of a sample actually obeys a single, constitutive torque equation describing damped, precessional motion, and (5) Nonlinear coherent motion, in which the coupling between different modes of excitation is no longer amenable to simple harmonic analysis. The type of reversal process that a given material is expected to exhibit strongly influences the choice of measurement technique to be used for characterization. Given such a wide variety of magnetic reversal modes, it should not come as a surprise that a variety of different measurement techniques to characterize magnetization dynamics have been developed over the years. Each technique is generally optimized to study dynamics in one of the five regimes outlined above. In some cases, the measurement can be achieved by use of a commercially available magnetometer. In other, extreme cases, a linear particle accelerator has been pressed into service as characterization tools. While a wide spectrum of techniques and instrumentation are available, it should be emphasized that no one characterization method stands out as the best tool for the magnetodynamic study of all magnetic materials. In our treatment of this subject, we will emphasize the advantages and disadvantages of any given method so that the reader can make an informed decision as to the best approach to be used for a given class of materials.

4.1.1 Magnetic Viscosity and Thermal Effects

Magnetic viscosity (also referred to as magnetic after-effect) is typically described as a dependence of the measured hysteresis on the field sweep-rate of the magnetometer. Néel [1] was the first to recognize that such an effect must be the result of random processes acting to bring a magnetic material into thermal equilibrium with its environs. It is generally assumed that such a material is sufficiently inhomogeneous in its magnetic structure that a statistical approach to the problem is appropriate. Polycrystalline thin films typically used as recording media are excellent examples of materials exhibiting magnetic viscosity.

Perhaps the easiest way to observe viscous effects with a conventional magnetometer is to simply apply a magnetic field close to the coercivity and measure the magnetic moment as a function of time. One typically observes a slow relaxation of the magnetization into the direction of the applied field. If one plots the magnetization against the logarithm of time, it is often found that the curve is roughly linear. The reason for the quasi-logarithmic time-dependence is usually ascribed to an inhomogeneous distribution of switching parameters for an ensemble of switching volumes [2] (e.g., exchange-decoupled grains in polycrystalline thin films). In plotting such data, one must be careful in the assignment of $t = 0$, as an unwise choice for the origin in a semi-log plot can cause a misleading curve shape at the shortest timescales. A reasonable choice for $t = 0$ is the moment the field source has stabilized. For measurements of thermal relaxation of magnetization, the amount of time during which thermal relaxation can occur is the dwell time of the applied magnetic field at the maximum field magnitude. The rise and fall time of the field can be ignored since thermal relaxation is significant only when the applied field is held above a threshold value [3]. In other words, the magnetization response during the rise-time of the applied field can and should be ignored if one is interested in assessing the thermal relaxation of the magnetization. Such a choice of time-scale is justifiable because the rate of thermal relaxation is an exponential function of the applied field.

4.1.2 Domain Wall Motion

In many situations where the applied field has a significant component anti-parallel to the initial magnetization direction, magnetic reversal will proceed by a process of domain nucleation and growth. This process is particularly important in materials that display a large degree of inhomogeneity in magnetic structure. Local regions of the sample can be driven into full magnetic reversal while the bulk of the material still resists switching. The nucleation process remains poorly understood, in part because we currently lack experimental methods for the clear observation of the nucleation event and in part because stochastic processes drive the event. The lower bound on the size of a nucleated domain for a given material should be roughly equal to the domain wall width L_d, or

$$L_d = \sqrt{\frac{A}{K_u}} , \tag{4.1}$$

where A is the exchange constant with units of energy per unit length and K_u is the anisotropy energy density with units of energy per unit volume. For most ferromagnetic materials, 10^9 J/m $< A <$ 4×10^9 J/m. K_u can range from 10^{22} J/m^3 to 3×10^{25} J/m^3. Thus, we can expect the characteristic size of a nucleation event to range anywhere from 0.6 µm to 7 nm. Nevertheless, it should be clear that detection and study of the nucleation process requires high-resolution magnetic imaging techniques that range in resolution from micrometer to nanometer scales.

The process of domain wall motion is much better understood [4,5]. It is believed that the damping of gyromagnetic precession ultimately limits the velocity of domain wall motion. Domain wall velocity v_d is typically a linear function of applied field H for fields in excess of the pinning field H_0, with the relation

$$v_d = \mu(H - H_0) , \tag{4.2}$$

Values of mobility μ typically range from 0.1 to 1 m^2/(A s) (10 to 100 m/(Oe s)).

Domain wall velocity is often described as a slower mechanism for magnetic reversal. One should be careful in ascribing a specific time regime to switching via domain wall motion because both the wall velocity and the characteristic distance a wall must move to fully reverse a sample will set the time scale. For example, it is plausible that a field pulse of only 8 kA/m (25 Oe) could switch a 0.25 μm magnetic particle *in only* 100 ps by means of domain wall motion. Thus, conclusions that switching is driven by domain wall motion should be supported with experimental evidence that the process is indeed characterized by separate stages of domain nucleation and growth.

4.1.3 Incoherent Rotation

If a magnetic field is slowly applied along the hard axis of an ideal uniaxial sample, i.e., orthogonal to the anisotropy axis, then the magnetization rotates adiabatically into the direction of the applied field. From Stoner-Wohlfarth theory [6] for a single-domain object with anisotropy energy U_k, the angle of rotation θ is given by

$$\theta = \arcsin\left(\frac{H}{H_k}\right) , \tag{4.3}$$

where H_k is the anisotropy field, defined by $H_k = 2U_k/\mu_0 M_s$ ($\mu_0 = 4\pi \times 10^{-7}$ H/m, M_s = saturation magnetization). It is understood, however, that most materials of practical importance for data storage applications suffer from some dispersion in the anisotropy properties throughout a given sample. The dispersion in the anisotropy gives rise to a spatial nonuniformity of the magnetization, or what is generally called ripple [7,8]. As such, (4.3) represents a spatially averaged response that can grossly deviate from the actual microscopic response at the nanometer length scale. In a dynamical situation where the applied field is altered faster than the relaxation rate of the magnetization, it is then possible for the switching process to produce nonuniformities in the magnetization distribution that are not in the thermodynamic ground state [8]. If the nonuniformity in the magnetization state is of a degree sufficient to significantly *impede* the switching process, then one speaks of an incoherent switching process [8]. It is believed that the impediment to switching is the result of significant internal dipole fields that are generated during the reversal process [8]. In any case, the instigation of highly non-uniform states during the dynamic response of magnetization to an applied field is an extremely complex process that remains poorly understood.

4.1.4 Coherent Rotation

In the limit of an ideal ferromagnet with uniaxial anisotropy, the response of the magnetization to an applied field is presumed to be highly coherent. The coherent regime is typified by highly uniform dynamics throughout the excited volume and very reproducible behavior. In addition, the speed of reversal is at the fundamental limit, as determined by the precessional frequency of the constitutive spins. Indeed, it is the phase coherence of the spin precession in a given volume that uniquely distinguishes this switching mode from any of the others that are described above. The Landau–Lifshitz–Gilbert (LLG) equation (See Sect. 4.2.1) is only rigorously applicable for the case of coherent rotation of a given magnetic volume. Thus, use of LLG for micromagnetic simulations presumes that coherence is maintained for length scales smaller than the discretization level of the simulation. While the existence of such a length scale for arbitrary experimental parameters has yet to be established, there are experimental results to confirm that coherent rotation does indeed occur under certain limiting conditions. For example, if a field pulse is applied along the hard axis of a well-aligned uniaxial sample, coherent response is typically observed [9]. Coherent response is preserved even if the field pulse is misaligned from the hard axis by a moderate angle. It is only as the field pulse approaches the easy axis direction that coherent response is replaced by reversal modes that involve local instabilities in the magnetization orientation [10]. Such instabilities result in highly stochastic response that confounds any measurement technique based upon the averaging of reproducible phenomena.

4.1.5 Small Signal vs. Switching

When studying the response of a magnetic material to small-amplitude excitations (as in a permeability measurement), one should keep in mind that magnetic systems are extremely susceptible to nonlinear interactions that can lead to unstable behavior at large amplitudes [11,12]. One should never presume that the simplifying assumptions (which lead to the usual equations of magnetodynamic response) are applicable for large excitations. This warning is not to be taken lightly, because the magnetic recording process is undoubtedly a process involving very large excitations. Efforts to experimentally measure the bandwidth of materials relevant to data storage applications should involve some attempts to duplicate the amplitudes and conditions applicable to the magnetic recording process. Unfortunately, it is not always possible experimentally to fully duplicate the recording process and still extract data that are amenable to theoretical understanding. Thus, the reader should consider the following outline of methods and results as a summary of tools to uncover the underlying physics of the high-speed recording process, as opposed to being simply "revved-up" recording simulations. While the latter has utility in the testing and evaluation of recording methods and components, the former can be used to enhance our basic understanding of the magnetic switching process, and it is upon the foundation of solid, physical understanding that technological innovation may be built.

4.2 Inductive Methods

Inductive techniques are perhaps the simplest class of measurements to determine the bandwidth of a magnetic material. A high-frequency magnetic field is applied to a specimen using a waveguide structure and the magnetic response is then determined by measuring an inductive voltage induced in a pick-up coil. The general approach is equivalent to a high-frequency measurement of inductance. The excitation sources can be either broadband (impulse and/or step pulses) or narrowband (swept rf signal.) Broadband signal sources are complementary with time-resolved measurement methods, such as a high-speed sampling oscilloscope, whereas the narrowband sources are more compatible with frequency-resolved measurements using a network analyzer. Both sources have various advantages and disadvantageous that will be discussed below in more detail.

Electrical sampling methods use a high-speed sample-and-hold circuit to measure the voltage in a given waveform at an exact moment in time relative to the activation of a trigger circuit. While the bandwidth of the sample-and-hold circuit is very broad, the output circuit (typically an analog-to-digital converter) can be quite slow, facilitating a very high signal-to-noise ratio (SNR). The disadvantage is the inability to measure single-shot processes: Each time a trigger pulse is sent to the oscilloscope, the oscilloscope actually measures only one point of the waveform. A complete trace is obtained after multiple measurements with varying delay between the arrival of the trigger pulse and the activation of the sample-and-hold circuit.

4.2.1 Narrowband Measurements

To understand what response we should expect from the high permeability magnetic materials typically used for data storage applications, we need to review the fundamentals of ferromagnetic resonance (FMR) [13]. The fundamental response of electron spins to a magnetic field H is precession. The frequency of precession is given by the Larmor equation,

$$\omega = \gamma \mu_0 H \ , \tag{4.4}$$

where γ is the gyromagnetic ratio (typically $2\pi \times 28$ GHz/T) and μ_0 is the permeability of free space ($4\pi \times 10^{-7}$ H/m). However, because of the high density of spins in a ferromagnet, dipole coupling cannot be neglected. Solving the gyromagnetic equation of motion for the case of a thin film geometry, and including intrinsic uniaxial anisotropy, we have the important result

$$\omega = \gamma \mu_0 \sqrt{M_s (H_k + H)} \ , \tag{4.5}$$

where H is applied either parallel (+) or orthogonal (−) to the easy axis, H_k is the anisotropy field, and M_s is the saturation magnetization. For a material such as Permalloy ($Ni_{81}Fe_{19}$), with $M_s = 800$ kA/m (800 emu/cm^3) and $H_k = 400$ A/m (5 Oe), the theoretical FMR frequency is 600 MHz for $H = 0$. The resonance frequency is higher for materials with larger moment (or anisotropy) than Permalloy. Therefore, any instrument for the assessment of the bandwidth of head materials should have a bandwidth in excess of several GHz.

The Landau–Lifshitz–Gilbert (LLG) equation is the usual starting place for the analysis of high-frequency response of ferromagnets [5]. The equation is typically written as

$$\frac{d\vec{M}}{dt} = -\gamma \vec{T} + \frac{\alpha}{M_s} \left(\vec{M} \times \frac{d\vec{M}}{dt} \right) \ , \tag{4.6}$$

where we use a generalized torque as the source of precessional motion. Here, α is the phenomenological damping parameter. The second term in (4.6) describes the eventual relaxation of the magnetization into the equilibrium direction defined by $\vec{T} = 0$.

The generalized torque allows us to include anisotropy effects for a ferromagnetic slab in the $\theta = \pi/2$ plane using a free energy description of the magnetic state (in spherical coordinates):

$$\vec{T} = -\vec{r} \times \vec{\nabla} U(\theta, \phi) = \frac{\partial U}{\partial \theta} \vec{\phi} - \frac{1}{\sin\theta} \frac{\partial U}{\partial \phi} \vec{\theta} \ , \tag{4.7}$$

where the film lies in the $\theta = 0$ plane. In the limit of a sample with large magnetic moment and small motion of the magnetization out of the film plane, we have the following approximate form for LLG:

$$\frac{d^2\phi}{dt^2} + \left(\frac{\alpha}{1+\alpha^2}\right)\gamma\mu_0 M_s \frac{d\phi}{dt} - \frac{\gamma^2}{(1+\alpha^2)}\mu_0 \frac{\partial U}{\partial \phi} = 0 \,. \tag{4.8}$$

For a single domain magnet with uniaxial anisotropy (easy axis along $\phi = 0, \pi$), the free energy is

$$U(\phi) = \mu_0 M_s \left(\frac{H_k}{2}\sin^2(\phi) - H\cos(\phi - \beta)\right) \,, \tag{4.9}$$

where β is the angle of an in-plane applied field H relative to the easy axis. Assuming small angle motion of the magnetization, we can reduce (4.9) to that of a harmonic oscillator:

$$\frac{d^2\phi}{dt^2} + \frac{2}{\tau}\frac{d\phi}{dt} + \omega_0^2(\phi - \phi_0) = 0 \,, \tag{4.10}$$

where

$$\tau = 2(1+\alpha^2)/\alpha\gamma\mu_0 M_s \,, \tag{4.11}$$

and

$$\omega_0 = \frac{\gamma\mu_0}{\sqrt{1+\alpha^2}}\sqrt{M_s(H\cos(\phi_0 - \beta) + H_k\cos 2\phi_0)} \,, \tag{4.12}$$

where ϕ_0 is the static equilibrium angle for the magnetization. Applying a Fourier transform to (4.10) and solving for the normalized response function $\Phi(\omega)$, we have the usual result for a harmonic oscillator,

$$\Phi(\omega_0, \omega) = \frac{\omega_0^2}{(\omega_0^2 - \omega^2) - i\frac{2\omega}{\tau}} \,. \tag{4.13}$$

Two applied field geometries merit special attention: 1) longitudinal bias ($\beta = 0$), and 2) transverse bias ($\beta = \pi/2$). In the case of longitudinal bias, we have

$$\omega_2 = \frac{\gamma\mu_0}{\sqrt{1+\alpha^2}}\sqrt{M_s(H + H_k)} \,. \tag{4.14}$$

Eq. (4.14) is almost identical to (4.5), except for the appearance of a factor of $1/(1+\alpha^2)^{1/2}$. This factor is not present when using a Landau–Lifshitz formulation for ferromagnetic dynamics [14], where the precessional and damping torque are rigorously orthogonal. Physically, this factor accounts for the contribution of the damping torque to the precessional motion of the magnetization. The damping torque has a component that opposes the precessional motion in the LLG formulation. This component results in a reduction in the precessional frequency ω_0. In addition, the damping torque is proportional to the rate of magnetization change, such that a reduced precessional frequency also results in a reduced relaxation rate, $1/\tau$. This is manifest in (4.11). Both rate reductions are negligible when $\alpha \ll 1$, but they should have a dramatic effect for materials where $\alpha \to 1$ or higher.

For the case of transverse bias, we have two possible forms for the resonance frequency,

$$
\begin{aligned}
\omega_0 &= \frac{\gamma\mu_0}{\sqrt{1+\alpha^2}}\sqrt{M_s(H_k(1-(H/H_k)^2))}\ ; \quad H \le H_k \\
&= \frac{\gamma\mu_0}{\sqrt{1+\alpha^2}}\sqrt{M_s(H-H_k)}\ ; \quad H \ge H_k
\end{aligned}
\tag{4.15}
$$

We can now use (4.13) to calculate the small-angle susceptibility spectrum for a magnetic thin film, where the susceptibility is defined by $M = \chi H$. The real (or reactive) part of the spectrum is simply

$$
\chi'(\omega) = \frac{M_s}{(H+H_k)}\left[1-(\omega/\omega_0)^2\right] / \left(\left[1-(\omega/\omega_0)^2\right]^2 + \frac{1}{Q^2}(\omega/\omega_0)^2\right)
\tag{4.16}
$$

and the imaginary (or lossy) part is

$$
\chi''(\omega) = \frac{M_s}{(H+H_k)}\frac{1}{Q}(\omega/\omega_0) / \left(\left[1-(\omega/\omega_0)^2\right]^2 + \frac{1}{Q^2}(\omega/\omega_0)^2\right) , \tag{4.17}
$$

where $Q = 2\omega_0\tau$ is the "quality" factor of the resonance. Fitting of a measured permeability spectrum to (4.16) and (4.17) permit the determination of resonance frequency ω_0 and damping time τ. The resonance frequency may be approximated from either the zero-crossing of or the peak of χ' or the peak of χ''. If, however, the damping is not small enough to ignore, the measured precessional frequency ω_p will be slightly less than ω_0. The difference is given by

$$
\omega_p^2 = \omega_0^2 - (1/\tau)^2 \ . \tag{4.18}
$$

For a material like Permalloy with $H_k \sim 400$ A/m (5 Oe) and with a damping parameter $\alpha = 0.015$, the effect of the natural resonance-frequency shift for $H = 0$ is equivalent to an anisotropy shift of 40 A/m, or 10% of H_k. However, if the damping is 0.03, the frequency shift is equivalent to an anisotropy reduction of 160 A/m (2 Oe), or 40% of H_k. Thus, simply using the peak frequency in χ'' to determine the anisotropy of an arbitrary material can be seriously misleading, even for moderate damping.

The damping time can be estimated using $\tau = 1/4\pi\Delta f$ where Δf is the full-width-at-half-maximum of the peak in χ''. If a bias field is applied, then the dependence of the resonance frequency on H (found in (4.14) and (4.15)) can be used to extract the anisotropy and gyromagnetic ratio, if M_s is already known for the film in question.

For bias fields much less than the magnetization of the film, the magnetic motion is highly elliptical. Nevertheless, the motion of the magnetization out of the film plane is inescapable in the limit of precessional motion. To estimate the ellipticity ε of the motion, one need consider only the ratio of the perpendicular and in-plane restoring forces:

$$\varepsilon = \sqrt{\frac{M_s}{H + H_k}} \, . \tag{4.19}$$

For Permalloy, the ellipticity in the absence of any applied field is 45. Thus, a dynamic response of 45 degrees in the film plane is accompanied by a maximum arc of only one degree out of the film plane.

Several approaches may be used to measure permeability with continuous wave (cw) microwave excitations [15–18]. All of these methods use a broadband network analyzer both as the source of cw signal and the instrument for signal analysis. The basic idea is to determine the impedance change in a microstrip waveguide when loaded with a magnetic sample. Magnetic effects are isolated from all other origins of impedance change by using external field coils to saturate the sample for a background measurement that is subtracted from subsequent data. The microstrip employed is typically on the order of 1 cm wide, generating fields on the order of 1 A/m (13 mOe) for a 250 mV signal and 50 Ω characteristic impedance. Bandwidths as high as 2–3 GHz have been demonstrated. Such an instrument may be used to quickly determine the microwave response at the low excitation amplitudes of many magnetic materials, without the need for any sample preparation. The simplicity of the measurement system is appealing for those wishing to quickly characterize a variety of materials. Comparison of the data (such as shown in Fig. 4.1) with the linear response model depicted in (4.16) and (4.17) can reasonably test the degree to which a given material is limited purely by gyromagnetic processes, or further constrained by additional limitations, such as eddy current losses or inhomogeneous effects.

4.2.2 Broadband Measurements

The first broadband measurements were accomplished shortly after the invention of the high-speed electrical sampling oscilloscope. Dietrich, Proebster, and Wolf [19] used a homemade sampling oscilloscope to measure the switching speed of Permalloy films intended for thin-film memory applications. A modern variation on the inductive method has been developed recently by Silva et al. [20] (see Fig. 4.2). This new approach uses a waveguide as a pulse-delivery mechanism, but avoids the complication of an inductive, broadband, pickup coil by relying on the same waveguide for signal detection. Small coplanar waveguide structures are employed to facilitate the use of

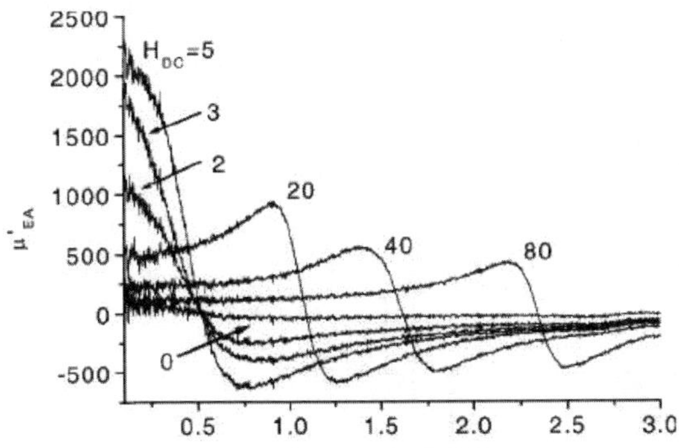

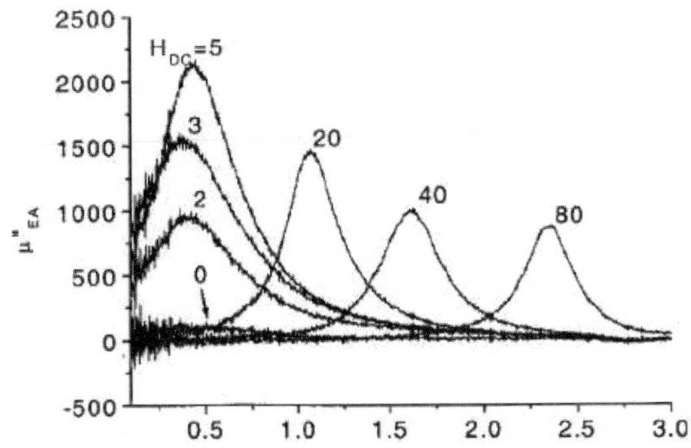

Fig. 4.1. Real and imaginary permeability spectra measured by narrow-band technique. Bias field H_{DC} is applied along hard axis. Sample is Mo-doped Permalloy. From [18], used with permission from © The AIP

solid-state pulse generators. A coplanar geometry has the advantage that the width of the center conductor w can be modified gradually without changing the characteristic impedance of the waveguide. A waveguide with a 50 µm center conductor width was demonstrated in [20]. A sample is placed on top of the waveguide, where the transverse magnetic field H_p of the mode structure excites the dynamic response. Since H_p scales as $I/2w$ directly on top of the center conductor, fields as large as 2 kA/m (25 Oe) are generated by a 50 µm waveguide with 50 Ω characteristic impedance and a 10 V pulse generator [20].

Since both the excitation pulse and the inductive signal of interest are measured with the sampling oscilloscope, care must be taken to avoid damaging the oscilloscope with large field pulses. It may be necessary to attenuate the signal entering the oscilloscope if large voltage pulses are employed.

Time traces are acquired *with* and *without* a saturating magnetic field applied orthogonal to the waveguide direction. These two traces are then

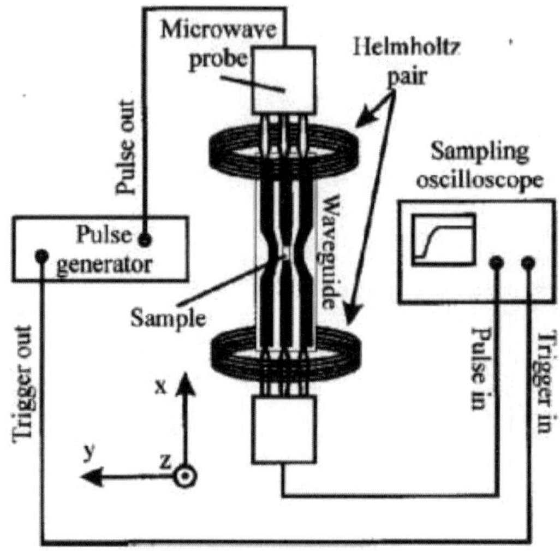

Fig. 4.2. Diagrammatic sketch of modern inductive measurement system. From [20], used with permission from © The AIP

subtracted, with the inductive signal from the sample remaining. With sufficient averaging, the signals are obtained with 16 bits of dynamic range. For example, if a sampling oscilloscope has a signal range of 800 mV, then the noise floor is only 12 μV with maximum averaging (typically 4096 averages). Thus, a 1 mV inductive signal can be measured with a SNR of 38 dB, even with a superimposed 800 mV excitation pulse. If a pulse generator of greater voltage is employed, the attenuators that are required at the front end of the oscilloscope reduce the effective dynamic range.

An example of measured precessional ringing is shown in Fig. 4.3. For $H_b = 0$, the magnetization rotation is in excess of 90 degrees. For increasing bias field, the rotation is reduced, but the precessional frequency increases in accordance with (4.5).

The magnitude of the inductive signal is proportional to the thickness and length of the sample placed upon the waveguide, as well as the spacing between the magnetic film and the waveguide. In addition, since the signal is inductive in nature, it is proportional to the characteristic frequency of the magnetic motion. To estimate the magnitude of the inductive signal, we can use

$$|V_{ind}|_{p-p} = \frac{\mu_0 L \delta \chi_0 \omega_0}{4} \left(\frac{V_{step}}{2 Z_0 w} \right) , \tag{4.20}$$

assuming negligible spacing between the sample and waveguide and $\delta/w \to 0$. The permeability of free space is μ_0, the characteristic angular frequency of oscillation is ω_0, V_{step} is the amplitude of the voltage step, the film thickness

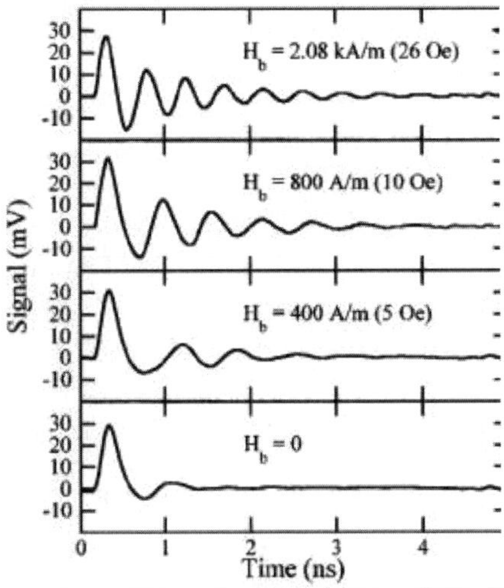

Fig. 4.3. Precessional ringing for thin film Permalloy obtained by an inductive technique. Sample is driven by a 50 ps, 10 V step excitation. Bias field H_b is applied along the easy axis of sample. From [20]. Used with permission from © The AIP

is δ, the length of the sample is L, the waveguide impedance is Z_0, the waveguide width is w, and the DC susceptibility of the sample is χ_0. The factor of 1/4 in (4.20) accounts for two geometrical effects that reduce the measured inductive voltage: First, only half of the flux produced by a sample can thread around the center conductor of a waveguide, inducing a voltage. Second, the induced voltage pulse is actually bimodal, with half the pulse traveling in the forward direction, and half travelling back toward the pulse generator. Only the forward pulse is detected with the sampling oscilloscope. For a Permalloy film ($M_s \sim 800$ kA/m) that is 1 mm long and 50 nm thick, with a characteristic frequency of 1 GHz and full saturation response, the inductive signal is estimated to be 80 mV. Such a signal can be measured with a SNR of 36 dB if a 10 V pulse is used.

Spacing effects can be minimized by placing the sample in direct contact with the waveguide. Shorting of the waveguide is avoided by coating the sample with insulating photoresist before measurement. If the spacing between the waveguide and the sample is less than 10 µm for a waveguide 500 µm in width, then the effect of spacing is less than 5.4 dB of signal loss.

Sample shape can affect the amplitude and resonance frequency obtained from such an instrument [20]. One must consider the dipole fields produced when the width of a sample is comparable to the thickness of the sample times the ratio of the saturation magnetization to the intrinsic anisotropy. The dipole fields are an additional source of anisotropy that will reduce the magnitude of the magnetic excitation and increase the resonance frequency.

For a sample of Permalloy, 50 nm in thickness, this shape-induced anisotropy is significant for waveguide widths below 1 mm.

We can use spin wave theory to calculate the effect of shape anisotropy for the case of a sample that is much wider than the waveguide [21,22]. Under such circumstances, the waveguide is effectively an antenna that couples to long wavelength spin waves in the sample. In the limit where the film thickness is much less than the antenna width, we can use the Damon-Eshbach (DE) theory for magnetostatic (long wavelength) spin waves [23]. The DE theory is applicable for wavelengths sufficiently long that exchange energy plays a negligible role compared to the internal demagnetizing fields. For a sample biased such that $\vec{M}$ is oriented along the axis of the waveguide, the field pulses delivered by the waveguide couple predominantly to the so-called DE surface modes. The DE surface modes propagate in a direction orthogonal to the average magnetization direction. In general, a continuous spectrum of such modes can be generated, with a dispersion relation

$$\omega_0(k) = \gamma\mu_0\sqrt{(M_s)\left(\frac{1}{2}M_s\delta k \pm H_k + H\right)}\,, \tag{4.21}$$

where $\delta k \ll 1$ and $H + H_k \ll M_s$ [22]. The sign for K_k in (4.21) depends on whether a longitudinal (+) or transverse bias (−) is employed. Under linear conditions, each mode has an associated susceptibility for excitation by waveguide field components of the same k-value:

$$\chi_0(k) = \frac{\chi_0}{2(\omega_0(k)/\omega_0(0)) - 1}\,, \tag{4.22}$$

where χ_0 is the DC susceptibility of the uniform mode (i.e. $k = 0$), given by $\chi_0 = M_s/(H + H_k)$. Using elementary linear response theory and the reciprocity theorem, we can calculate the spectrum of the induced voltage [24], which includes the contribution of all the excited modes:

$$V(\omega) = \frac{\mu_0 L\delta V_{step}}{8\pi Z_0}\int_0^\infty [\sin^2(kw/2)/(kw/2)^2]e^{-2kd}\chi_0(k)\Phi(\omega_0(k),\omega)dk\,, \tag{4.23}$$

where w is the width of the waveguide, Z_0 is the characteristic impedance of the waveguide, and d is the spacing between the sample and the waveguide. The derivation of (4.23) assumes a perfect step excitation and ignores the propagation of the DE surface modes away from the waveguide. The principal consequence of the DE mode excitation for a wide waveguide (i.e., $\delta/w \ll (4/\pi\chi_0)$) is a shift of the measured resonance to a higher frequency than would occur for a uniform excitation.

In the limit of a wide waveguide and small waveguide-sample spacing ($2\pi d/w \ll 1$), we can estimate the measured frequency shift Δf. To good approximation, the shift is given by [24]

$$\Delta f = \frac{\omega_0(0)}{4\pi}\left[\frac{\bar{\chi}}{\ln(1+\bar{\chi})} - 1\right] , \tag{4.24}$$

where $\bar{\chi} = (\pi/2)(\delta/w)\chi_0$. For example, suppose one measured a 50 nm film of Permalloy ($M_s = 800$ kA/m (800 emu/cm^3), $H_k = 400$ A/m (5 Oe)) with a waveguide of dimensions $w = 500$ µm and $d = 10$ µm. The resultant frequency shift from (4.24) is $\Delta f = 47$ MHz. Numerical integration of (4.23) gives a frequency shift of $\Delta f = 48$ MHz, in good agreement with the approximate value. Such a frequency shift is equivalent to an additional anisotropy component of 63 A/m (0.8 Oe). Indeed, it may be shown [24] that the role of film thickness is analogous to that of shape anisotropy, with a resultant anisotropy contribution:

$$H_{k,DE} \approx \frac{\pi}{4}\frac{M_s\delta}{w}; \qquad \frac{\delta\chi_0}{\pi w} \leq 1 . \tag{4.25}$$

In the limit of linear response ($\sin\phi \sim \phi$), the magnetodynamics should obey (4.10). For a time-domain measurement, a Laplace transform solution to (4.10) can be used for either step or impulse excitations. The step response solution is [20]

$$\phi(t) = \frac{H_p}{(H_k + H)}\left(1 - \frac{\exp(-t/\tau)}{\cos(\beta_s)}[\cos(\omega_p t + \beta_s)]\right) u(t) , \tag{4.26}$$

where H_p is the field amplitude of the step excitation, $u(t)$ is the Heaviside function, and the phase angle is defined by $\tan(\beta_s) = 1/(\omega_p\tau)$. The impulse response solution is found by simply taking the first derivative of (4.26):

$$\phi(t) = \frac{H_p}{(H_k + H)}\frac{(\omega_p\Delta t)}{\cos(\beta_i)}\exp(-t/\tau)\sin(\omega_p t + \beta_i)u(t) , \tag{4.27}$$

where the area of the impulse used to excite the magnetization is $H_p\Delta t$, and the phase angle β_i is given by $\tan(\beta_i) = 2/(\omega_p\tau)$.

The integrated signal from a time-resolved inductive measurement may be fitted with (4.26) and/or (4.27) to determine the bandwidth and damping for a given material. Alternatively, the raw inductive waveform for a step excitation may be subjected to a Fourier transform in order to directly obtain the permeability spectrum [25]. Since the inductive signal is proportional to the time-derivative of the magnetic response, and the step response is time-integrated relative to an impulse excitation, the two effects (integration due to step response, differentiation due to inductive response) cancel each other to result in the impulse response waveform. Fourier transform of the waveform results in a spectrum comparable to (4.16) and (4.17). Care must be taken to account for the phase of the time-response waveform when applying a fast-Fourier-transform (FFT) algorithm. In essence, one is using the FFT to approximate the Laplace transform. For example, if one wishes to obtain a positive-only imaginary spectrum, it may be necessary to multiply the resultant spectrum by a phase factor of the form $\exp(i\omega t_0)$, where t_0 is a suitable time shift.

The inductive coupling between the waveguide and sample can act as an additional source of damping. The voltage oscillations induced in the waveguide at the precessional frequency of the sample carry energy away from the sample in the form of microwave radiation. It is straightforward to estimate this source of loss. Using the Stoner-Wohlfarth equation [6] for the free energy of a single-domain entity, the precessional energy stored in the sample immediately after application of the field pulse, H_p, is equal to the difference in the free energy immediately before and immediately after the field pulse is applied:

$$U = \frac{\mu_0}{2} \Delta M H_p \delta L w = \frac{\mu_0}{2} \frac{M_s A w}{(H_k + H)} H_p^2 , \tag{4.28}$$

where $A = \delta L$ and $\Delta M = \mu H p$ is the magnitude of the magnetization that rotates into the field pulse direction. The power transmitted into the waveguide immediately after application of the pulse is simply

$$P = 4\frac{VV^*}{Z_0} = \frac{(\omega \mu_0 \Delta M \delta L)^2}{4Z_0} , \tag{4.29}$$

where the factor of 4 accounts for the fact that both forward and backward-traveling inductive pulses generated during the precessional oscillations carry energy away from the sample. The time scale τ_{rad} for the radiative loss is then the ratio of the energy stored to the energy dissipation rate

$$\tau_{rad} = \frac{U}{P} = \frac{2wZ_0}{(\gamma\mu_0)^2 \mu_0 M_s^2 A} = \frac{2wc}{A\omega_M^2 \sqrt{n_{\text{eff}}}} , \tag{4.30}$$

where c is the speed of light, n_{eff} is the effective refractive index of the waveguide, and $\omega_M = \gamma\mu_0 M_s$. This source of precessional decay is predominantly a function of the measurement geometry and the saturation moment of the sample. For example, if one is measuring a 1 cm $\times$ 1 cm sample of 50 nm thick Permalloy with a 50 Ω waveguide that has a 0.5 mm wide center conductor, the associated decay time in the absence of any other damping mechanism would be $\tau_{rad} = 2.5$ ns. This is equivalent to a damping parameter of $\alpha_{rad} = 2/\tau_{rad}\omega_0 = 0.004$. With this example, we see that the radiative losses are not necessarily negligible. To correct for this additional damping in an attempt to determine the sample-intrinsic damping, α_{rad} must be subtracted from any measured values of α. If the sample properties are such that radiation is the dominant source of precessional decay, one can increase the width of the waveguide to decrease the radiative damping, albeit at the expense of the measured inductive signal because of the reduced field pulse amplitude.

We have seen that several corrections must be made to (4.5) in order to correctly determine γ and H_k from precessional frequency data. To summarize, first, there is a "red shift" in the spectrum due to the mixing of damping and precessional torque in the LLG equation (4.14). To correct for this effect, one should multiply the fitted value of γ by $\sqrt{1+\alpha^2}$ to obtain the gyromagnetic ratio for LLG. Second, there is a "blue shift" due to the coupling to finite-wavelength DE spin wave modes in the sample (4.25). Here, the correction is an adjustment to the fitted value of H_k. Finally, there is a "red shift" due to the bandwidth limiting effect of nonzero damping (4.18). This requires one to shift the square of the measured precessional frequency in accordance with (4.18). This, too, most stronly affects the fitted value for H_k. The sign of the first and last corrections to the anisotropy are independent of whether traverse or longitundinal bias is used. Thus, if one measures the frequency as a function of bias field for both longitudinal and transverse bias geometries and fits the data to extract the anisotropy for the two geometries, the true anisotropy may be accurately determined by taking the average result for H_k [18].

Both the time-resolved and frequency-resolved methods rely upon careful design and fabrication of the waveguide used to couple rf radiation to the magnetic sample. The final result should be a structure that is properly impedance-matched to the coaxial line leading to and from the waveguide. If a narrow conductor width is desired for the measurement in order to determine shape effects or to increase the magnetic component of the rf fields, one should use a coplanar structure, since the impedance of a coplanar waveguide is a function of the ratio of the width of the center conductor to the separation between the ground planes [26,27]. A wide center conductor may be used to launch the wave in the guide, followed by a tapered section along which the width-to-separation ratio is kept constant. If, on the other hand, maximum magnetic field amplitude is required, but at macroscopic dimensions to avoid any possible shape effects in the precessional dynamics, a microstrip waveguide is perhaps better suited to the task. If the sample is placed within the dielectric space between the conductive strip and the ground plane, a magnetic field almost twice that of the coplanar structure may be produced in the sample. The increased magnetic field results from an image current distribution produced in the ground plane of equal but opposite direction to that in the conductive strip.

Time-domain reflectivity (TDR) is a powerful means of diagnosing the impedance properties of a waveguide structure [28]. With TDR, a fast rise-time step pulse is emitted from the front end of a high speed sampling oscilloscope. If the pulse encounters any mismatch in impedance, the reflected pulse is captured by the oscilloscope with a time delay indicative of the spatial location of the impedance mismatch. This allows one not only to assess whether an impedance mismatch is present, but also to determine the location of the mismatch. For example, a poor interconnect joining the waveguide to the coaxial line may be found quickly by such a method.

4.3 Magneto-Optical Methods

Ultrafast laser systems have been used for many years to characterize high-speed electrical signals in microwave circuits [29]. The technique was extended to study high-speed magnetic phenomenon using the magneto-optic Kerr effect (MOKE) [30]. Bandwidths as high as 200 GHz have been demonstrated with an insulating ferromagnetic sample using picosecond laser systems [31]. Both electrical and magneto-optic techniques rely upon sampling methods to obtain such high bandwidths. The sampling method is described in detail in the previous section. Lasers capable of producing 50 fs optical pulses using mode-locked Ti:sapphire architectures are now available commercially. Such lasers have a potential bandwidth of 7 THz. In practice, most magnetic phenomena of interest for data storage applications are restricted to a bandwidth of no more than 10 GHz, thus limiting the required pulse duration to only 35 ps. Indeed, a diode laser with the capability to produce pulses of 10 ps duration has been demonstrated as a cost-effective alternative to mode-locked solid state lasers [32].

As with the time-resolved inductive methods outlined in the previous section, samples are excited with the stray magnetic fields produced by a waveguide structure transmitting a microwave pulse. Placing the sample between the conductive strips of a coplanar waveguide produces magnetic fields perpendicular to the film plane. On the other hand, the region directly over the conductive strips produces a longitudinal magnetic field. Both geometries have been used effectively to excite precessional dynamics in ferromagnetic samples. To generate the magnetic pulse, part of the laser pulse is diverted with a beam splitter toward a photoconductive switch, consisting of metallic electrodes deposited on a semiconductor substrate. A diagram of such a system is shown in Fig. 4.4. Such switches can produce pulses with rise times as fast as 15 ps [33]. A less exotic alternative that has proven effective for magnetodynamic studies is to use a reverse-biased photodiode with a high-bandwidth choke to isolate the DC voltage from the microwave electronics. A commercial InGaAs diode has been used to produce ~1 V pulses as short as 200 ps in duration [20].

The photo-induced pulses may be used to directly excite the sample, or they may be used to trigger a solid-state pulse generator. Pulse generators have the advantage of large amplitudes and controlled pulse duration. On the other hand, pulse generators can operate at only somewhat limited repetition frequencies to avoid damaging their circuitry. Since most mode-locked high-speed lasers operate at a repetition frequency on the order of 100 MHz, direct triggering of the pulse generator with the laser is not possible. "Pulse picker" technology must be used to reduce the repetition rate of the laser in the case of

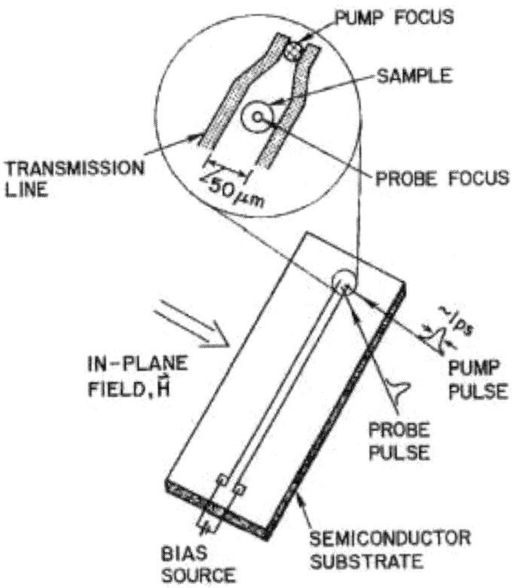

Fig. 4.4. Photoconductive switch and coplanar charge line used to excite magnetic sample with picosecond field pulse. Pump pulse creates an electrical short at the end of the charged waveguide due to the generation of photo-excited carriers in the semiconductor substrate. The resultant electrical disturbance propagates back from the short to the sample. From [31]. Used with permission from © The AIP

conventional mode-locked lasers. A pulse picker is a high-speed electro-optic beam deflection system that acts as a fast shutter to permit the unimpeded passage of optical pulses at a user-defined rate. If used in conjunction with countdown electronics, pulse pickers can reduce the repetition frequency to an arbitrarily small number.

Pulse pickers do not attenuate the rejected pulses with 100% efficiency. They leak power in the rejected pulses with an extinction ratio usually on the order of 200:1 to 400:1. This means that the leakage power in 200–400 of the rejected pulses is equivalent to one accepted pulse. If the output of the pulse picker is only 1 MHz and a 100-MHz repetition-rate laser is being used, then 20–30% of the *average* optical signal is due to rejected pulses. This leakage signal must be accounted for in a time-resolved measurement if one wishes to calibrate the MOKE signal relative to the magnetic state of the sample. For example, if the MOKE signal for a sample in an ambient magnetic state is ϕ_0, and the instantaneous MOKE signal in the excited state is $\phi_0 + \Delta\phi$, then the actual measured signal in the excited state (assuming a 300:1 extinction ratio and a 1 MHz final repetition rate with a 100 MHz mode-locked laser) will be (leakage + ambient + signal) $= (1/300) \times 99 \times \phi_0 + \phi_0 + \Delta\phi = 1.32\phi_0 + \Delta\phi)$. If $\Delta\phi$ is on the order of ϕ_0 (to be expected for a weak contrast signal), then the leakage contribution will constitute a serious error in the accuracy of a calibrated measurement. This is particularly important if one wishes to make a vectorial measurement with such a system.

The choice of electronic delay can seriously affect such a time-resolved measurement. The jitter produced by many commercial delay generators will greatly reduce the effective bandwidth of the measurement system. Analog pulse generators typically produce the lowest measured jitter for the 10–20 ns delays usually required.

Accurate electrical trigger pulses can be derived from the optical output of a laser using a reverse-biased small-area photodiode. Surprisingly large bias voltages (~30–40 V) can be used if the optical pulses are of low total energy (<1 nJ) without damaging many PIN photodiodes. At such high-bias voltages, the bandwidth of a diode is much higher than specified. Electrical pulses of 1 ns duration can be produced when using a 0.3 mm diameter PIN diode with a 45 V reverse bias and optical pulses with 0.5 nJ of energy. The amplitude of such a pulse, assuming quantum efficiency ε, output impedance Z_0, and electrical pulse duration t_{50}, will be given by

$$V_{pulse} = \frac{\varepsilon e Z_0 \lambda}{t_{50}} \frac{U_{pulse}}{hc} , \tag{4.31}$$

where U_{pulse} is the total energy of the optical pulse, h is Planck's constant, and λ is the laser wavelength. Thus, a diode with 20% quantum efficiency connected to a 50 Ω load will produce −5 V pulses *without any amplification.* Assuming a laser with 0.1% rms noise over a 10 MHz bandwidth, the intrinsic jitter for the rising edge of such a pulse is only (5 mV/10 MHz) × (0.35/1 ns)/ (5 V/1 ns) = 35 ps. Such pulses are an excellent means of triggering any ancillary electrical equipment used for the measurements.

If a commercial pulse generator is used with a mode-locked laser, one must accommodate the intrinsic insertion delay associated with commercial delay- and pulse-generators. Typical insertion delays are on the order of 30–50 ns. Since the laser is used as the base timing source for the entire experiment, one is not at liberty to delay the output of a laser pulse until after the insertion delay has transpired. It is therefore necessary to either delay the optical pulse by a time comparable to the insertion delay, or delay the electrical pulse by a time comparable to the pulse repetition rate *after* the pulse picker. Adequate delay of the optical pulse requires an additional spatial propagation distance for the laser beam of 15 m after the launch of the electrical timing pulse to account for 50 ns of insertion delay. This requires a total of 5 round-trip passes of the optical beam across the length of a 1 m × 3 m optical table. During the course of such a delay, a laser beam with 0.6 mrad of beam divergence (diffractive spread) will grow from 1 mm diameter to 1 cm, making the beam more cumbersome to manipulate. In addition, the passage of dust through the optical beam will add to the modulation noise of the measurement. Alternatively, the delay of the electrical pulse is inversely proportional to the repetition rate of the pulse picker. This way, the magnetic pulse is generated by the optical pulse previous to the one used to measure the magnetic state

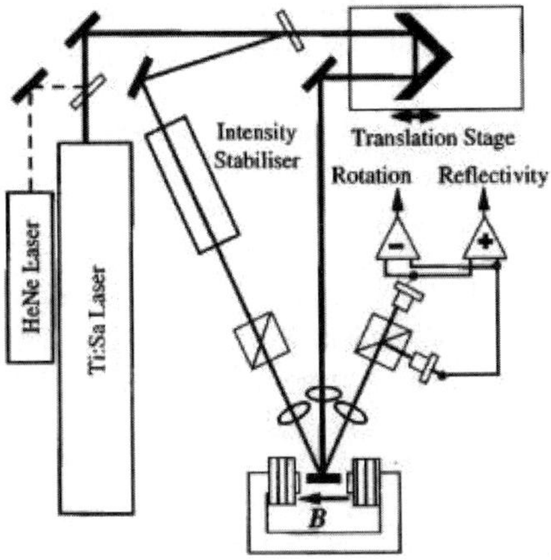

Fig. 4.5. Time-resolved MOKE system for detection of polar and longitudinal magnetization components. From [34]. Used with permission from © The AIP

of the sample. If the operational rate is 1 MHz, then an electrical delay of ~1 μs is necessary. This delay can be accomplished with a length of RG 174 coaxial cable of approximately (10^8 m/s) × (1 μs) = 100 m. It is necessary to restore the trigger pulse after propagation through such a long length of cable. Simple amplification with a 100 MHz bandwidth amplifier will usually suffice.

Detection of the magnetic state of a material is inferred from the polarization state of the reflected light when using MOKE. Oblique orientation of the incident beam with respect to the sample will make the measurement sensitive to the longitudinal magnetization component. An example of how one might make such a measurement is shown in Fig. 4.5. A complication arises from the similarity of the signal induced by the polar and longitudinal magnetization components (i.e., $\vec{M}$ perpendicular to the film plane and $\vec{M}$ in the film plane, albeit parallel to the plane of incidence.) Both the polar and longitudinal magnetization components cause a polarization rotation of reflected light for oblique incidence. Compounding the difficulty, the polar magnetic contrast is typically orders of magnitude greater than the longitudinal contrast. Thus, if the excitation pulse is sufficiently fast, then the precessional dynamics instigated in the sample will cause a significant polar component of the magnetization, which will be detected with similar sensitivity to the longitudinal component [34]. To isolate the two signals, techniques can be employed that use the nonreciprocal nature of MOKE: The helicity of rotation for the plane of polarization is strictly a function of the magnetization direction, independent of the direction of light propagation. Techniques that exploit this property of MOKE have been used effectively for static imaging of magnetic domains [35,36]. Only recently has this approach been used to isolate the signal due to polar and longitudinal magnetization components under dynamic excitation [37]. Alternatively, one may use the transverse form of MOKE to measure magnetization change in the direction perpendicular to the optical plane of incidence. This effect is completely insensitive to the perpendicular magnetization component [38].

Several methods may be used to measure the very slight changes in polarization state of the reflected light. If the reflected light is passed through a polarizing beam splitter such as a Wollaston prism, then the two resulting beams may be measured with two photodiodes in a differential detection scheme [39,40]. Such a method is immune to additive common-mode noise such as fluctuations in the voltage of a shared power supply for both photodiodes. More importantly, division of the differential signal by the summed signal from the two diodes can remove much of the common-mode modulatory noise resulting from noise in the laser intensity and drift in the gain of any electrical amplifiers. The weakness of such a scheme is the need to balance the sensitivity of the two detectors to achieve a high degree of common mode rejection. Rotation of the beam splitter with a precision stage can balance the two detection arms. Any DC offset at the output of the transimpedance amplifiers for the individual diodes could upset the balance if the offset drifts with time.

AC detection schemes offer the advantage that a single detector is used for all measurements, eliminating noise due to the divergent behavior of separate detectors. A popular method for modulating the optical signal is the photoelastic modulator (PEM) [41]. The PEM affects the ellipticity of the optical beam. Time-varying strain along one axis of the PEM optical element results in time-varying retardation of the phase for light polarized along the strained axis. If the polarization is oriented either perfectly parallel or perfectly perpendicular to the PEM axis that produces a time-varying phase retardation, there is no effect on the polarization of the optical beam. If, however, there is some misalignment of the plane of polarization relative to the axis of AC retardation, then time-varying ellipticity is induced in the transmitted light. The time varying state is measured by subsequent passage of the beam through a polarizer, followed by a detector. The AC components of the resultant electrical signal are extracted using a lock-in amplifier. The signal at the modulator frequency is proportional to the ellipticity of the original beam. The signal at twice the modulator frequency is proportional to the polarization of the original beam. Both signals are immune to any additive noise associated with the detector by virtue of the lock-in detection method. Division of the AC components by the DC component (measured with a low pass filter) achieves the same immunity to modulation noise components as that of differential detection.

The MOKE technique is particularly useful for measuring the characteristic response time in thin-film recording heads [42–44]. If the illumination light is projected onto the surface of the head through a microscope objec-

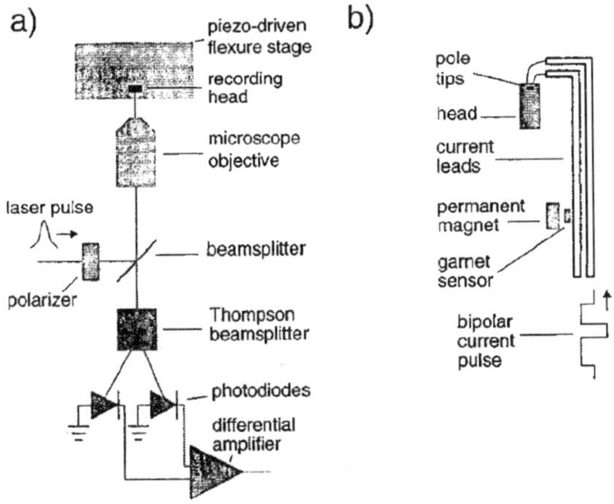

Fig. 4.6. Optical (**a**) and electrical (**b**) schematic of time-resolved microscope for determination of recording head rise-time. From [44]. Used with permission from © The AIP

tive lens, the spot size of the beam on the sample surface may be as small as 0.3 µm. The reflected light is then captured with the same objective lens and directed to polarization detection optics. A diagram outlining the optical and electrical arrangement of such a system is shown in Fig. 4.6. Usually, use of a laser as an illumination source will result in deleterious interference effects referred to as "speckle" [45]. Fortunately, the coherence length for the light emanating from a mode-locked laser system scales linearly with the pulse duration. For example, the light from a Ti:sapphire laser capable of generating 50 fs pulses will have a coherence length of only $(3 \times 10^{10}$ cm/s$) \times$ (50 fs) $=$ 15 µm. This means that light-scattering objects in the microscope optics (dust, scratches, finger prints) cannot produce interference effects with any other scattering objects spaced more that 15 µm apart. This effectively cancels most speckle effects.

Also shown in Fig. 4.6 is a method for the measurement of the current pulse used to drive the recording head [46]. As the head presents a substantial nonlinear impedance discontinuity as a load, it is not possible to assume that the current flow in the head is linearly proportional to the voltage waveform. To make such a measurement, it is possible to use a soft ferromagnetic material in proximity to a waveguide structure that acts as an interconnect between the recording head and the pulse generator (i.e., a "write driver"). The stray fields from the waveguide rotate the magnetization in the ferromagnetic sense material, which is then measured with the same probe beam usually used to interrogate the recording head. Application of a DC bias field improves the bandwidth of the sense material in accordance with (4.5). Use of a material with large magneto-optic activity such as yttrium iron garnet (YIG) makes the measurement simple, given the large polarization

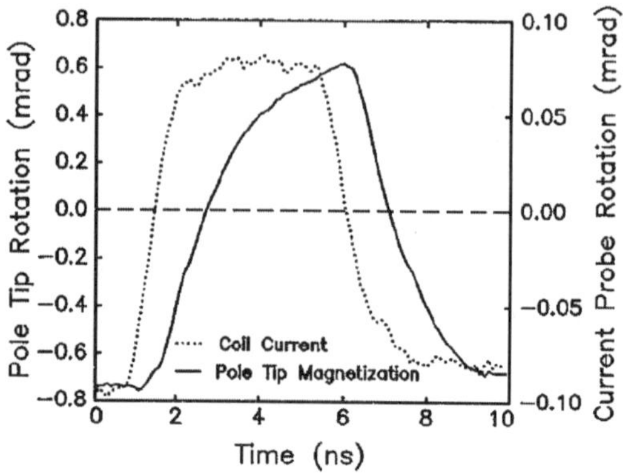

Fig. 4.7. Comparison of current pulse and magnetization response for 10-turn thin-film recording head, as obtained by time-resolved Kerr microscopy. From [44]. Used with permission from © The AIP

rotations induced by such a material. The system is calibrated by passing a DC current through the coplanar structure and measuring the polarization rotation induced in an optical beam passing through the YIG. By comparing the response time of the current sense layer with that of the recording head, as is shown in Fig. 4.7, one gains an immediate determination of how the head materials and/or design can act as a bandwidth limit to the recording channel.

The orientation of the illuminating beam onto the sample surface determines the contrast mechanism giving rise to magnetic contrast. If the beam is directed exactly along the optical axis of the objective lens, then only polar MOKE contrast is obtained. This is typically the only contrast used for the microscopic inspection of recording heads along the air-bearing surface; the magnetization component perpendicular to this surface is believed to be proportional to the recording field [46].

The resolution of a microscope is limited by diffraction and determined by both the wavelength λ of illuminating light and the numerical aperture NA of the objective lens. Using the Rayleigh criterion for resolution, the smallest distance between two objects that can be discerned is $d = 0.61\ \lambda/\mathrm{NA}$ [47]. The NA is defined as $\mathrm{NA} = n \sin\theta$, where n is the refractive index of the medium directly above the specimen's surface and θ is the maximum half angle of the oblique rays from the sample that are subtended by the objective lens. For 400 nm light and an oil-immersion objective lens with $n = 1.4$, the best resolution is $d = 0.17$ μm. It is not obvious that the resolution of a MOKE microscope is equal to this absolute limit. Only a very small fraction of the objective lens aperture is actually used in order to maximize the MOKE contrast, so the NA is reduced. This acts to reduce the resolution of the microscope [48]. One route to better resolution involves the use of a solid immersion lens (SIL) [49]. Such a lens replaces the oil used with a conventional immersion lens, enhancing the resolution by increasing the refractive index of the medium directly above the sample. (Increasing the refractive index decreases the wavelength of the illuminating light, resulting in enhanced resolution.) Fabrication of SIL structures with high-refractive-index materials has been described in detail [50]. The true advantage of the SIL approach is the potential to use materials with a very high index of refraction for the solid immersion lens, thereby exceeding the resolution gains of conventional oil-immersion objective lenses. A lens made of LaSF6-type glass ($n = 1.89$ at $\lambda = 442$ nm) has been successfully implemented with a demonstrated resolution of $\lambda/2.2$ [50]. This resolution is equivalent to that obtained with a conventional oil immersion lens with NA = 1.34.

A very different approach to time-resolved magneto-optics utilizes the recently discovered second-harmonic magneto-optic Kerr effect (SHMOKE) [51–56]. This technique makes use of the second-harmonic generation of light that occurs at the surface of a centrosymmetric material at sufficiently high optical power densities. Commercial Ti:sapphire lasers produce optical pulses of 10 nJ energies and of 50 fs duration. The resultant peak power is 200 kW. The focusing of such an optical pulse down to a spot of 5 µm diameter produces staggering peak power densities of 1 TW/cm^2. Such extreme power densities would normally vaporize many materials if illumination were sustained. However, the sample is insulated from damage by both the very brief duration of the pulses and the low energy of the pulses. Nevertheless, these power densities are sufficient to produce measurable second-harmonic output from the surface of a transition-metal film. The second-harmonic yield is not copious: typical efficiencies are only 10^{-3}–10^{-4} second harmonic photons per optical pulse. For a laser repetition rate of 100 MHz, this is a photon yield of $\sim 10^5$ counts-per-second. Such photon yields are more than sufficient for time-resolved studies [9]. An example of a SHMOKE-based instrument for magnetodynamic measurements is shown in Fig. 4.8 [57].

While such signal levels are acceptable for precision photometry measurements, it would hardly seem worthwhile to encumber the difficulties associated with photon counting for magnetodynamic measurements. The strength of this method is apparent in the unusual nature of magnetic contrast obtained with SH-MOKE. For a transverse geometry, where the magnetization changes in a direction perpendicular to the optical plane of incidence, the second-harmonic yield can vary by as much as 60% [55]. Such contrast is three orders of magnitude larger than for the transverse linear MOKE. At the same time, the longitudinal geometry results in polarization rotations of 5–10 degrees when the magnetization change is in the plane of incidence [56,58,59]. Thus, one finds in practice that the SHMOKE signal is of sufficiently large contrast to afford true, shot-noise-limited measurements of the magnetic state in a sample. Such precision is possible because intensity drift in modern mode-locked lasers is sufficiently small that $1/f$ noise is constrained to only 1% of the average output power of the laser. Thus, to estimate the

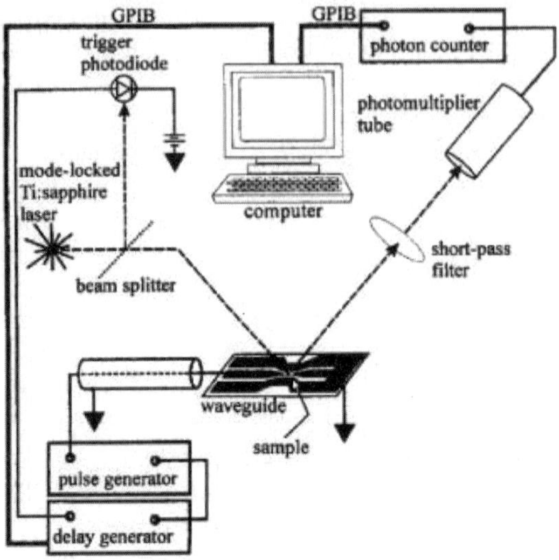

Fig. 4.8. Diagram of a time-resolved SHMOKE system. From [57]. Used with permission from © The IEEE

SNR for such a measurement, one need only keep track of the photon count N. The SNR for an intensity measurement is simply $N^{1/2}$.

Oddly, SHMOKE is not particularly sensitive to the polar magnetization component for air-exposed magnetic films. The reason for this insensitivity is not well understood. This is in contrast with conventional MOKE, where the polar effect is usually more sensitive than the longitudinal effect by an order of magnitude or more.

Another unique feature of SHMOKE is its sensitivity to surface and interfacial magnetization [53,59]. The surface sensitivity is derived from the symmetry properties that can give rise to a second-order non-linearity [54]. This makes SHMOKE a powerful tool for the assessment of surface magnetodynamics. Since eddy-current effects are believed to impede the bulk response of micrometer-thick magnetic films at microwave frequencies, the surface response stands as the true measure of the ultimate magnetic switching speed possible in a film of a given thickness.

The large contrast for longitudinal and transverse SHMOKE makes it possible to measure the planar vector magnetization [61]. The weak sensitivity to polar magnetization removes any mixing of the polar and longitudinal effects that complicate linear MOKE measurements [34]. Vector measurements are useful for determining the coherence of the response for large rotations and/or arbitrary initial conditions. It is well established that high speed magnetic switching is often incoherent [62]. Intrinsic nonuniformities in the magnetization distribution are believed to nucleate inhomogeneous dynamics [63,64]. Such nonuniform processes would be manifested as a transitory reduction of

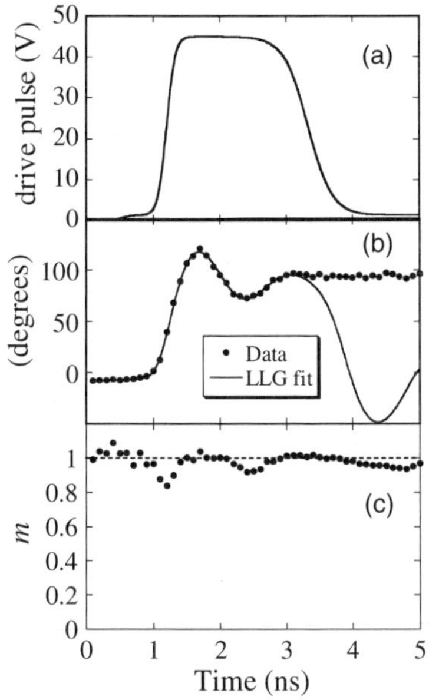

Fig. 4.9. Vectorial magnetodynamics for 50 nm Permalloy film under step excitation with field pulse along the hard axis. The excitation pulse is shown in **a**. The angle response is shown in **b**. along with a fit to LLG, shown as the *solid curve*. $\theta \sim 0$ is along the easy axis of sample. The normalized magnitude of the magnetization is presented in **c**. The *dashed line* is the presumed magnitude within the context of LLG phenomenology. From [66] and [67]. Used with permission from © The IEEE

the net volume-averaged magnetic moment [65]. A vector-sensitive technique can detect such a reduction in the magnetic moment. An example of data obtained with vectorial, time-resolved SHMOKE for 50 nm Permalloy is shown in Fig. 4.9. Here, $\left|\vec{M}\right|$ is approximately constant during the entire switching process for a hard-axis field pulse of 1 kA/m (13 Oe) [66]. Such data illustrate the possibility of coherent switching, even for large excitations. However, the magnetization remains pinned along the hard axis direction upon termination of the field pulse. This is an example of a transient metastable state, to be covered in more detail in Sect. 4.4 [67].

An electrical short circuit at the end of the waveguide near the sample may be used to some advantage for optical measurements. The short reflects the incident pulse and thereby doubles the amplitude of the applied field pulse. The voltage of the reflected pulse is equal and opposite to the incident pulse, reducing the voltage at the short to zero. The current is doubled since the direction of current flow changes sign with the reversal of the propagation direction. However, the forward- and backward-propagating pulses are coincident only at the sample location if the sample is not spaced too far from the short. The spacing l for coincident arrival must be less than $l < \frac{1}{2}(c/n_{eff})t_r$, where c is the speed of light, n_{eff} is the effective refractive index of the waveguide (typically $n_{eff} \sim (\varepsilon_r^{1/2})/2$ for a coplanar waveguide, where ε_r is the dielectric constant for the waveguide substrate) and t_r is the rise-time of the voltage pulse. If a pulse generator with $t_r = 50$ ps is used with a coplanar waveguide fabricated on sapphire, then the sample should be kept within 4 mm of the short to keep from broadening the rise-time of the excitation pulse by more than 50%.

On the other hand, if the sample is placed sufficiently far from the short, interesting effects can be observed resulting from the coherent interference of the precessional oscillations instigated in a sample. The staggering of the incident and reflected pulses constitutes a primitive form of pulse shaping that can cause very rapid switching without the usual precessional oscillations observed in many soft ferromagnetic materials [68]. Such suppression of precessional ringing has also been observed by conventional MOKE methods by tuning the duration of an applied field pulse during an impulse measurement [69].

4.4 Transport Methods

To determine the dynamic response for sub-micrometer patterned samples, one must turn to other methods that employ high speed transport measurements. Waveguide interconnects can transmit fast electrical pulses to small patterned elements. The magnetic state of the elements is then measured in real time using the giant magnetoresistance (GMR) effect. Such methods have been put to good use in assessing bandwidth limitations to the switching performance of magnetic random access memories [70,71].

Single-shot switching speeds as fast as 250 ps have been measured in 1 μm × 5 μm GMR devices [72]. Electrical contact is made with a microwave probe station equipped with three-axis micromanipulators and 40 GHz bandwidth probes. The probes connect with contacts on a lithographically patterned test structure. The test structures were constructed on high resistivity Si (>2 kΩ cm) with a thermal oxide surface. This reduced parasitic losses due to the absorption of microwave power by the substrate. The coplanar structures were ~100 μm in dimension, simplifying alignment of the probes. The waveguides tapered to 4 μm widths, at which point they contacted microstrip lines that were separated from a ground-plane with several hundred nanometers of sputtered SiO insulator. The microstrip lines led over the GMR devices, exposing the device to the combined field produced by the current in the conductive strip and the image current in the ground-plane. The additional field from the ground plane is problematic, insofar as the exact current distribution is not well known. The field without the ground plane contribution is calculated to be as large as 10 kA/m (125 Oe) when excited with 4 V pulses. The pinned layer in the GMR structure had a pinning field in excess of

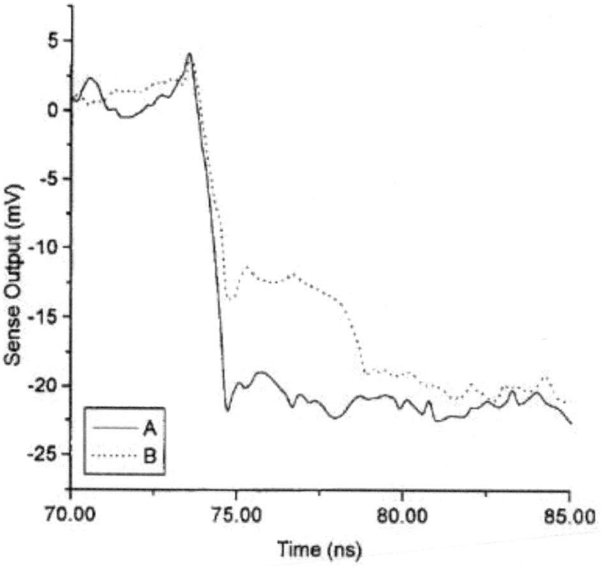

Fig. 4.10. Two single-shot measurements of switching in a GMR device in response to 6 kA/m (75 Oe) field pulses of 1 ns duration. All conditions for the two measurements are nominally identical. From [72]. Used with permission from © The AIP

80 kA/m (1 kOe), making it immune to the field pulses. The resistive change in the GMR device is measured with a DC bias voltage applied through a microstrip sense line. The bias voltage is isolated from the dynamic signal response with a high-bandwidth commercial bias tee. The change in voltage due to switching of the free layer was then measured using a 1.5 GHz bandwidth digitizing oscilloscope. The excitation lines also extended to coplanar contact pads that allowed for measurement of the transmitted pulse amplitudes and rise time. Such a measurement indicated an excitation bandwidth of 1.4 GHz. The layer composition for the GMR devices was Ta(5 nm)–$Ni_{0.8}Fe_{0.2}$(2 nm)–Co(1 nm)–Cu(3 nm)–Co(2 nm)–Ru(0.6 nm)–Co(1.5 nm)–FeMn(10 nm)–Ta(5 nm). Only the 3 nm free layer is presumed to contribute to the observed dynamic response. An example of data obtained with this test structure is shown in Fig. 4.10.

Patterned GMR devices exhibit a simple relationship between the pulse duration and the pulse amplitude required to reverse the magnetization. The data are fitted to the switching speed equation

$$\frac{1}{T} = \frac{1}{S_w}(H_p - H_0) , \tag{4.32}$$

yielding a switching speed parameter of $S_w = 1.2$ kA·ns/m (15 Oe·ns) and the threshold field H_0 of 4.8 kA/m (60 Oe). Thus, a pulse of 1 ns duration is required to switch a device for a pulse amplitude of only 4.8 kA/m + (1.2 kA·ns/m)/(1 ns) = 6 kA/m (75 Oe). This result required the application of an 800 A/m (10 Oe) transverse bias field to initiate switching [72]. It was also found that a full-switch event occurred only with a ~90% probability, even for field pulses as large as 18 kA/m (230 Oe). Instead of a complete reversal, a device would usually switch into an intermediate metastable state, indicative of partial reversal. The occurrence of such metastable states is presumed to be the result of inhomogeneous reversal mechanisms [72]. Micromagnetic simulations suggest that reversal of a device is initiated by the nucleation of reversed domains at the ends of an elongated device. The nucleation sites then grow through a process analogous to domain wall motion [72]. Since the devices are of length $L = 1$ µm, and nucleation of reversal is presumed equally likely at the two ends of the device, then the domain wall mobility μ required to explain the observed switching speed is $\mu = L/(2S_w) = 0.4\ \mathrm{m^2/(A{\cdot}s)}$ (33 m/(Oe·s)). Such a value for μ is quite reasonable for thin-film Permalloy. Middelhoek [73] obtained values for mobility of 0.13–1 $\mathrm{m^2/(A{\cdot}s)}$ (10–80 m/(Oe·s)) for 50 nm Permalloy in the early 1960's using single-shot pulsed waveguide techniques.

The theory for domain wall motion presumes that a constant speed of motion is achieved when the viscous damping torque is equal to the net torque applied to the spins in the middle of the wall structure, where the magnetization is oriented orthogonal to the anisotropy direction. In such a viscous limit, the time required for the spins to rotate by 90 degrees is [74]

$$\tau_{90} = \frac{\alpha}{\gamma\mu_0(H - H_k)} . \tag{4.33}$$

The wall velocity v is the ratio of the domain wall width w to τ_{90}, or

$$\nu = \frac{w}{\tau_{90}} = \frac{w\gamma\mu_0(H - H_k)}{\alpha} . \tag{4.34}$$

So, the switching-speed parameter for a device presumed to reverse through a mechanism of domain wall motion is given by

$$S_w = \frac{L(H - H_k)}{2\nu} = \left(\frac{L}{w}\right)\frac{\alpha}{2\gamma\mu_0} . \tag{4.35}$$

The domain wall width from micromagnetic simulations is w ~150 nm [75]. Solving for α using the measured value for S_w obtained in [75], we find $\alpha = 0.08$. The extracted value of damping from the measured switching speed is about five times larger than the damping measured by precessional methods [75]. Such discrepancy between the damping associated with large-angle switching and small-angle free induction decay has been a long-standing issue [14]. It is generally believed that the enhanced damping in the case of large-angle switching is the result of the increased spin-wave generation (i.e. inhomogeneous precessional dynamics) when the magnetization is driven through the hard axis direction with a fast field pulse along the easy axis [75]. It is not yet clear whether the excitation of short-wavelength spin oscillations is intrinsic to the large-angle switching process or is simply the result of the domain wall motion.

Since these measurements detect the instantaneous state of a single magnetization component, one can determine the *actual* magnetic switching time for a single event. A device subject to a 1 ns, 6 kA/m (75 Oe) pulse would actually switch in less than 1 ns [72]. This result has the implication that the time interval required to actually nucleate the reversal process was indiscernibly short. In addition, multiple single-shot measurements revealed that the degree of reversal was a highly stochastic process. Often, switching would initially occur on these very short time scales, but would achieve only a 60% reversal. Full reversal would occur only after some additional dwell time, on the order of several tens of nanoseconds. This is compelling evidence for the importance of real-time measurement techniques that reveal the detailed nature of the switching physics. A measurement made with a time-averaged technique (such as sampling) would obtain a switching time longer than observed in a single-shot measurement. It may be concluded that the domain wall motion that drives the magnetization reversal occurs in a much more staccato fashion, with periods of rapid motion separated by intervals where the wall is pinned, with its motion impeded. Thus, we should use the extracted damping parameters of the previous paragraph cautiously, insofar as they reflect a spatially averaged response of the magnetization as the wall moves through multiple pinning sites. The coupling of domain wall motion to the imperfections of the device effectively enhances the viscous forces impeding the domain wall motion, resulting in a larger apparent damping value.

Similar switching speeds have been measured with 0.8 µm × 1.6 µm magnetic tunnel junctions (MTJs) using a sampling oscilloscope [71]. The free layer thickness was only 5 nm. A synthetic antiferromagnet structure was used for the pinned layer. While less is known about the bandwidth limitations of the tunnel junctions used for these measurements, results very similar to those in [72] were obtained. In particular, a switching-speed determination for a transverse bias field of 4 kA/m (50 Oe) yielded $S_w = 1.1$ kA·ns/m (14 Oe·ns). For these measurements, the actual rise time of the MTJ resistance was approximately equal to the minimum field-pulse duration required to switch a device, even though a sampling oscilloscope was used for the measurement, requiring an average over multiple pulse excitations. An example of the data obtained by such a measurement is shown in Fig. 4.11.

The fact that these devices were three times shorter than for the GMR structures would seem to imply that the actual domain wall velocity was three times slower. The result is an extracted damping parameter that is three times larger, or $\alpha \approx 0.2$–0.3. This is 20–30 times larger than obtained by free-induction decay measurements made with the same structures [71]. This is perhaps the result of a large density of defects, resulting in enhanced viscous forces that slow the domain growth. Subsequent measurements of the switching probability for pulse amplitudes smaller than that required to insure a full switch show a peak in the switching probability as a function of pulse duration [76]. Such a peak is the result of an integral number of inter-

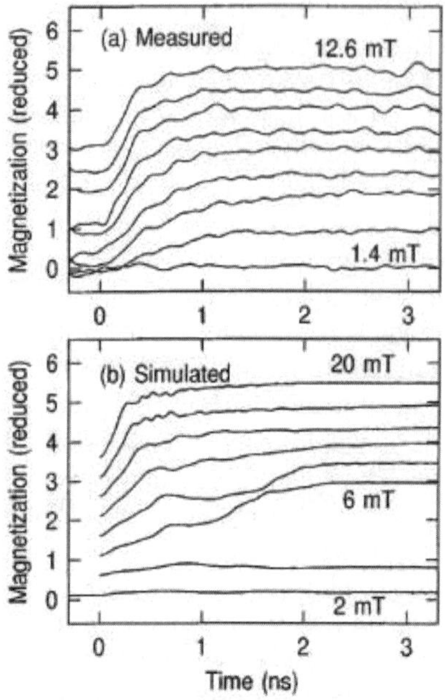

Fig. 4.11. a Average switching response in a MTJ test structure. Each curve is for a different pulse amplitude, 1–10 kA/m (13–125 Oe). **b** Results of a micromagnetic simulation for a nominally identical magnetic device. From [71]. Used with permission from © The Physical Review

mediate states that must be traversed to fully switch a device. In essence, the reversal occurs by a process of domain wall motion through multiple pinning sites that impede the motion [76]. If the pulse amplitude is too small to deterministically break the wall away from a pinning site, then switching is contingent on the occurrence of a thermal fluctuation sufficient to overcome the pinning energy of the defect. The peak in the switching probability occurs at the average total time associated with thermally assisted motion through all N pinning sites for a given pulse amplitude. Fitting of the data to this model yields $N \sim 4$, or an average distance of 0.2 μm between each pinning site. If one instead assumes that the domain wall need only move far enough that a simple majority of the spins in a device must switch during the field pulse to insure full eventual reversal, then the average distance between pinning sites is 0.1 μm. This distance is comparable to the domain wall size of 0.1 μm, as calculated using micromagnetics [76].

An additional assumption of the model described above is that intermediate multi-domain states are only metastable and do not lead to magnetization reversal if the field pulse is terminated while a device is in such a state. In other words, the magnetization state of the device reverts to its initial condition (the state it was in before the field pulse is applied) if left in a metastable multi-domain state at the end of a field pulse. A more detailed study of such metastable states finds that they are associated with any condition where a magnetic system is left with a large fraction of the spin system in a state of unstable equilibrium and oriented along the hard axis direction after application of a pulse [69]. If the total volume of a domain wall (where the spins orient along the hard axis) is a significant fraction of the total volume of the free layer, then a pinned wall will tend to drift over long time intervals toward the state of lowest free energy. The remaining "unswitched" volume of the device acts as an effective bias field source, pushing the device state back to the initial magnetic orientation. This assumes that the total switched volume is less than the unswitched volume. Otherwise, the same mechanism leads to the eventual reversal of the device, but on a time scale much longer than the applied field pulse duration. Such metastable states leading to the switched state have been observed in high-speed switching studies [72].

References

1. L. Néel, Ann. Geophys. **5**, 99 (1949)
2. R. Street and J. C. Wooley, Proc. Phys. Soc. London, A **62**, 562 (1949)
3. T. J. Silva and N. D. Rizzo, unpublished
4. K. J. Sixtus and L. Tonks, Phys. Rev. **34**, 1441 (1929)
5. L. Landau and E. Lifshitz, Phys. Z. Sowjetunion **8**, 154 (1935)
6. E. C. Stoner and E. P. Wohlfarth, Philos. Trans. R. Soc. London, Ser. A **240**, 599 (1948)
7. K. J. Harte, J. Appl. Phys. **31**, 283S (1960)
8. H. Hoffmann and M. H. Kryder, IEEE Trans. Magn. **9**, 554 (1973)
9. T. M. Crawford, T. J. Silva, C. W. Teplin, and C. T. Rogers, Appl. Phys. Lett. **74**, 3396 (1999)
10. F. B. Humphrey, J. Appl. Phys. **29**, 284 (1958)
11. N. Bloembergen and S. Wang, Phys. Rev. **93**, 72 (1954)
12. H. Suhl, J. Phys. Chem. Solids. **1**, 209 (1957)
13. C. Kittel, Introduction to Solid State Physics, Sixth Edition, (John Wiley, New York, 1986) 479 ff.
14. D. O. Smith, J. Appl. Phys. **29**, 264 (1958)
15. T. Kawazu, M. Yamaguchi and K. I. Arai, IEEE Trans. Magn. **30**, 4641 (1994)
16. V. Korenivski, R. B. van Dover, P. M. Mankiewich, Z.-X. Ma, A. J. Becker, P. A. Polakos, and V. J. Fratello, IEEE Trans. Magn. **32**, 4905 (1996)
17. P. Queffelec, P. Gelin, J. Gieraltowski, and J. Loaec, IEEE Trans. Magn. **30**, 224 (1994)
18. T. J. Klemmer, K. A. Ellis, and B. van Dover, J. Appl. Phys. **87**, 5846 (2000)
19. W. Dietrich, W. E. Proebster, and P. Wolf, IBM J. Res. Dev. **4**, 189 (1960)
20. T. J. Silva, C. S. Lee, T. M. Crawford, and C. T. Rogers, J. Appl. Phys. **85**, 7849 (1999)
21. M. Sparks, Phys. Rev. B **1**, 3831 (1970)
22. K. J. Harte, MIT Lincoln Laboratory Technical Note 1965-31 (MIT Lincoln Laboratory, Lexington MA, 1965) 1–3
23. R. W. Damon and J. R. Eshbach, J. Phys. Chem Solids. **19**, 308 (1961)

24. A. B. Kos, T. J. Silva, and P. Kabos, unpublished
25. C. Alexander, Jr., J. Rantschler, T. J. Silva, and P. Kabos, J. Appl. Phys. **87**, 6633 (2000)
26. Reinmut K. Hoffmann, Handbook of Microwave Integrated Circuits (Artech House, Norwood MA, 1987) 351–375
27. K. C. Gupta, R. Garg, I. Bahl, P. Bhartia, Microstrip Lines and Slotlines, 2nd ed. (Artech House, Norwood MA, 1996) 375–456
28. Agilent Technologies application note 1304-2
29. J. A. Valdmanis, G. Mourou, and C. W. Gabel, Appl. Phys. Lett. **41**, 211 (1982)
30. M. R. Freeman, R. R. Ruf, and R. J. Gambino, IEEE Trans. Magn. **27**, 4840 (1991)
31. M. R. Freeman, M. J. Brady, and J. Smyth, Appl. Phys. Lett. **60**, 2555 (1992)
32. M. Bauer, R. Lopusnik, J. Fassbender, and B. Hillebrands, Appl. Phys. Lett. **76**, 2758 (2000)
33. A. Y. Elezzabi and M. R. Freeman, Appl. Phys. Lett. **68**, 3546 (1996)
34. R. J. Hicken and J. Wu, J. Appl. Phys. **85**, 4580 (1999)
35. W. W. Clegg, N. A. E. Heyes, E. W. Hill, and C. D. Wright, J. Magn. Magn Mater. **95**, 49 (1991)
36. T. J. Silva and A. B. Kos, J. Appl. Phys. **81**, 5015 (1997)
37. G. E. Ballentine, W. K. Hiebert, A. Stankiewicz, and M. R. Freeman, J. Appl. Phys. **87**, 6830 (2000)
38. Y. Acremann, C. H. Back, M. Buess, O. Portmann, A. Vaterlaus, D. Pescia, and H. Melchior, Science **290**, 492 (2000)
39. P. Kasiraj, R. M. Shelby, J. S. Best, and D. E. Horne, IEEE Trans. Magn. **22**, 837 (1986)
40. S. A. Crooker, D. D. Awschalom, and N. Samarth, IEEE J. Sel. Topics Quant. Electron. **1**, 1082 (1995)
41. P. Q. J. Nederpel and J. W. D. Martens, Rev. Sci. Instrum., 687 (1985)
42. M. E. Re and M. H. Kryder, J. Appl. Phys. **55**, 2245 (1984)
43. B. Petek, P. L. Trouilloud, and B. E. Argyle, IEEE Trans. Magn. **26**, 1328 (1990)
44. M. R. Freeman and J. F. Smyth, J. Appl. Phys. **79**, 5898 (1996)
45. B. E. Argyle, Proceedings of the Symposium on Magnetic Materials, Processes and Devices **90-8**, 85 (1990)
46. A. Y. Elezzabi and M. R. Freeman, Appl. Phys. Lett. 68, 3546 (1996)
47. Max Born and Emil Wolf, Principles of Optics, 6th edition (Pergamon Press, New York, 1980) 418–419
48. D. Treves, J. Appl. Phys. **32**, 358 (1961)
49. B. D. Terris, H. J. Mamin, D. Rugar, W. R. Studenmund and G. S. Kino, Appl. Phys. Lett. **65**, 388 (1994)
50. J. A. H. Stotz and M. R. Freeman, Rev. Sci. Instrum. **68**, 4468 (1997)
51. R.-P. Pan, H. D. Wei, and Y. R. Shen, Phys. Rev. B **39**, 1229 (1989)
52. W. Huebner and K.-H. Bennemann, Phys. Rev. B, **40**, 5973 (1989)
53. J. Reif, J. C. Zink, C.-M. Schneider, and J. Kirschner, Phys. Rev. Lett. **67**, 2878 (1991)
54. B. Koopmans, M. G. Koerkamp, Th. Rasing, and H. v. d. Berg, Phys. Rev. Lett. **74**, 3692 (1995)
55. T. M. Crawford, C. T. Rogers, T. J. Silva, and Y. K. Kim, Appl. Phys. Lett. **68**, 1573 (1996)

56. K. Boehmer, J. Hohlfeld, and E. Matthias, Appl. Phys. A **60**, 203 (1995)
57. T. J. Silva and T. M. Crawford, IEEE Trans. Magn. **35**, 671 (1999)
58. T. M. Crawford, C. T. Rogers, T. J. Silva, and Y. K. Kim, IEEE Trans. Magn. **32**, 4087 (1996)
59. H. A. Wierenga, W. de Jong, M. W. J. Prins, Th. Rasing, R. Vollmer, A. Kirilyuk, H. Schwabe, and J. Kirschner, Phys. Rev. Lett. **74**, 1462 (1995)
60. R.-P. Pan, H. D. Wei, and Y. R. Shen, Phys. Rev. B **39**, 1229 (1989)
61. P. Kabos, A. B. Kos, and T. J. Silva, J. Appl. Phys. **87**, 5980 (2000)
62. F. B. Humphrey and E. M. Gyorgy, J. Appl. Phys. **30**, 935 (1959)
63. C. D. Olson and A. V. Pohm, J. Appl. Phys. **29**, 274 (1958)
64. F. B. Humphrey, J. Appl. Phys. **29**, 284 (1958)
65. H. B. Callen, J. Phys. Chem. Solids. **4**, 256 (1958)
66. T. J. Silva, P. Kabos, and M. R. Pufall, unpublished
67. P. Kabos, S. Kaka, and T. J. Silva, IEEE Trans. Magn. **36**, 3050 (2000)
68. T. M. Crawford, P. Kabos, and T. J. Silva, Appl. Phys. Lett. **76**, 2113 (2000)
69. M. Bauer, R. Lopusnik, J. Fassbender, and B. Hillebrands, Appl. Phys. Lett. **76**, 2758 (2000)
70. S. E. Russek, J. O. Oti, S. Kaka, and E. Y. Chen, J. Appl. Phys. **85**, 4773 (1999)
71. R. H. Koch, J. G. Deak, D. W. Abraham, P. L. Trouilloud, R. A. Altman, Y. Lu, W. J. Gallagher, R. E. Scheuerlein, K. P. Roche, and S. S. P. Parkin, Phys. Rev. Lett. **81**, 4512 (1998)
72. S. Kaka and S. Russek, J. Appl. Phys. **87**, 6391 (2000)
73. S. Middelhoek, IBM J. Res. Dev., 351 (1966)
74. J. F. Dillon, Jr., in: *Magnetism*, ed. by G. T. Rado and H. Suhl (Academic Press, New York, 1963), vol. 3, p. 415
75. S. E. Russek, S. Kaka, and M. J. Donahue, J. Appl. Phys. **87**, 7070 (2000)
76. R. H. Koch, G. Grinstein, G. A. Keefe, Y. Lu, P. L. Trouilloud, W. J. Gallagher, and S. S. P. Parkin, Phys. Rev. Lett. **84**, 5419 (2000)

5 Thermal Effects in High-Density Recording Media

Andreas Moser and Dieter Weller

Magnetic recording is an efficient method to record, store and retrieve vast amounts of digital information. In such a storage system, instability of the stored data or data loss cannot be tolerated. This chapter reviews the effects of thermal excitations on the stability of stored data in magnetic recording media. In order to understand the effect of thermal excitations we first consider energy barriers that prevent the system from spontaneous transitions between different states. In granular magnetic recording media, as used today, these energy barriers are governed by the product of the anisotropy energy density and the grain volume. As head and media parameters are scaled to smaller dimensions, which has been the main path toward achieving higher areal densities, grain volumes are reduced in order to keep the number of grains per bit cell approximately constant. As a result, energy barriers become smaller and thermal excitations may give rise to superparamagnetic effects, which in turn may lead to signal loss over time. In order to avoid superparamagnetic effects and to ensure data stability one typically compensates for the reduction in grain volume by increasing the anisotropy energy density. This approach, however, results in higher write coercivities and is naturally limited by the available write field of the recording head, which traces the thermal stability problem in magnetic recording media more to a switching field and writing problem than to an intrinsic media grain stability problem. Note also, that the write coercivity, as discussed in this chapter, typically increases with increasing write frequency, which aggravates the write field problem.

In 1997, Charap et al. [1] considered the thermal stability of recorded bit transitions in longitudinal magnetic recording using micromagnetic modeling. They arrived at an areal density limit of about 40 $\mathrm{GBit/in^2}$ using the following scaling assumptions: A minimum of 100 grains per bit cell are necessary to maintain an adequate signal-to-noise ratio (SNR), a bit aspect ratio (ratio of width and length of a bit cell) BAR $= 12$ is necessary to maintain tracking; the maximum available write field is $H_{\max} \approx 5100$ Oe limiting the anisotropy density of the medium and hence the minimal stable grain size. Since then, new paths to higher storage densities of 100 $\mathrm{GBit/in^2}$ and beyond have been proposed by various authors. Improved signal-to-noise ratio (SNR) was found in models under idealized conditions when the bit aspect ratio is lowered to about 4 [2]. Fewer grains per bit, improved coding schemes and a transition to perpendicular recording may eventually lead to $\mathrm{TBit/in^2}$ recording [3].

Currently, the experimentally demonstrated areal density record stands at 63.2 GBit/in^2 at BAR = 5.7 [4]. Compound annual growth rates in both, laboratory demonstrations and products have been approximately 130% since 1997. The question arises, as to how long the magnetic recording industry will be able to sustain this areal density growth and what the ultimate achievable areal density is going to be? We leave it to others to answer this question [5]. However, we point out, that from the thermal stability perspective alone it is conceivable to fabricate stable media with grain diameters in the 3 nm range, which hypothetically leads up to areal densities in the 10–50 TBit/in^2 regime.

The main focus of this chapter is to present an experimental study of thermal stability in magnetic recording media and to discuss the data and relevant parameters in basic models of superparamagnetism. We start by discussing magnetization reversal in an isolated particle model [6,7] induced by a magnetic field and by thermal fluctuations in Sect. 5.1. We then present measurements of the interbit demagnetizing field, which tends to destabilize recorded bits (Sect. 5.2). Measured signal decay strongly depends on temperature and sample thickness, as discussed in Sect. 5.3. Pulse-width dependent coercivity and signal decay rate measurements are used to estimate the stability requirement for data storage over a period of 10 years. The chapter concludes with brief discussions of antiferromagnetically coupled recording media and alternative materials candidates in Sect. 5.4.

5.1 Isolated, Single Domain Particle Model

5.1.1 Magnetization Reversal by Magnetic Field

In a first approximation, recording media can be modeled as assemblies of single-domain, isolated or weakly coupled particles with uniaxial anisotropy, which reverse magnetization by coherent rotation. These so-called Stoner-Wohlfarth particles have two stable states I and II, separated by energy barriers E_B^+ and E_B^-, respectively. The magnetization in either of these states is aligned parallel to the anisotropy axis, as depicted in Fig. 5.1. In the absence of an external field H the activation energy barriers for magnetization reversal become equal and are solely determined by the anisotropy energy density K_u and the particle volume V [8]

$$E_B^0 = K_u V \,. \tag{5.1}$$

In the presence of an external field H, e.g. the demagnetizing field H_d across a magnetization transition, the activation barrier may be approximated by [9]:

$$E_B^{\pm}(H,V) = E_B^0 \cdot (1 \mp H/H_0)^m \,. \tag{5.2}$$

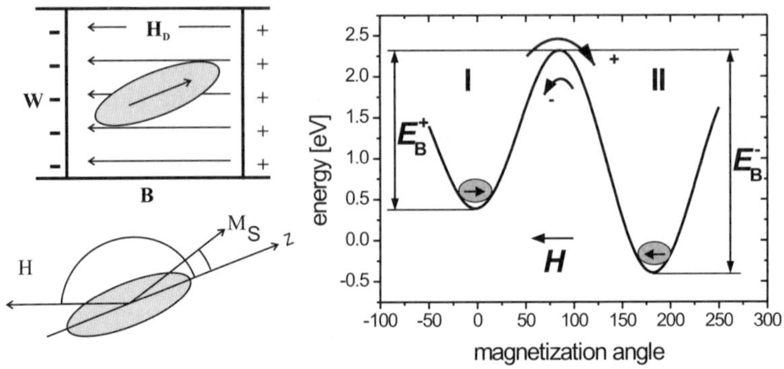

Fig. 5.1. Calculated energy of an isolated Stoner-Wohlfarth particle exposed to a magnetic field H pointing from right to left, such as a demagnetizing field in a bit cell of length B and width W. In state I the magnetic field is opposite to the magnetization of the particle. The tilt of the particle is 21°, representing an average angle in a 2D random system

H_0 is the intrinsic switching field, which is dependent on the angle Θ between the applied magnetic field and the anisotropy axis. It can be expressed by

$$H_0 = x(\Theta) \cdot H_K - N_{eff} M_s = x(\Theta) \cdot \frac{2K_u}{M_s} - N_{eff} M_s \tag{5.3}$$

H_K is the anisotropy field and $x(\Theta) = (cos^{2/3}(\Theta) + sin^{2/3}(\Theta))^{-3/2}$ is the so-called Stoner-Wohlfarth astroid [8]. For a 2D random array of non-interacting particles, which we use here to describe longitudinal recording media, the average field angle is $\Theta = 21°$ and x becomes $x(21°) = 0.567$ [10,11]. Exchange generally reduces x. The exponent m in (5.2) may be approximated by $m \approx 0.86+1.14 \cdot x(\Theta)$ (Pfeiffer approximation [9]) and becomes $m = 3/2$ [12] for the 2D random case, $x(21°) = 0.567$. N_{eff} is an effective demagnetizing field factor, which for a uniform system without exchange interactions was modeled to be $N_{\text{eff}} = 1$ [1]. M_s is the saturation magnetization. Without the presence of thermal activation, which will be discussed in the next chapter, only particles with $E_B^{\pm}(H,V) \leq 0$ $(H \geq H_0)$ can move into the low-energy state.

In recording media, particles have a size or diameter distribution. An example is given in Figs. 5.2 and 5.3, where a transmission electron microscopy (TEM) image of the IBM 35 GBit/in^2 demonstration media and the corresponding grain diameter distribution are shown [13]. Individual Co-rich grains (dark) are clearly separated by Cr and B rich (light) areas, in which the exchange coupling is supposedly reduced. In sputtered thin films, the distribution of the grain diameter D (Fig. 5.3) is usually approximated by a lognormal function $G_{\text{lognormal}}$ (Gaussian distribution in $\ln(D)$) with di-

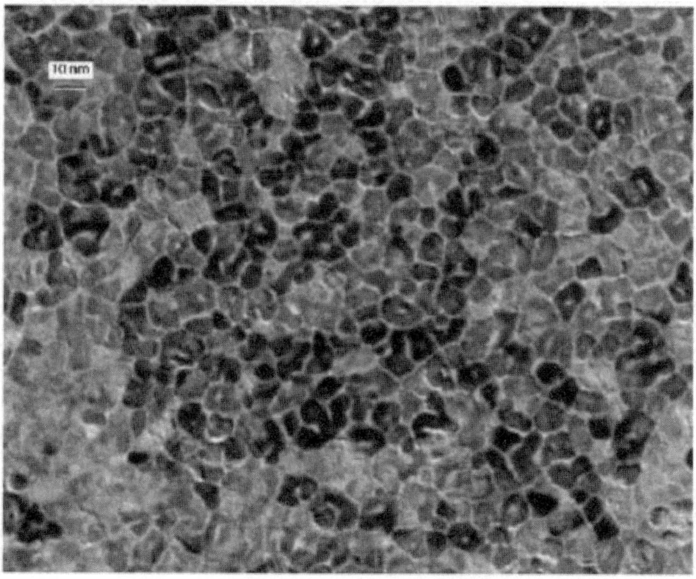

Fig. 5.2. Transmission electron microcopy (TEM) image a CoPtCrB recording media used for the 35 $GBit/in^2$ demonstration [13]

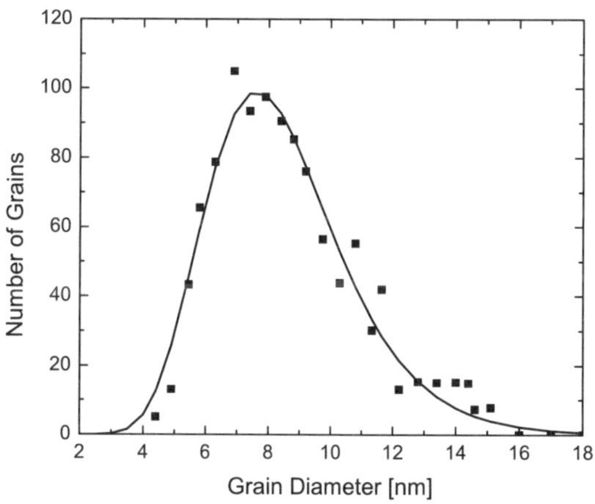

Fig. 5.3. Grain diameter distribution of the TEM image in Fig. 5.2 [13]. The distribution is centered at 7.6 nm and has standard distribution of $\sigma_D = 0.27$

mensionless width σ_D [14]:

$$G_{\text{lognormal}}(D) = \frac{1}{\sqrt{2\pi}\sigma_D} \exp\left[-\frac{1}{2}\left(\frac{\ln(D/D_c)}{\sigma_D}\right)^2\right] \tag{5.4}$$

and $\int_0^\infty G(D')dD' = 1$, where $G(D)dD$ is the fraction of particles with a diameter between D and $D + dD$. Fitting the grain diameter distribution to (5.4) results in $D_c = 7.6$ nm and $\sigma_D = 0.27$.

The grain diameter distribution translates directly into a grain volume distribution via the relationship $V = \pi(D/2)^2 \cdot \delta$, which assumes that the medium can be approximated by cylindrical grains of constant thickness δ. Such a volume distribution results in a distribution of activation energy barriers with center energy $E_C = K_u V_C$. In principle, both particle volume and anisotropy distributions need to be considered [15]. Here, we take the simplified approach of neglecting anisotropy variations, as done by others [16,17].

The full details of magnetization reversal processes in magnetic recording media can not be described by a simple independent particle model. Interactions between the grains of a recording medium and the magnetic field generated by the transitions generally enhance the complexity and dimensionality of the energy surface. Furthermore, particles with small energy barriers might be more stable as a consequence of coupling to larger neighboring particles [18,19]. Such processes are typically investigated in more complex micromagnetic models [20].

5.1.2 Magnetization Reversal by Thermal Excitation

Magnetization reversal by thermal excitations is a manifestation of a magnetic aftereffect, which has been studied for more than 100 years [21,22]. Experiments to measure the time-dependence of the magnetization are usually carried out by measuring the total sample moment as a function of time typically in a reverse field H. Here, we first consider a single Stoner-Wohlfarth particle in the presence of a magnetic field H and then continue with a more realistic case of an assembly of particles with an energy barrier distribution.

The energy barrier of a single particle is calculated using (5.2) and the thermally activated switching is characterized by a time constant following the Néel–Arrhenius law

$$\tau_\pm^{-1} = f_0 \cdot \exp\left(-E_B^\pm / k_B T\right) . \tag{5.5}$$

Here, the attempt frequency f_0 sets the frequency scale for thermally activated magnetization reversal and is of order $10^9 - 10^{12}$ Hz [23,24]. It is dependent on material parameters, such as the anisotropy field, the damping constant and the energy barrier, and external parameters, such as the applied magnetic field and the temperature [7]. $k_B = 1.38 \cdot 10^{-14}$ erg/K is Boltzmann's constant and T is the absolute temperature in Kelvin. Thermal equilibrium is reached when the net flow across the energy barrier vanishes. In the presence of a magnetic field H or a demagnetizing field H_d the activation barrier is asymmetric as shown in Fig. 5.1. In this case the time constants τ_+ and τ_-, which characterize switching in opposite directions across the activation barrier, are different (Fig. 5.1). The so-called "stability ratio"

$$C^{-1} = E_B^{\pm}/k_B T \tag{5.6}$$

is a key parameter characterizing thermally activated magnetization reversal. Note that in general, $E_B^{\pm}$ and therefore thermal stability are also a function of temperature as a result of the temperature dependence of K_u.

In an assembly of isolated ferromagnetic particles, the time evolution towards thermal equilibrium is determined by statistical switching of individual particles. The probability $P(t)$ that a particle resides in state I (see Fig. 5.1) with the activation energy barrier E_B^+ is a solution of the probability equation of motion $dP/dt = -P/\tau_+ + (1-P)/\tau_-$ [7,16]:

$$P(t) = \tau/\tau_- + (1 - \tau/\tau_-) \exp\left[-t/\tau\right] . \tag{5.7}$$

Here, $P(t=0) = 1$ characterizes the initial remanence state I, and $\tau^{-1} = \tau_+^{-1} + \tau_-^{-1}$. If switching occurs only in one direction, i.e. for strongly assymmetric barriers $E_B^+ \gg E_B^-$ and $\tau_- \to \infty$, the solution of (5.7) simplifies to the well-known form: $P(t) = \exp\left[-t/\tau_+\right]$. If all particles have the same magnetic moment density, the time dependent statistically averaged magnetization becomes

$$M(t) = M(0)\left[2P(t) - 1\right] . \tag{5.8}$$

If a particle size distribution $G(V)$ such as a lognormal distribution, (5.4), is included, the time dependent magnetization becomes

$$M(t) = \frac{1}{V} \int_0^{\infty} G(V') \cdot V' \cdot M(t, V') \cdot dV' . \tag{5.9}$$

In the special case of $H = \text{const}$ and a uniform distribution $G(V')V' = \text{const}$ within some limits of $V_1 < V_2$, (5.9) can be integrated [25] leading to the frequently observed logarithmic time dependence [22]

$$M(t) = M(t_0) + S(H,T) \cdot \log(t/t_0) \tag{5.10}$$

which is valid only in a limited time range $\tau_1(V_1) \ll t \ll \tau_2(V_2)$. The slope parameter $S(H,T)$ in (5.10) is known as the magnetic viscosity coefficient and it characterizes magnetization decay in such systems.

Next we need to apply the above model to calculate a typical magnetic recording situation, e.g. a decaying di-bit as depicted in Fig. 5.1. In this problem one has to incorporate the time-dependence of the demagnetizing field $H_d(t)$ (see discussion of demagnetizing field in Sect. 5.2), which is the result of the decaying magnetization. Accordingly, $\tau_\pm$ [(5.2) and (5.5)] are not constant in time and the differential equation for $P(t)$ has to be integrated numerically.

Viscosity measurements of magnetic recording media are typcially very time consuming. By design, a recording medium is "stable", hence its magnetic viscosity effect is expected to be negligible. As a consequence, decay data have to be collected over many decades in time in order to arrive at statistically meaningful viscosity coefficients. This problem is usually solved by accelerating decay by either performing the measurements in a reverse field, or by heating the sample or by reducing the thickness, i.e. average grain volume. Examples will be discussed in the experimental Sect. 5.3. In all of these experimental situations, the energy barrier $E_B(H)$ and the switching time constant τ (5.5) are reduced. A particularly important experimental method is based on controlling the time during which a reverse field is applied, in addition to controlling its amplitude. In these pulse-width dependent remanence measurements, the reverse magnetic field is applied for a specific duration t_p, followed by a remanent magnetization measurement after a fixed waiting time. The result is a set of pulse-width dependent dc-demagnetization (DCD) curves or remance curves, which can be inspected to extract the switching field distribution or even the energy barrier distribution as first pointed out in [26] (see detailed discussion in Chap. 6).

The remanent coercivity H_{CR} is extracted from such a remanence curve as the field where the remanent magnetization in (5.8) vanishes: $M_r(t_p) = 0$. At this point, half the sample volume has switched, provided that each grain has the same saturation magnetization. For $\tau_- \gg \tau_+$, it is straightforward to calculate the remanent coercivity as a function of the pulse width by combining (5.2), (5.5) and (5.8) and setting $t = t_p$ [27] and $n = 1/m$:

$$\frac{H_{CR}}{H_0} = 1 - \left[\frac{k_B T}{K_u V_m} \ln\left(\frac{f_0 t_p}{\ln 2}\right)\right]^n . \tag{5.11}$$

$K_u V_m / k_B T = C_m^{-1}$ is a median stability ratio. The intrinsic switching field H_0 is reached at write times $t_p = \ln 2 / f_0$. Note however, that (5.11) is only valid in the limit of large barriers, i.e. conditions where $t_p \gg f_0^{-1}$. At short times, precessional effects become dominant and (5.11) no longer applies.

Higher or lower energy barriers may be studied by probing different states $M_r \neq 0$ in time-dependent coercivity measurements or by applying a reverse magnetic field in decay measurements. Quantitative relationships between both measurements have been suggested and experimentally verified [15].

5.2 Demagnetizing Fields

5.2.1 Single Transition

Following (5.2), the demagnetizing field $H = H_d$ in longitudinal recording media may considerably reduce the energy barrier for magnetization reversal. As a result, switching rates increase and significant decay (see Chap. 5.3.2), as well as transition broadening and percolation may occur. These effects

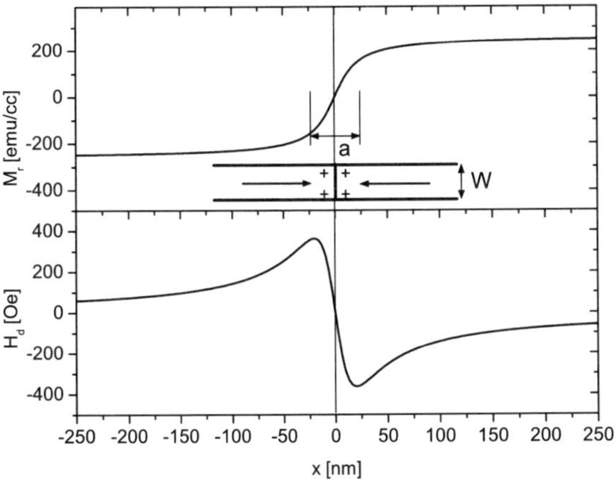

Fig. 5.4. M_r (top) and H_d (bottom) of a single arctangent transition, calculated in the center of the medium using (5.13) and typical 35 GBit/in^2 demonstration medium parameters [29]

may ultimately limit the achievable areal density in longitudinal magnetic recording.

Let us first consider the extreme case of an infinitely sharp bit transition in longitudinal recording media and assume the condition $\delta \ll B$, where δ is the medium thickness and B is the transition spacing. The demagnetizing field H_d in the center of the bit cell is then proportional to the linear transition density $LD = B^{-1}$. For $W \gg B$, where W is the track width, H_d may be approximated by [2]

$$H_d(D) \approx k \cdot M_r\delta \cdot LD = s \cdot LD\,. \tag{5.12}$$

$M_r\delta$ is the areal moment density. The prefactor k is 16 for the case of an isolated di-bit [2] and becomes 4π for a square wave pattern [28], which is considered here.

For higher transition densities the finite width of the transitions, characterized by the transition parameter a (Fig. 5.4), has to be incorporated explicitly into the calculations and deviations from the linear dependence of the demagnetizing field in (5.12) are expected. If the magnetization change across a transition assumes an arctangent form, the demagnetizing field along the center line of the recording medium may be calculated analytically [29]:

$$H_d(\Delta x) = -8M_r\left[\tan^{-1}\left(\frac{a+\delta/2}{\Delta x}\right) - \tan^{-1}\left(\frac{a}{\Delta x}\right)\right]\,, \tag{5.13}$$

where Δx is the distance from the center of the transition. An example of calculated demagnetizing fields is illustrated in Fig. 5.4, in which a single transition in the IBM 35 GBit/in^2 demonstration medium is considered.

The above model is fairly crude and does not take into account the actual granular structure of the media. It is therefore important to devise experiments to determine bit demagnetizing fields quantitatively, as discussed next.

5.2.2 Experiments

Qualitative measurements of the demagnetizing field in longitudinal magnetic recording media have been reported in [32] using spin-stand testing. They investigated tracks with square wave bit patterns at different linear densities and measured the DC write current I_c that is necessary to reduce the signal amplitude of the read-back signal in half. The total field is the sum of the field produced by the write head H_w and the internal demagnetizing field H_d. A measurement of the current I_c as a function of the linear density LD should therefore produce the density dependent demagnetizing field $H_d(LD)$ via the relation $H_d(LD) = H_{CR} - H_w(I_c)$, H_{CR} being constant. Indeed, it was found that I_c is reduced for tracks written at higher linear densities, suggesting that the demagnetizing field increases with LD.

A similar, but quantitative approach is used in [33]. The medium is sputtered onto a coplanar waveguide [34]. Launching a current pulse through the waveguide results in a characteristic magnetic field pulse at the magnetic layer with a controlled field amplitude. It is determined by the geometry of the sample and the amplitude of the current pulse and can be accurately calculated using a Karlquist model [35].

To determine the average demagnetizing field in a bit cell, bit tracks are written into the recording medium using a contact write/read tester [30]. A schematic drawing of this instrument is shown in Fig. 5.5. The read-back signal amplitude is measured before and after application of a 10 ns magnetic field pulse by the waveguide. For such short pulses, nucleation of reversal processes occurs only in the regions where the total magnetic field exceeds the local switching field, and the effects of thermal excitation may be neglected. The process is repeated for different pulse amplitudes and for different linear densities D. The effective remanent coercivity

$$H^*_{CR}(LD) = H_{CR} - H_d(LD) \tag{5.14}$$

is then extracted as the reverse pulse field, which is required to reduce the read-back signal in half (inset to Fig. 5.6).

The investigated medium is a 25 nm thick $CoCr_{10}Ta_4$ film on a 25 nm thick Cr underlayer, both sputtered onto a coplanar waveguide [34], and it has an in-plane easy-axis orientation. The magnetic $CoCr_{10}Ta_4$ layer has a remanent coercivity of $H_{CR} = 1040$ Oe (measured at ≈ 1 s pulse width) and an areal moment density of $M_r\delta = 1.2$ memu/cm^2. The high areal moment density results in a transition parameter of $a \approx 74$ nm (Williams-Comstock model with head media spacing of ≈ 80 nm).

The effective remanent coercivity $H^*_{CR}(LD)$ is plotted as a function of linear density LD in Fig. 5.6. At low transition densities ($LD = 500$ fc/mm, flux

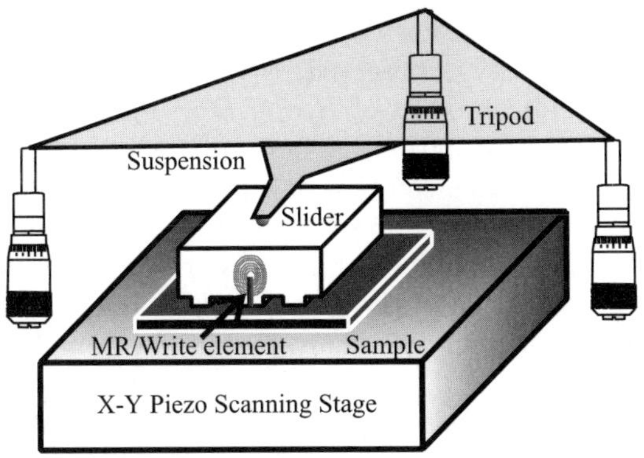

Fig. 5.5. Contact write/read tester or Scanning Magneto-Resistive Microscope [30,31]. This instrument uses a conventional write/read head, which is in contact with the sample. The sample is attached to a piexo-electric x-y stage with a scan range of 100 µm * 100 µm and can be heated to 390 K using a resistive heater. The suspension of the head is attached to a tripod for precise adjustment of pitch and roll angles. The instrument is used to write transitions and image stray fields from such transitions

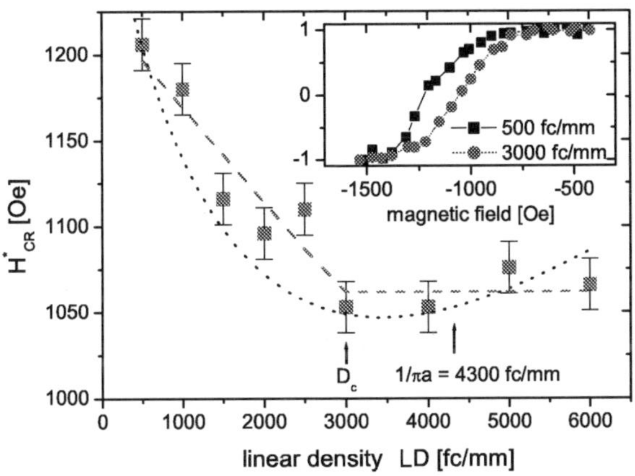

Fig. 5.6. Effective remanent coercivity $H_{CR}{}^*$ as a function of linear density LD according to (5.14). The dashed line is a fit to the experimental $H_{CR}{}^*$ data assuming infinitely sharp transitions. The dotted line represents the demagnetization field averaged across the bit cell using a finite transition parameter, $a = 74$ nm [percolation density $(1/\pi a) \simeq 4300$ fc/mm]. The inset shows two representative remanence loops at $LD = 500$ and 3000 fc/mm

changes per mm), H^*_{CR} agrees well with the intrinsic or short time switching field of 1230 Oe, measured by an indpendent pulse-width dependent coercivity method (see Chap. 5.3.3). For $LD < LD_c$, where $D_c \approx 3000$ fc/mm is the position of the shoulder in Fig. 5.6, H^*_{CR} decreases roughly linearly with LD as a result of the additional, density-dependent demagnetizing field

$H_d(LD) \approx s \cdot LD$: $H^*_{CR} = H_{CR} - s \cdot LD$. For $LD > LD_c$, no further decrease of the effective remanent coercivity is observed: H^*_{CR} = const. A linear fit in the regime $LD < LD_c$ yields an offset of $H_{CR}(LD = 0$ fc/mm$)$ = 1224 ± 20 Oe, a slope $s = (-5.5 \pm 1.0) \times 10^{-3}$ Oe cm and a shoulder at $LD_c = 2940 \pm 370$ fc/mm. From slope s one can now calculate the prefactor in (5.12) to be $k = s/M_r\delta = 4.6 \pm 0.8$. This analysis shows that the demagnetizing fields are considerably reduced if the bit transitions are spread over a width a. For the present case, a reduction of the prefactor k by about 2–3 times ($k \approx 4.6$) compared to the case of infinitely sharp transitions ($k = 4\pi$) is found.

The observed shoulder in Fig. 5.6 occurs at ~ 70% of the estimated percolation density $\frac{1}{\pi a} \simeq 4300$ fc/mm. A more accurate fit to the data may be obtained by incorporating the finite transition width parameter into the calculation of H_d. The dotted line in Fig. 5.6 represents such a calculation of the averaged demagnetizing field between two neighboring transitions (bit cell), which is straightforward to carry out by superimposing demagnetizing fields according to (5.13).

In modern recording systems, transition parameters are much smaller than in the discussed example. As a result, k is larger than 4.6 and tends toward 4π for infinitely sharp transitions. Using $LD_c \approx 0.7 \cdot \frac{1}{\pi a}$ one estimates

$$(H_d)_{\max} = k \cdot LD_c \cdot M_r\delta \approx 0.7k \cdot \frac{M_r\delta}{\pi \cdot a} . \tag{5.15}$$

This amounts to a maximal demagnetizing field of about 700 Oe for the 35 GBit/in^2 IBM demonstration medium ($M_r\delta$ = 0.39 memu/cm^2, a = 16 nm, $k = 4\pi$) [13].

5.3 Measurements of Thermal Effects

5.3.1 Samples

In this section, experiments on thermal effects of two sets of samples will be presented. The structural and magnetic properties of both sets are summarized in Table 5.1.

Table 5.1. Structural and magnetic parameters a set of $CoCr_{22}Pt_{14}$ (I–IV) media provided by NSIC, and a set of CoCrPtB media (A–D) from IBM internal sources

Sample #	I	II	III	IV	A	B	C	D
δ [nm]	13.0	10.2	7.5	5.5	17.7	15.6	13.9	11.5
$M_r\delta$[memu/cm^2]	0.39	0.31	0.22	0.17	0.46	0.40	0.36	0.30
H_{CR} [Oe]	2420	1700	1210	340	4800	4600	4250	3750

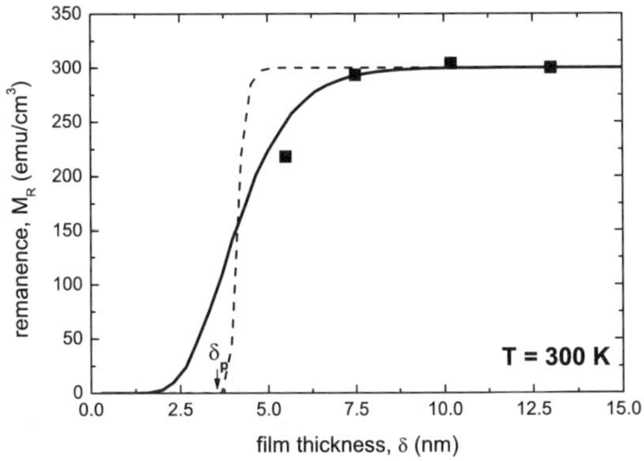

Fig. 5.7. Thickness dependent remanent magnetization of samples I–IV. Calculations with (solid line) and without (dotted line) lognormal energy barrier distribution using (5.9) [26]

The first set consists of $CoCr_{22}Pt_{14}$ longitudinal recording media with thicknesses ranging from 5.5 nm to 13.0 nm. The magnetic layers are deposited by magnetron sputtering on a combination of 20 nm thick CrV_{20} intermediate and 60 nm thick NiAl underlayers. Substrates are NiP-plated Al-Mg (see refs. [36,37]). The samples were specifically designed to exhibit different degrees of superparamagnetic effects ranging from thermally stable (I and II) to thermally unstable (IV).

The second set of samples (A–D) uses a similar magnetic alloy to the one used in the IBM 35 GBit/in^2 recording demonstration [13]. The magnetic CoPtCrB-alloy layers are deposited on glass substrates using a different seed structure compared to the one in samples I–IV. The layers are $(11\bar{2}0)$ textured and have a median grain diameter of about 8 nm.

Figure 5.7 shows the remanent magnetization of samples I–IV from Table 5.1. The magnetization remains roughly constant as the superparamagnetic transition thickness δ_p [26] is approached and starts dropping off only for the thinnest sample, which is a result of thermally activated switching processes. This behavior is quite well reproduced by a fit using (5.9) with a lognormal energy barrier distribution $\sigma_E = 0.3$ and a time constant of 100 s (solid line). The dotted line represents a fit without an energy barrier distribution.

5.3.2 Signal Decay

Signal decay of recorded data tracks is one of the most commonly used measurements of thermal excitations in recording media. In contrast to conventional magnetometry measurements, signal decay measurements fully capture the complexity of decaying bit patterns. Here, effects of the time-dependent and spatially varying demagnetizing field as well as thermally activated transition broadening are directly included in the measurement.

Signal decay measurements are usually carried out using a spin-stand. The read-back amplitude of a square wave bit pattern is either measured using a spectrum analyzer or an analog signal peak detector. In order to correct for tester instabilities (drift), e.g. induced by time dependent variations of the read head sensitivity or the flying height, the read-back amplitude is usually normalized with respect to a read-back signal of a freshly written track [38] or an aged control track [39]. The aged track barely decays during the experiment as a result of the logarithmic nature of the decay process and provides an almost constant reference. In recent implementations of thermal decay measurement, the control track and the track of interest occupy different sectors of the same track [40,41]. This method is discussed in detail by Richter in Chap. 6.

The decay measurements discussed in the following have been carried out using a contact write/read tester [30,36], as introduced in Sect. 5.2.2. The investigated track is 80 μm long and the observation period is about 1 day (86400 s). Similar to spin-stand measurements, care has to be taken to eliminate amplitude changes from sources other than thermal magnetization decay. Here, the track is re-recorded at the end of each experiment and measured to check for read head sensitivity variations and tester drifts. Typically, the initial amplitude is reproduced within 1%.

Within this experimental error, an observable decay in the read back signal is then attributable to the decay of the magnetization toward thermal equilibrium (see Chap. 5.1.2). In analogy to the magnetic viscosity parameter $S(H,T)$, which was introduced in (5.10), we define the signal decay rate, in units of %/decade, as

$$R(H_d, T) = 100 \cdot \frac{A(t) - A(t_0)}{A(t_0)\log(t/t_0)}, \tag{5.16}$$

where $A(t)$ is the read-back signal amplitude and t is the elapsed time after the track has been written.

Linear Density Dependence. As already introduced in Sect. 5.2 the demagnetizing fields present in the bit cells lower the energy barriers for magnetization reversal (substitute $H = H_d$ in (5.2), resulting in higher switching rates (5.5). In longitudinal media, the demagnetizing fields increase with higher linear densities (5.12), reducing the energy barriers further. Accordingly, the measured decay rate is expected to strongly depend on the linear density. This trend is experimentally observed in signal decay measurements as e.g. shown in Fig. 5.8 for sample IV of sample set one. Note that the demagnetizing fields have the opposite effect in perpendicular recording. There, the demagnetizing fields are generally reduced at higher linear densities, and the written transitions are expected to exhibit improved thermal stability. The

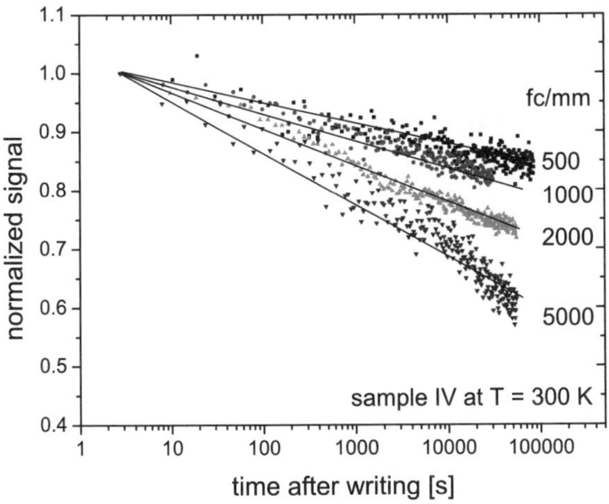

Fig. 5.8. Signal decay measurement of sample IV ($\delta = 5.5$ nm) at 500, 1000, 2000 and 5000 fc/mm at T = 300 K

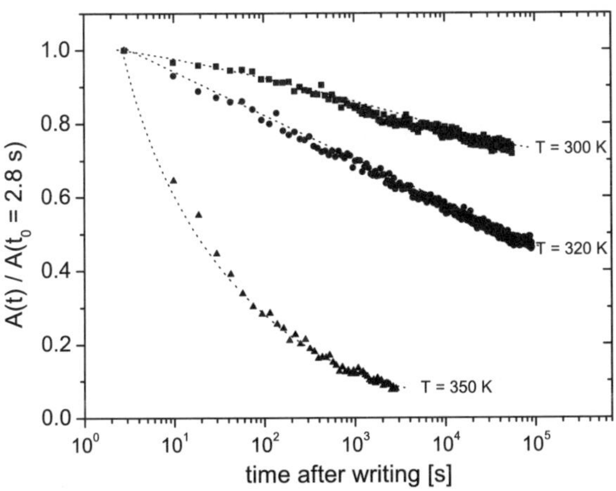

Fig. 5.9. Signal decay measurement at 1000 fc/mm of sample IV ($\delta = 5.5$ nm) at T = 300 K, 320 K and 350 K. The measurements are normalized to the first data point at $t = 2.8$ s. The lines are to guide the eye [36]

latter conclusion, of course, is valid only if the recorded patterns are DC-free. Perpendicular recording is discussed in detail by Victora in Chap. 7.

Temperature Dependence. Magnetic harddrives typically operate at 40–70 K above room temperature. At these elevated temperatures the Néel–Arrhenius switching rates according to (5.5) increase and enhanced thermal decay is expected. This effect is indeed observed as illustrated in Fig. 5.9 for sample IV ($\delta = 5.5$ nm). Even at room temperature, this sample exhibits

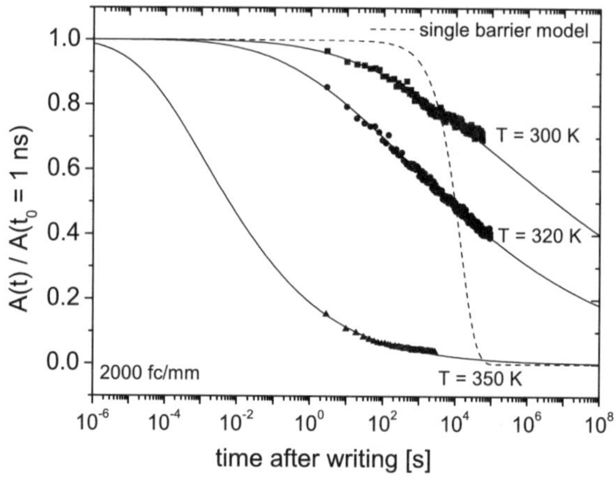

Fig. 5.10. Signal decay measurements at 2000 fc/mm of sample IV ($\delta = 5.5$ nm at T = 300 K, 320 K and 350 K. The data are normalized at $t_0 = 1$ ns and fit to decay model (solid lines) with lognormal barrier distribution. For comparison, the decay of a single barrier media is modeled (dashed line) [36]

considerable signal decay of $\approx 6\%$/decade. As shown further in Fig. 5.9 the decay accelerates as the temperature increases to 320 K and 350 K. The respective decay rates according to (5.16) are $\approx 11\%$/decade at 320 K and $\approx 30\%$/decade at 350 K. As expected from a lognormal energy barrier distribution, the decay data are generally not linear in $\log(t/t_0)$ over the entire time range of interest from recording ($\sim$ ns) to storage (10 years). Spin-stand measurements have shown similar deviations from pure logarithmic decay [17].

When normalizing to the first measured data point, as done in (5.16), one neglects the signal decay before the measurement is started. This is illustrated in simulations in Fig. 5.10, which are based on the independent particle model with a lognormal energy barrier distribution with $\sigma_E = 0.3$ at $H = 0$ Oe. They correctly reproduce the different shapes of the experimental data (dotted lines in Fig. 5.10) and clearly demonstrate that the experimental data are measured at completely different phases of the decay process.

Discussion. Energy barrier distributions and their effect on magnetization decay are more closely considered in Fig. 5.11. A lognormal distribution with a $\sigma_E = 0.3$ illustrates the situation in the IBM 35 GBit/in^2 demonstration medium (solid line). The energy barriers are reduced in a reverse magnetic field (dotted line) or at higher temperatures. In comparison, a wider distribution with $\sigma = 0.5$ (dashed line) has more grains with smaller energy barriers.

The relation of the energy barrier distribution to signal decay may be understood by first considering the decay of a hypothetical recording medium

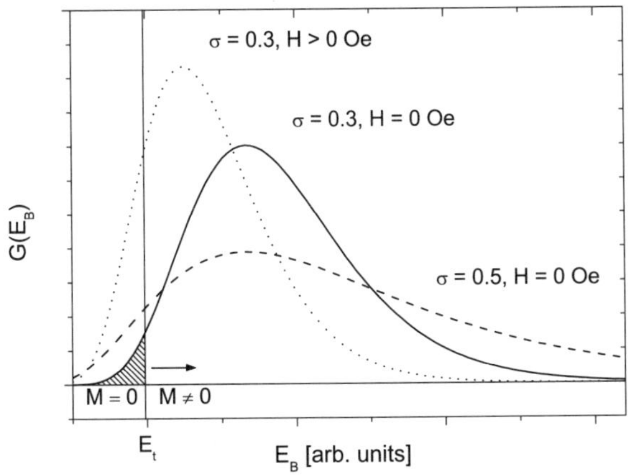

Fig. 5.11. Lognormal energy barrier distribution (solid line). The dashed line shows a distribution with the same median energy barrier and a broader distribution width. The energy barriers are lower in the presence of a reverse a magnetic field (dotted line) or at higher temperatures

compromised of monodisperse Stoner-Wohlfarth particles. Signal decay in such a medium occurs in a very narrow range (1-2 decades in time), as indicated by the simulated dashed line in Fig. 5.10. Such a decay process may be approximated by a step function. In a medium with a finite distribution width, only grains within a narrow range of energy barriers decay at a given time in this simple model. These energy barriers are given by $E_B = E_t \simeq \ln(f_0 t) \cdot k_B T$ (5.5), as illustrated by the solid vertical line in Fig. 5.11. Grains with larger energy barriers have not decayed yet and are still fully magnetized. In contrast, grains with smaller energy barriers are superparamagnetic on the considered time-scales and to first order have a vanishing averaged magnetic moment. As time progresses, grains with larger energy barriers start to decay and the solid vertical line shifts logarithmically to the right.

In theory, thermal decay experiments provide all relevant information about the energy barrier distribution, since the signal decay rate is roughly proportional to the energy barrier density $G(E_B)$, where $E_B \simeq \ln(f_0 t) \cdot k_B T$. In practice, however, the measurements are limited by the time available for the experiment and the quality of the experimental data. In the following we take into account the limited available time and consider only the decay of grains, for which $E_B < E_t$, i.e. to the left of the solid vertical line in Fig. 5.11. In particular, in thermally stable media, only the decay of particles with small energy barriers may be observed. The total integrated decay, which is proportional to the shaded area underneath the distribution curve, is then only of the order of a few percent.

Several experiments to characterize a wider portion of the energy barrier distribution within the limited amount of observation time have been suggested and carried out. One involves measurements at elevated temperatures, where grains with higher energy barriers decay within the observation time. However, intrinsic material parameters, such as the anisotropy energy density, are temperature-dependent, and may affect the resulting decay. Also wider energy barrier distributions exhibit a larger decay (e.g. dashed line in Fig. 5.11 with $\sigma = 0.5$ and $H = 0$ Oe), as such recording media have more grains with small energy barriers.

Alternatively, in order to accelerate thermal decay, one can lower energy barriers by applying a reverse magnetic field, such as the demagnetizing field or a magnetic field pulse, as discussed in the Sect. 5.1.2. In both cases, the energy barrier distribution shifts to lower energy barriers (dotted line in Fig. 5.11) and a larger sample volume becomes superparamagnetic. Experiments are shown in the next section.

5.3.3 Pulse-width Dependent Coercivity

Pulse-width dependent coercivity measurements were first used to study particulate magnetic tape [42]. Recently, strip-line [43,44], spin-stand [45,46] and contact write/read tester [26,30,36] experiments have been employed to measure pulse-width dependent coercivities in thin film media.

The following experiments were performed using a contact write/read tester (Fig. 5.5). A write head is statically placed in contact with the sample and used to generate a magnetic field pulse. The resulting di-bit is then read back by scanning the read element across the di-bit transition. These steps are repeated for different field amplitudes resulting in M-H remanence curves

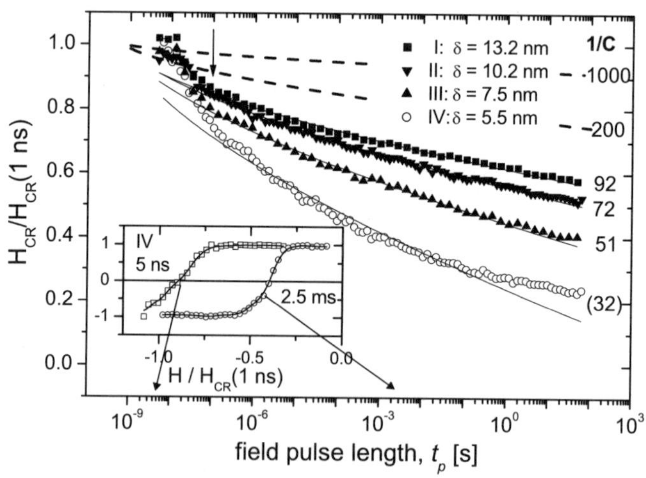

Fig. 5.12. Pulse-width dependent coercivity measurements over 11 decades in time. The fitted stability ratio is indicated to the right. The inset shows remanence magnetization loops at magnetic field pulse widths of 5 ns and 2.5 ms [26]

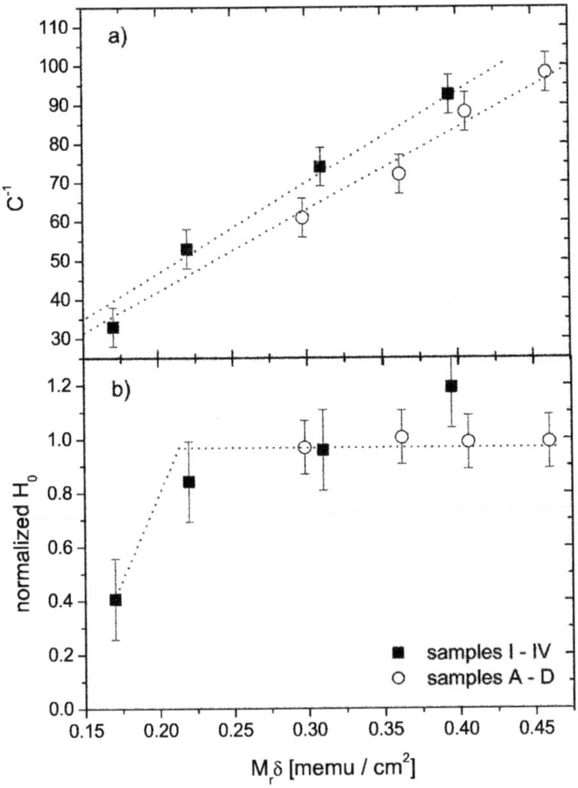

Fig. 5.13. Stability ratio as a function of $M_r\delta$ with linear fits in **a** and the normalized intrinsic switching field in **b**. The dotted line in **b** is to guide the eye

(see inset to Fig. 5.12). The remanent coercivity is then extracted and plotted as a function of the magnetic field pulse width. The measurement procedure is explained in detail in references [30,36,26]. Figure 5.12 shows examples of remanent coercivity measurements for a pulse width range from 5 ns to 65 s (10 decades in time).

Thickness Dependence. Strong pulse-width dependent coercivity effects (Fig. 5.12) are observed in all samples with a variation as large as 50%. The stability ratio $1/C$ and the switching field parameter H_0 are obtained by fitting the data to (5.11) and are plotted in Fig. 5.13. The exponent was set to $n = 2/3$ [9,12], and f_0 was assumed to be 10^9 Hz. At short pulse widths, deviations from the simple thermal model occur, which are believed to be due to the energy barrier dependence of f_0 [47] (see also Chap. 3 by Safonov and Bertram) and precessional switching effects. These data are generally excluded from the present fitting procedure and the analysis range is confined to pulse widths 10^{-6} s $< t_p < 1$ s.

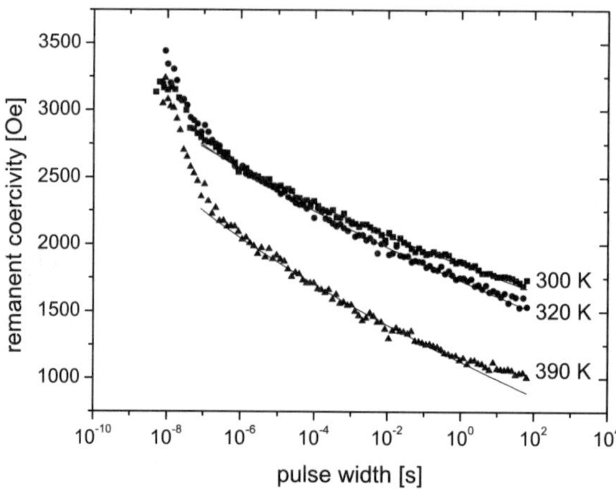

Fig. 5.14. Pulse-width dependent coercivity measurements of the 10.2 nm thick CoCrPt media (sample II) at 300 K, 320 K and 390 K

Both sets of samples show similar trends with the stability ratio dropping off approximately linearly with the areal moment density $M_r\delta$ (Fig. 5.13a). This behavior can be easily understood if one rewrites $C^{-1} = (K_u V_m)/(k_B T)$ in (5.11) by substituting $K_u = 1/2 H_K M_s$ and assuming a cylindrical grain volume with median diameter D_m according to $V_m = \pi/4 D_m^2 \delta$. The stability ratio may then be written $C^{-1} = a \cdot M_r\delta$, with $a = (H_K \cdot D_m^2 \cdot \pi)/(8 k_B T S)$. $S \simeq 0.8$ is the remanence squareness, and H_K is the anisotropy field (5.3). For granular samples, H_0 should be independent of the sample thickness. This behavior is found for samples A–D and I–III in Table 5.1 (Fig. 5.13b). The strong reduction of H_0 in sample IV is an indication that the magnetic properties are different, which is presumably due to microstructural changes in the thinnest sample (see Chap. 2 for in depth discussion of structural and microstructural properties of recording media).

Temperature Dependence. Examples of pulse width dependent coercivity measurements of sample II are plotted in Fig. 5.14 for three different temperatures. Clearly, the remanent coercivity at short pulse widths, corresponding to the intrinsic switching field H_0, is almost independent of temperature. By contrast, the slope C of the curves representing the inverse of the stability ratio C^{-1} in (5.11) increases, reflecting stronger thermal excitations at higher temperatures. As a consequence, the signal decay rate increases at elevated temperatures as already discussed in conjunction with Figs. 5.9 and 5.10 in Chap. 5.3.2 [39,36]. Similar trends for H_0 and C are found for all samples I–IV.

Figure 5.15 shows the temperature dependence of the normalized stability ratio $C^{-1}(T)/C^{-1}(T = 300\ \mathrm{K})$, where $C^{-1}(T) = E_B/k_B T = K_u V_m/k_B T$. Interestingly, the stability ratio of all samples is reduced at roughly the same

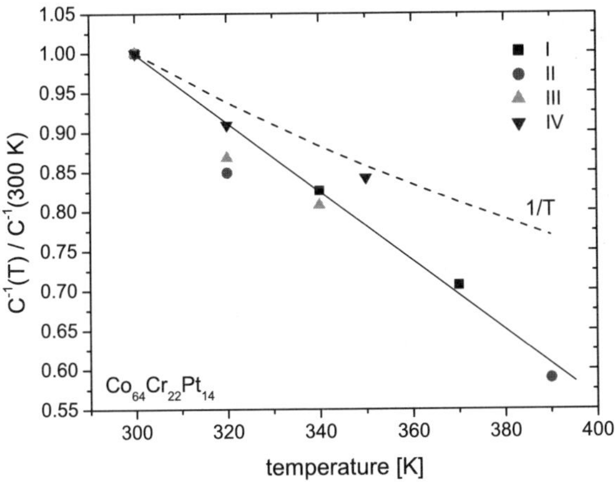

Fig. 5.15. Temperature dependence of the stability ratio $1/C$ normalized to $1/C(300\text{ K})$. The dashed line illustrates the $1/T$ bahavior expected for temperature-independent energy barriers. The solid line is to guide the eye

rate of about 0.5%/K. Clearly, the observed stability ratios drop off faster than the simple $1/T$ behavior (dashed line), showing that the energy barriers must be temperature-dependent. The additional reduction due to the temperature dependence of K_u, $C^{-1}(T) = K_u(T)V_m/k_BT$, amounts to $\sim 10\%$ at typical drive temperatures. By performing a linear fit to all data, a temperature coefficient of $-0.2\%/\text{K}$ is obtained for the anisotropy energy density.

5.3.4 Stability Cliff

The stability limit of the characterized $Co_{64}Cr_{22}Pt_{14}$ media may be deduced by correlating signal decay and time dependent coercivity measurements. Figure 5.16 shows signal decay rates at two linear densities of 500 fc/mm and 2000 fc/mm as a function of the measured stability ratio C^{-1}. Different measurement conditions (temperature) and sample parameters (thickness) result in stability ratios varying over a wide range of values between 27 and 93. Clearly, substantial signal decay rates occur for stability ratios below the "cliff" $(1/C)_0 = 35$ for a measurement time of 1 day. A simple theoretical estimate which does not include particle size distributions is in remarkable agreement with the experimental minimal stability ratio, $(1/C)_0$. It can be derived from (5.5) and (5.7) using a signal retention criterion of $P(t = 10\text{ years})/P(t = 0\text{ s}) = 0.95$ as suggested in [1]: $C^{-1} = \ln\left[-t * f_0/\ln(0.95)\right] = 35$ with t=86400 s (1 day) and $f_0 = 10^9$ Hz.

The onset of decay shifts to higher stability ratios with increasing linear density. This is a direct consequence of the effect of increasing demagnetizing fields H_d, which reduce the reversal barrier and destabilize the bit transitions.

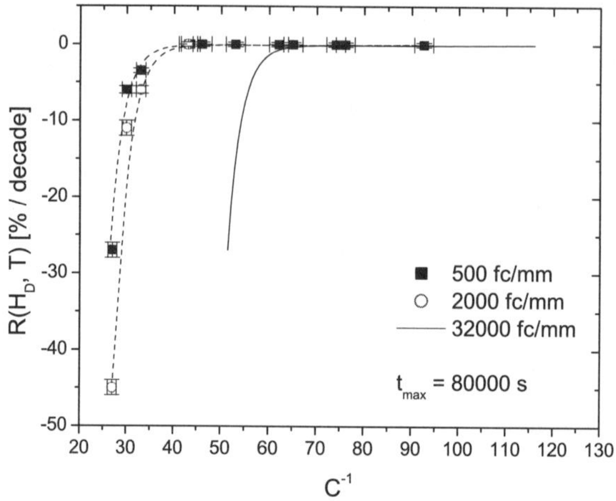

Fig. 5.16. Signal decay measurements at 500 fc/mm and 2000 fc/mm are plotted as a function of the stability constant obtained from pulse-width dependent coercivity measurements. The solid line shows an extrapolation to a linear density of 32000 fc/mm. The dashed lines are to guide the eye [28]

An extrapolation to higher linear densities with sufficient thermal stability for 10 years data storage is readily done in two steps. First, the minimum stability ratio is increased to account for the extension of the laboratory time scale of 1 day by 3.5 decades to 10 years. From simulations using the single particle model with an energy barrier distribution $\sigma_E = 0.33$ we estimate $\Delta((1/C)_0) \approx 4$/decade in time, resulting in a minimal stability ratio ("cliff") of $(1/C)_0 \approx 35 + 3.5 \cdot 4 = 49$ [26]. Second, the larger demagnetizing fields at higher linear densities need to be accounted for. A combination of (5.2) with (5.15) yields for the minimal stability ratio

$$(1/C)_0 = 49 \cdot \left(1 - \frac{k \cdot M_r\delta \cdot LD_c}{H_0}\right)^{-3/2} . \tag{5.17}$$

Table 5.2 gives some examples of the application of (5.17). The minimal estimated stability ratio of the 40 GBit/in^2 simulation is higher than the stability criterion theoretically derived in [1] for linear densities of 32000 fc/mm (814 kfci). This finding suggests, that the effect of the demagnetizing fields is considerably overestimated, as expected when using $k = 4\pi$. A 100 GBit/in^2 scenario, as discussed in [2], with a reduced bit-aspect ratio BAR $\simeq 4$, requires a similar $(1/C)_0$. In [13], the IBM 35 GBit/in^2 demonstration medium was reported to have a stability ratio between 70 and 80 and no thermal decay was detected. This ratio is well above the estimated limit $(1/C)_0 = 54$, which explains the good thermal stability.

Table 5.2. Comparison stability ratio requirements for micromagnetic simulations at 40 and 100 GBit/in^2 and the 35 GBit/in^2 demonstration assuming infinitely sharp transitions $k = 4\pi$

Example	$M_r\delta$ [memu/cm^2]	H_0 [kOe]	a-param. [nm]	$(1/C)_0$ (5.17) $k = 4\pi$
40 GBit/in^2 [1] model	0.5	5.1	16	74
100 GBit/in^2 [2] model	0.13	4.0	9.2	57
35 GBit/in^2 [13] IBM demonstration	0.39	9.0	16	57

5.3.5 Discussion

The results of the investigated alloy systems, which have been presented in this chapter, provide guidance for the development of future magnetic recording media. The stability ratio is reduced at elevated temperatures resulting in enhanced thermal decay. By contrast H_0 is found to be almost temperature-independent between room temperature and 390 K (see Fig. 5.14), i.e. write field requirements are not relaxed. Future recording schemes, e.g. thermally assisted recording, in which one attempts to lower H_0 temporally during the writing instant, need to take this into account. Special alloys with an increased temperature dependence are needed.

Experimentally, pulse-width dependent coercivity measurements are extremely sensitive to thermal effects. They are less time consuming and easier to perform than signal decay measurements. These measurements are well suited to compare the relative stability of recording media with similar structural and magnetic parameters, i.e. samples with similar energy barrier distribution width. Generally, however, both measurements are necessary in order to assess the median stability ratio and the width of the energy barrier distribution. This conclusion was illustrated in conjunction with Fig. 5.11 (Sect. 5.3.2).

5.4 Alternative Recording Media

5.4.1 Antiferromagnetically Coupled Media

Traditionally, recording systems at higher areal density are accomplished by scaling all relevant media and head parameters to smaller sizes. One of these parameters is the demagnetization length $M_r\delta/H_{CR}$, which is reduced to lower the transition parameter [29]. Recently, antiferromagnetically coupled (AFC) recording media [48] or longitudinal recording media with stabilization layers [49] have been suggested as an evolutionary path to improve thermal

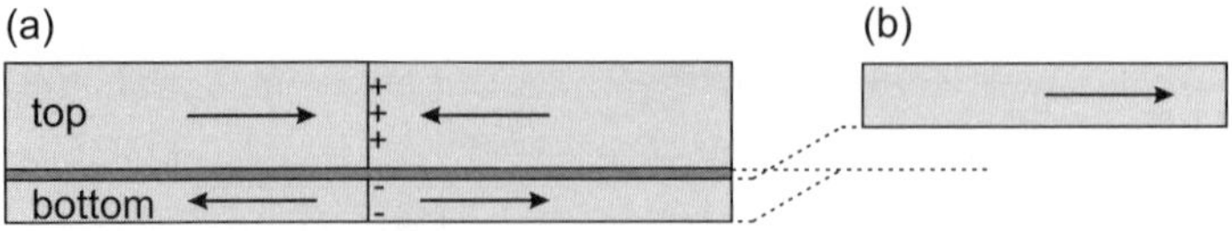

Fig. 5.17. (**a**) Schematic representation of a single transition in an AFC media with two ferromagnetic layers. (**b**) Single layer media with the same effective $M_r\delta = (M_r\delta)_{\text{top}} - (M_r\delta)_{\text{bottom}}$

stability without compromising the recording performance. In these laminated recording media, two or more ferromagnetic layers are separated by thin non-magnetic layers, which induce antiferromagnetic coupling between the ferromagnetic layers (as indicated in Fig. 5.17a). The exchange coupling between the layers and a reduction of the demagnetizing fields are expected to enhance thermal stability. Indeed, experiments presented in [48] and [49] demonstrate that AFC structures exhibit considerably less thermal decay than single layer structures with the same effective $M_r\delta$ (Fig. 5.17b)). In addition a reduction of the read-back pulse width PW_{50} and an increase in signal-to-noise ratio (SNR) have been observed.

Switching processes and thermal activation processes in AFC media are much more complex than in single layer media. In order to simplify the arguments, we consider in the following the basic AFC double layer structure, as shown in Fig. 5.17a. Both layers have equal anisotropy density K_u and saturation moment density M_s.

In the limit of strong exchange coupling between both layers, i.e. the exchange field is larger than the switching field of the storage layer, both layers switch together and the stability ratio is the sum of the stability ratios of the individual layers:

$$C^{-1} = \frac{(K_u V)_{\text{eff}}}{k_B T} = \frac{(K_u V)_{\text{top}} + (K_u V)_{\text{bottom}}}{k_B T} \,. \tag{5.18}$$

The switching field in this strong coupling case when neglecting demagnetizing fields is then calculated to be

$$H_0 \simeq 0.5 \left(\frac{2K_u}{M_s}\right)_{\text{eff}} = \frac{K_u(V_{\text{top}} - V_{\text{bottom}})}{M_s(V_{\text{top}} + V_{\text{bottom}})} \,. \tag{5.19}$$

Such a strongly coupled AFC structure does not exhibit any advantages, since the stability ratio and the switching field are identical to that of a single layer medium with an enhanced anisotropy density $(K_u)_{\text{single}} = K_u(V_{\text{top}} + V_{\text{bottom}})/(V_{\text{top}} - V_{\text{bottom}})$ and a reduced grain volume $V_{\text{single}} = V_{\text{top}} - V_{\text{bottom}}$ [50].

The situation is different for weakly coupled systems, which are currently considered for use as storage media [48,49]. Here, the exchange field is much smaller than the switching field of the storage layer, and in the remanent

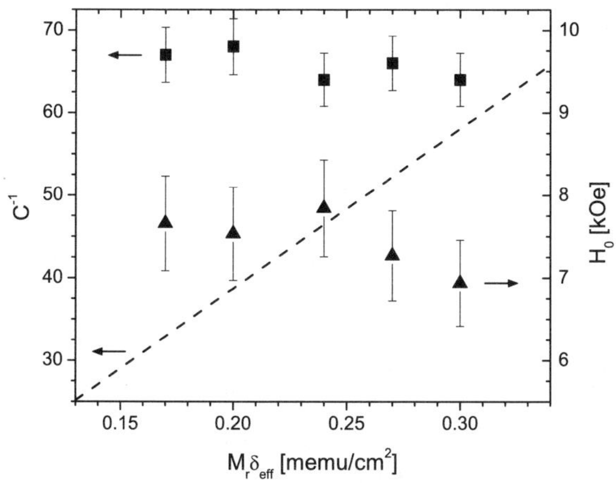

Fig. 5.18. Stability ratio (squares) and switching field (triangles) of an AFC media with constant top layer and variable bottom layer thickness as a function of the effective $M_r\delta_{\mathrm{eff}}$. The dashed line shows the calculated stability ratio of a single layer medium with $M_r\delta = M_r\delta_{\mathrm{eff}}$

state the bottom layer moves into an antiferromagnetic state relative to the top layer. As shown in the following experiments, such a structure takes full advantage of the enhanced thermal stability of the top layer, while barely affecting the writeability of the medium.

Pulse-width dependent coercivity measurements of a series of AFC media with a fixed top layer and a variable bottom layer thickness were performed, and both the stability ratio and the switching field were found to be constant within experimental error, as shown in Fig. 5.18 [51]. These results suggest that the stability ratio and the switching field, to first order, are determined by the top layer and the exchange field. Furthermore, $M_r\delta_{\mathrm{eff}}$ can be tailored over a wide range to achieve the required recording properties without compromising thermal stability or write field requirements. Moreover, $M_r\delta_{\mathrm{eff}}$ is smaller, resulting in reduced demagnetizing fields. In essence, AFC media combine two advantages: they are "thick" regarding thermal stability, and "thin", when magnetic properties, such as PW_{50} and demagnetizing fields, are considered.

The reduced $M_r\delta_{\mathrm{eff}}$ of the AFC media results in an areal density advantage in comparison to stable single layer media. Following [52] the achievable areal density is $AD \propto M_r\delta^{-2/3}$, when only thermal stability issues are considered. Using a rather aggressive thickness ratio of top and bottom layer of 3:1, we estimate an areal density gain of $[(3-1)/3]^{-2/3} \sim 1.3$ over a stable single layer medium with thickness 3. Improved recording performance of AFC media and reduced demagnetizing fields may increase the areal density gain further.

5.4.2 High-K_u Alternative Media

A dilemma arises, as one tries to reduce noise and maintain thermal stability at the same time. Compensating the loss of volume by increasing the anisotropy energy density K_u runs into the problem of increased write field, which must be sufficient to switch all grains. At write frequencies approaching the thermal attempt frequency f_0 (1–100 GHz), thermal excitations become less relevant and the write field approaches or exceeds the switching field in (5.3). Temperature assisted methods may eventually be needed to ease this problem [53]. However, new special alloys or multilayers with strong temperature coefficients will be needed (see Sect. 5.3.4).

Provided that the write field problem can be solved, high K_u-alloys are extremely attractive because of the enormous potential of drastically reduced "stable" grain diameters. This potential is briefly illustrated in the following.

Today's recording materials are based on hexagonal Co-alloys with an approximate maximum anisotropy constant of $K_u = 4.5 \cdot 10^6$ erg/cm^3 for pure metallic Co. The smallest thermally stable grain diameter is estimated using

$$D_p \simeq \left(60\frac{k_B T}{K_u}\right)^{1/3} = 1.35 \cdot 10^{-4}\left(\frac{1}{K_u}\right)^{1/3} \tag{5.20}$$

for $t_{max} = 10$ years and $T = 300$ K. Results are $D_p = 8$ nm for Co and $D_p = 9.1$ nm for the CoPtCrB media alloy used in the IBM 35 GBit/in^2 demonstration[1]. Grain sizes of these magnitudes are already achieved today. Table 5.3 summarizes intrinsic hard magnetic alloy properties, as compiled in [26].

Face centered tetragonal (fct) structures like FePt, FePd or CoPt ($L1_0$ phases) and rare-earth transition metals, such as $SmCo_5$ and $Fe_{14}Nd_2B$, have a much higher anisotropy energy density ranging from $1.7 \cdot 10^7$ erg/cm^3 to $2 \cdot 10^8$ erg/cm^3 [26,54]. Consider e.g. the $L1_0$-phase of the FePt system. With an anisotropy of about 10^8 erg/cm^3, grains with diameters as small as $D_p = 2.8$ nm should be thermally stable according to (5.20). Such small grain diameters may translate into a potential 10-fold areal density increase, assuming a $1/D_p^2$ areal density scaling. The write fields of such materials, however, providing that their properties survive grain engineering, would be in the vicinity of $H_0 = 50,000$–$100,000$ Oe, far beyond the capability of present days write heads.

Other materials rely on the high anisotropy generated by interfaces, such as Co/Pt or Co/Pd multilayers, which are currently investigated for usage in perpendicular recording systems [55]. Other systems are based on nanocomposite and nanostructured materials [56].

[1] Assuming cylindrical grains with area of 53.3 nm^2 and thickness 14.4 nm, the median grain volume for the IBM 35 GBit/in^2 demonstration media becomes 780 nm^3, which is larger but closer to than $D_p^3 = 753$ nm^3.

Table 5.3. Properties of magnetic materials capable of sustaining grain sizes $D_p \leq$ 10 nm over storage times of ten years, where D_p is the diameter of a sperical grain as calculated by (5.20) [26]

alloy system	**material**	K_u [10^7 erg/cm^3]	M_s [emu/cm^3]	H_K [kOe]	D_p [nm]
	CoCrPtB	0.33	320	20	8
Co-alloys	Co	0.45	1400	6.4	8
	hex Co_3Pt	2.0	1100	36	4.8
	FePd	1.8	1100	33	5.0
$L1_0$ phases	FePt	6.6 - 10	1140	116	2.8 - 3.3
	CoPt	4.9	800	123	3.6
	MnAl	1.7	560	69	5.1
RE-TM	$Nd_2Fe_{14}B$	4.6	1270	73	3.7
	Co_5Sm	11 - 20	910	240 - 400	2.2 - 2.7

Some of the problems common to hard magnet materials approaches are unfavorably high processing temperatures during either annealing or growth, which may be a disadvantage for achieving small grains.

Monodisperse chemically synthesized nanoparticles, as described in Chap. 9, are another attractive path to high-density media [57]. Their extremely small volume distribution promises reduced noise and reduced decay rate. Additionally, taking into account the high anisotropy of e.g. FePt, recording media with extremely small average grain volumes may be produced, which are close to the limit imposed by superparamagnetic effects.

5.5 Summary

For almost a decade, the thermal decay problem has been considered a "brick wall" in longitudinal magnetic recording. All predicted "limits" have been surpassed by engineering and innovation. Today, as discussed in this chapter, thermal decay gives rise to serious concerns for future high areal density improvements. New ideas and perhaps also new recording techniques are needed. Antiferromagnetically coupled media are an excellent example of an evolutionary approach to extend the limits of magnetic recording. However, systems based on perpendicular recording and thermally assisted recording, and perhaps recording on patterned media may soon be required.

References

1. S.H. Charap, Pu Ling Lu, and Yanjun He, "Thermal stability of recorded information at high densities," *IEEE Trans. Magn.*, vol. 33, pp. 978, 1997.
2. H.N. Bertram, H. Zhou, and R. Gustafson, "Signal to noise ratio scaling and density limit estimates in longitudinal magnetic recording," *IEEE Trans. Magn.*, vol. 34, pp. 1845, 1998.
3. R. Wood, "The feasibility of magnetic recording at 1 terabit per square inch," *IEEE Trans. Magn.*, vol. 36, pp. 36, 2000.
4. K. Stoev "Demonstration and characterization of greater than 50 GBit/in^2 reording systems," The 8th Joint MMM-Intermag Converence, San Antonio, TX, Jan. 7–11, 2001, paper EA-12 (IEEE Trans. Magn. (to be published)).
5. D.A. Thompson, and J.S. Best, *IBM J. Res. Develop.*, vol. 44, pp. 311, 2000.
6. L. Néel, "Théory du trainage magnétique," *Ann. Geophys*, vol. 5, pp. 99, 1949.
7. W.F. Brown, Jr, "Thermal fluctuations of a single-domain particle," *Phys. Rev.*, vol. 130, pp. 1677, 1963.
8. E.C. Stoner and E.P. Wohlfarth, "A mechanism of magnetic hysteresis in heterogeneous alloys," *Trans. Roy. Soc.*, vol. A240, pp. 599, 1948.
9. H. Pfeiffer, "Determination of anisotropy field distribution in particle assemblies taking into account thermal fluctuation," *Phys. Stat. Sol. A*, vol. 118, pp. 295, 1990.
10. H. Kronmüller and T. Reininger, "Micromagnetic background of magnetization processes in inhomogeneous ferromagnetic alloys," *J. Magn. Magn. Mater.*, vol. 112, pp. 1, 1992.
11. H.N. Bertram and H.J. Richter, "Arrhenius Néel thermal decay in polycrystalline thin film media," *J. Appl. Phys.*, vol. 85, pp. 4991, 1999.
12. R.H. Victora, "Predicted time dependence of the switching field for magnetic materials," *Phys. Rev. Lett.*, vol. 63, pp. 457, 1989.
13. M. Doerner, X. Bian, M. Madison, *et al.*, "Demonstration of 35 GB/in^2 using media on glass substrates," *IEEE Trans. Magn.* vol. 37, pp. 103, 2001.
14. W. Ciurzynska, G. Haneczok, and J. Zbroszczyk, "Magnetic after effect spectra analysis by applying the lognormal distribution of relaxation times," *J. Magn. Magn. Mater.*, vol. 189, pp. 384, 1998.
15. S.M. Stinnett, J.W. Harrell, A.F. Khapikov, and W.D. Doyle, "The relationship between the dynamic remanent coercivity and the viscosity and irreversible susceptibility in magnetic media," *IEEE Trans. Magn.*, vol. 36, pp. 148, 2000.
16. Cheng Yang, J.M. Sivertsen, and J.H. Judy, "Time decay of the remanent magnetization in longitudinal thin film recording media as a function of distributions of grain size and easy axis orientation," *IEEE Trans. Magn.*, vol. 34, pp. 1606, 1998.
17. Yun Zhang and H.N. Bertram, "Thermal decay in high density disk media," *IEEE Trans. Magn.*, vol. 34, pp. 3786, 1998.
18. B. Lengsfield, IBM Research, San Jose, CA, private communication, 2000.
19. H. Zhou, H.N. Bertram, A.F. Torabi, and M.L. Mallary, "The effect of intergranular exchange on thermal energy barrier distributions in longitudinal media," The 8th Joint MMM-Intermag Converence, San Antonio, TX, Jan. 7–11, 2001, paper EA-12 (IEEE Trans. Magn. (to be published)).

20. J. Fidler and T. Schrefl, "Micromagnetic modelling the current state of the art," *J. Phys. D, Appl. Phys.*, vol. 33, pp. R135, 2000.
21. J.A. Ewing, "On time-lag in magnetization of iron," *Proc. Roy. Soc.*, vol. 46, pp. 269, 1889.
22. R. Street and J.C. Woolley, "A study of magnetic viscosiy," *Proc. Phys. Soc.*, vol. A62, pp. 562, 1949.
23. P. Gaunt, "The frequency constant for thermal activation of a ferromagnetic domain wall," *J. Appl. Phys.*, vol. 78, pp. 3470, 1977.
24. W. Wernsdorfer, E.B. Orozco, K. Hasselbach, A. Benoit, B. Barbara, N. Demoncy, A. Loiseau, H. Pascard, and D. Mailly, "Experimental evidence of the neel brown model of magnetization reversal," *Phys. Rev. Lett.*, vol. 78, pp. 1791, 1997.
25. S. Chikazumi, *Physics of Magnetism, chapter 15*, John Wiley & Sons, Inc, New York, 1964.
26. D. Weller and A. Moser, "Thermal effect limits in ultrahigh density magnetic recording," *IEEE Trans. Magn.*, vol. 35, pp. 4423, 1999.
27. M.P. Sharrock, "Time-dependent magnetic phenomena and particle-size effects in recording media," *IEEE Trans. Magn.*, vol. 26, pp. 193, 1990.
28. A. Moser and D. Weller, "Thermal processes and stability of longitudinal magnetic recording media," *IEEE Trans. Magn.*, vol. 35, pp. 2808, 1999.
29. H.N. Bertram, *Theory of Magnetic Recording*, Cambridge University Press, U.K., 1994.
30. A. Moser, D. Weller, M.E. Best, and M.F. Doerner, "Dynamic coercivity measurements in thin film recording media using a contact write/read tester," *J. Appl. Phys.*, vol. 85, pp. 5018, 1999.
31. S.Y. Yamamoto and S. Schultz, "Scanning magnetoresistance microscopy," *Appl. Phys. Lett.*, vol. 69, pp. 3263, 1996.
32. R. Indeck, P. Dhagat, A. Jander, and M. Muller, "Effect of trackwidth and linear spacing on stability and noise in longitudinal and perpendicular recording," *J. Magn. Soc. Japan*, vol. 21(S2), pp. 205, 1997.
33. A. Moser, D. Weller, N.D. Rizzo, and T.J. Silva, "Determination of average demagnetizing fields in longitudinal magnetic recording using nanosecond field pulses," *Appl. Phys. Lett.*, vol. 77, pp. 1505, 2000.
34. N.D. Rizzo, T.J. Silva, and A.B. Kos, "Relaxation times for magnetization reversal in a high coercivity magnetic thin film," *Phys. Rev. Lett.*, vol. 83, pp. 4876, 1999.
35. T.J. Silva, C.S. Lee, T.M. Crawford, and C.T. Rogers, "Inductive measurement of ultrafast magnetization dynamics in thin film permalloy," *J. Appl. Phys.*, vol. 85, pp. 7849, 1999.
36. A. Moser, D. Weller, and M.F. Doerner, "Thermal stability of longitudinal magnetic recording media," *Appl. Phys. Lett.*, vol. 75, pp. 1604, 1999.
37. M. Yu, M.F. Doerner, and D.J. Sellmyer, "Thermal stability and nanostructure of cocrpt longitudinal recording media," *IEEE Trans. Magn.*, vol. 34, pp. 1534, 1998.
38. Y. Hosoe, I. Tamai, K. Tanahashi, Y. Takahashi, T. Yamamoto, T. Kanbe, and Y. Yajima, "Experimental study of thermal decay in high density magnetic recording media," *IEEE Trans. Magn.*, vol. 33, pp. 3028, 1997.
39. P. Dhagat, R.S. Indeck, and M.W. Muller, "Spin stand measurements of time and temperature dependence of magnetic recordings," *J. Appl. Phys.*, vol. 85, pp. 4994, 1999.

40. H.J. Richter, and E.A. Brockie, "Precise measurement of the linear density dependence of thermal decay of squarewave recordings," The 8th Joint MMM-Intermag Converence, San Antonio, TX, Jan. 7–11, 2001, paper EA-12 (IEEE Trans. Magn. (to be published)).
41. A. Moser *et al.*, "Signal decay measurements in longitudinal recording media at drive temperatures.," to be published.
42. L. He, W.D. Doyle, and H. Fujiwara, "High speed coherent switching below the stoner wohlfarth limit," *IEEE Trans. Magn.*, vol. 30, pp. 4086, 1994.
43. S.M. Stinnett, W.D. Doyle, P.J. Flanders, and C. Dawson, "High speed switching measurements in thin film disk media," *IEEE Trans. Magn.*, vol. 34, pp. 1828, 1998.
44. S.M. Stinnett, W.D. Doyle, O. Koshkina, and L. Zhang, "The relationship between high speed switching characteristics and the fluctuation field," *J. Appl. Phys.*, vol. 85, pp. 5009, 1999.
45. H.J. Richter, S.Z. Wu, and R. Malmhall, "Dynamic coercivity effects in thin film media," *IEEE Trans. Magn.*, vol. 34, pp. 1540, 1998.
46. K. Rubin, J. Goldberg, H. Rosen, E. Marinero, M. Doerner, and M. Schabes, "Dynamic coercivity and thermal decay of magnetic media," *Mat. Res. Soc. Proc.*, vol. 517, pp. 261, 1998.
47. H.N. Bertram, "Thermal dynamic reversal in thin films," The 8th Joint MMM-Intermag Converence, San Antonio, TX, Jan. 7–11, 2001, paper EA-12 (IEEE Trans. Magn. (to be published)).
48. E.E. Fullerton, D.T. Margulies, M.E. Schabes, M. Carey, B. Gurney, A. Moser, M. Best, G. Zeltzer, K. Rubin, H. Rosen, and M.A. Doerner, "Antiferromagnetically coupled magnetic media layers for thermally stable high density recording," *Appl. Phys. Lett.*, vol. 77, pp. 3806, 2000.
49. E.N. Abarra, A. Inomata, H. Sato, I. Okamoto, and Y. Mizoshita, "Longitudinal magnetic recording media with thermal stabilization layers," *Appl. Phys. Lett.*, vol. 77, pp. 2581, 2000.
50. E.E. Fullerton, D. Margulies, M. Schabes, and M.F. Doerner, "Antiferromagnetically coupled magnetic recording media," The 8th Joint MMM-Intermag Converence, San Antonio, TX, Jan. 7–11, 2001, paper EA-12 (IEEE Trans. Magn. (to be published)).
51. J. Lohau, A. Moser, E.E. Fullerton, D. Margulies, and M. Schabes, "Dynamic coercivity measurements of antiferromagnetically-coupled magnetic media layers," *Appl. Phys. Lett.*, vol. 78, pp. 2748, 2001.
52. T.C. Arnoldussen, M. Mirzamaani, M. Doerner, K. Tang, X. Bian, J. Feng, and M. Gatherwright, "Correlation of thermal stability and signal to noise ratio of thin film recording media," *IEEE Trans. Magn.*, vol. 36, pp. 92, 2000.
53. H. Katayama, M. Hamamoto, J. Sato, Y. Murakami, and K. Kojima, "New developments in laser assisted magnetic recording," *IEEE Trans. Magn.*, vol. 36, pp. 195, 2000.
54. D. Weller, A. Moser, L. Folks, M.E. Best, Wen Lee, M.F. Toney, M. Schwickert, J.U. Thiele, and M.F. Doerner, "High K_u materials approach to 100 Gbits/in^2," *IEEE Trans. Magn.*, vol. 36, pp. 10, 2000.
55. K. Takano, G. Zeltzer, D.K. Weller, and E.E. Fullerton, "$Co_{1-x}Cr_x$/Pt multilayers as perpendicular recording media," *J. Appl. Phys.*, vol. 87, pp. 6364, 2000.

56. M. Yu, Y. Liu, A. Moser, D. Weller, and D.J. Sellmyer, "Nanocomposite CoPt:C films for extremely high density recording," *Appl. Phys. Lett.*, vol. 75, pp. 3992, 1999.
57. S. Sun, C.B. Murray, D. Weller, L. Folks, and A. Moser, "Monodisperse FePt nanoparticles and ferromagnetic FePt nanocrystal superlattices," *Science*, vol. 287, pp. 1989, 2000.

6 Dynamic Effects in High-Density Recording Media

Hans Jürgen Richter

6.1 Introduction

Longitudinal magnetic recording systems are at present the basis of most information storage systems. During recent years, the data density stored on rigid disk media has increased at the astonishing rate of 60% per annum; more recently, the data density growth rate has even been increased to 100% per annum. Most of the gain in areal density has been achieved by scaling; that is, all dimensions of the recording process have been shrunk by the same amount.

For recording media, such as thin-film media, scaling means a reduction of grain size. In 1997, Charap et al. estimated that continued scaling in hard-disk recording would cause thermal instabilities at a density of about 56 bit/μm^2 (36 Gbit/in^2) [1]. The thermal energy strives to randomize any magnetic system and competes with the magnetic energy. For sufficiently large magnetic energy, as is the case for big grains, the probability that the thermal energy can randomly switch the magnetization of the grains is extremely small. For very small grains, however, the thermal energy becomes comparable to the magnetic energy and the probability of random magnetization switching can no longer be neglected (onset of superparamagnetism).

Since Charap's initial estimate, many areal density demonstrations have been publicized which exceed Charap's "limit" [2]. In this article, the theory of thermal activation is reviewed together with measurement data for phenomena related to thermal activation. In particular, decay phenomena of recorded information are reviewed in detail. A short section is devoted to the recently suggested anti-ferromagnetically coupled media and perpendicular recording.

6.2 Thermally Activated Magnetization Reversal

6.2.1 Magnetization Reversal in Single-Domain Particles

Magnetic recording media consist of very small grains, each of which has a strong uniaxial magnetic anisotropy. These magnetic particles are too small to support a domain structure and are therefore single domains. A good starting point for the theoretical description of the magnetization reversal, which also allows a simple inclusion of thermal effects, is the model of coherent rotation (Stoner–Wohlfarth model) [3]. The model of coherent rotation assumes that the magnetization in these single-domain particles is homogeneous. The model thus applies strictly to elliptical particles only. These particles show uniaxial anisotropy which is predominantly magnetocrystalline in nature and, to a much lesser extent, due to shape anisotropy. For simplicity, the symmetry axes of the two anisotropies are assumed to coincide. "*2D-isotropic*" longitudinal recording media have the magnetically easy axes of the grains oriented at random in the film plane. Many thin-film recording media have an additional anisotropy along the circumference ("*oriented media*"). The degree of orientation is only slight and the focus here is on 2D-isotropic media.

For an isolated single-domain particle with uniaxial magnetocrystalline anisotropy, the magnetic energy has two terms: the field energy and the anisotropy energy.

$$E\left(\vartheta\right) = -\mu_0 M_s H V cos\vartheta + K_1 V sin^2\left(\vartheta - \vartheta_0\right) \tag{6.1}$$

where:

$\mu_0 = 4\pi \times 10^{-7} \frac{V \cdot s}{A \cdot m}$ = permeability of free space
M_s = saturation magnetization
H = applied field
K_1 = (first order) magnetocrystalline anisotropy constant
V = volume of the particle.

As Fig. 6.1 illustrates, ϑ is the angle between the magnetization and the applied field, and ϑ_0 is the angle between the easy axis and the applied field. The magnetocrystalline anisotropy can be described by a fictitious anisotropy field, H_A, which is defined as: $H_A = 2K_1/\mu_0 M_s$. For convenience, the shape anisotropy, $\left(N_\perp - N_{||}\right) M_s$, where $N_\perp$ and $N_{||}$ are the demagnetizing factors

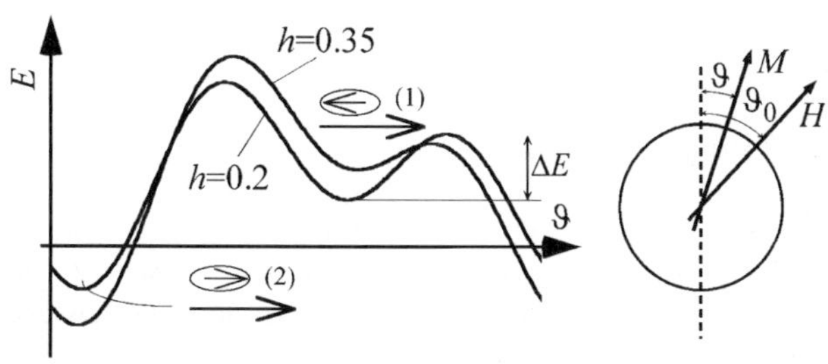

Fig. 6.1. Angle definitions and energy E as function of the angle between magnetization and applied field for a single-domain particle. The magnetization can be in state 1 in which it opposes the applied field, or in state 2 in which it is aligned with the field. The energy barrier, ΔE, keeps the magnetization in the energetically higher state 1. For thermally activated magnetization reversal, the energy barrier ΔE is provided thermally

perpendicular and parallel to the easy axis, respectively, is absorbed in the anisotropy field H_A.

The evaluation of (6.1) predicts magnetic hysteresis, whereby the angle of the magnetization, ϑ, has to be considered as a free parameter. Figure 6.1 shows the energy according to (6.1) as a function of ϑ. It is convenient to normalize equation (6.1) by $\mu_0 M_s H_A V$ and to introduce a normalized field: $h = H/H_A$.

Depending on the field h, there can be either one or two (local) energy minima (see Fig. 6.1). If there are two energy minima, two values for the magnetization can be assigned to one field value, i.e., there is hysteresis. Figure 6.1 illustrates that increasing the applied field makes the minimum associated with the higher energy shallower until it finally vanishes completely. At this point, the magnetization switches irreversibly. Finding the energy minima of (6.1) thus determines the hysteresis loop. A necessary condition for an energy minimum is $\frac{\partial E}{\partial \vartheta} = 0$, which reads in normalized form:

$$\frac{\partial e}{\partial \vartheta} = \mathrm{hsin}\vartheta + \frac{1}{2} sin\left(2\left(\vartheta - \vartheta_0\right)\right) = 0 \tag{6.2}$$

The stability of the magnetic state requires that $\frac{\partial^2 e}{\partial \vartheta^2} > 0$:

$$\frac{\partial^2 e}{\partial \vartheta^2} = \mathrm{hcos}\vartheta + \cos\left(2\left(\vartheta - \vartheta_0\right)\right) > 0 \tag{6.3}$$

The switching field, that is, the field at which the magnetization irreversibly switches, can be found by simultaneously solving (6.2) and (6.3). The result is:

$$h_{S,SW} = \frac{-1}{\left(\cos^{2/3}\vartheta_0 + \sin^{2/3}\vartheta_0\right)^{3/2}} \tag{6.4}$$

The angle $\varphi_{S,SW} = \vartheta_{S,SW} - \vartheta_0$ between the magnetization vector and the easy axis just before the magnetization switches is given by:

$$\varphi_{S,SW} = \arctan\left(\sqrt[3]{\tan\vartheta_0}\right) , \tag{6.5}$$

These considerations do not take thermal activation effects into account. It should also be mentioned that interactions between the grains are not considered. Since recording media consist of interacting particle systems, the theory outlined needs to be applied with caution.

6.2.2 Thermally Activated Magnetization Reversal: Arrhenius–Néel Approach

Thermally activated magnetization reversal is best discussed on the basis of the model of coherent rotation as described above. Consider state 1, as indicated in Fig. 6.1, in which the magnetization is opposing the applied field. Although a lower energy state is available, the magnetization remains in state 1, because it has to overcome an energy barrier at reversal. In the absence of thermal energy, the magnetization can reverse only if the applied field reduces the barrier to zero. If thermal energy is present, however, the activation energy can be provided thermally and the magnetization switches at smaller fields than predicted by the Stoner–Wohlfarth theory. The thermal energy strives to "disorder" the magnetic system; that is, it drives the system to zero magnetization. This competition between the magnetic energy and the thermal energy is reflected in the (statistical) relaxation rate:

$$r(h) = f_0 \exp\left(-\frac{\Delta E(h)}{k_B T}\right) , \tag{6.6}$$

where $\Delta E(h)$ is the (field dependent) energy barrier to be overcome, k_B is Boltzmann's constant (1.38×10^{-23} J/K) and T is the temperature in Kelvin. The frequency factor f_0 has the order of magnitude of 100 GHz for advanced recording media and is discussed more in detail later. Equation (6.6) was first introduced by Néel [4]; later W. F. Brown provided a more rigorous derivation [5]. The outstanding feature of (6.6) is that a relatively small change in particle volume (and thus in energy barrier) causes the relaxation rate to change dramatically. Once the magnetization has reached the energetically lowest state (state 2 as indicated in Fig. 6.1), it can again be activated thermally and switch back to state 1. This process can also be described by equation (6.6) with the appropriate energy barrier [5]. For very small grains, the lifetime of the magnetization in either state is short and the thermal energy quickly drives the ensemble magnetization to zero. This behavior is identical to that of a paramagnet. The only difference from a paramagnet is the magnitude of the magnetic moments, which has lent *superparamagnetism* its name [6].

A superparamagnetic grain ensemble is obviously useless for information storage. Moreover, a stable remanent magnetization is a necessary, but not a sufficient requirement for storage media. Kneller and Luborsky [7] reported that the onset of superparamagnetism is first observed as a reduction of coercivity. Using a two-state rate equation with switching rates r_{12} and r_{21} towards and away from the applied field direction respectively, the fraction of non-reversed grains, n, is:

$$\frac{\mathrm{d}n}{\mathrm{d}t} + n \cdot r_{tot} = r_{21} , \quad r_{tot} = r_{12} + r_{21} . \tag{6.7}$$

In thermal equilibrium $\frac{\mathrm{d}n}{\mathrm{d}t} = 0$ and the number of grains in either state is constant. For identical energy barriers, $r_{12} = r_{21}$, the two states are equally populated, i.e. the magnetization is zero. As soon as the energy barriers ΔE_{12} and ΔE_{21} are moderately different (without loss of generality, $\Delta E_{12} < \Delta E_{21}$), due to the exponential dependence, the relaxation process over the smaller energy barrier dominates and the backward switch can be neglected. In this case, the total switching rate is $r_{tot} \cong r_{12}$. The following discussion neglects the backward switch, which is a reasonable assumption for most purposes. Then the reduced remanent magnetization, $m_r(t)$ is:

$$m_r(h,t) = \frac{M_r(t)}{M_{r,\max}} = 2\exp\left[-r_{12}(h)\cdot t\right] - 1 \; , \tag{6.8}$$

where $M_{r,\max}$ is the maximum possible remanent magnetization. By definition, the remanent coercivity corresponds to the field that makes m_r zero; and thus it is *time dependent.* In order to evaluate the time dependence of coercivity, the energy barrier entering into (6.6) must be known. For the simplest case, namely that of alignment of the easy axis with the applied field, $\vartheta_0 = 0$, the energy barrier is:

$$\Delta E = K_1 V (1+h)^2 \tag{6.9}$$

Using $m_r(h_r, t) = 0$, the time dependent coercivity is [8] (see also Chap. 3):

$$|H_r(\vartheta_0 = 0, t)| = H_A\left(1 - \sqrt{\frac{k_B T}{K_1 V}\ln\frac{f_0 t}{\ln 2}}\right) \tag{6.10}$$

For the general case that the field is applied at an angle to the easy axis, $\vartheta_0 \neq 0$, the field dependence of the energy barrier is more complex and has to be evaluated numerically. Victora gave the following approximation [9]:

$$\begin{aligned}\Delta E(\vartheta_0, h) &\cong 2K_1 V\left(\frac{2}{3}\right)^{3/2}\left(1 + \frac{h}{h_{S,SW}(\vartheta_0)}\right)^{3/2} \sin\left(2\varphi_{S,SW}(\vartheta_0)\right) \\ &= K_1 V \cdot f(\vartheta_0, h)\end{aligned} \tag{6.11}$$

Although (6.11) breaks down for ϑ_0 near 0 and $\pi/2$ respectively, it generally matches the exact energy barriers very well. Analogous to the previous case, solving (6.8) and (6.11) for $h_r(\vartheta_0, t)$ yields an equation similar to (6.10). For thin-film media, the grains are oriented at random in the film plane, so that the case of a magnetically easy axis aligned with the field is highly exceptional. The calculation of the remanent coercivity for thin-film media therefore requires averaging over all grain orientation angles. It has been shown that the time dependent coercivity of a grain with field orientation angle $\vartheta_0 \cong 0.369$ (21.1°) can approximate the coercivity of the entire ensemble very well [10,11]. The result is:

$$\left|H_r^{2D-iso}(t)\right| \cong 0.566 H_A\left(1 - 0.977\left[\frac{k_B T}{K_1 V}\ln\frac{f_0 t}{\ln 2}\right]^{2/3}\right) \tag{6.12}$$

There has been much discussion on the exponent to be used for the time dependent coercivity. Equation (6.12) clarifies that the correct exponent is

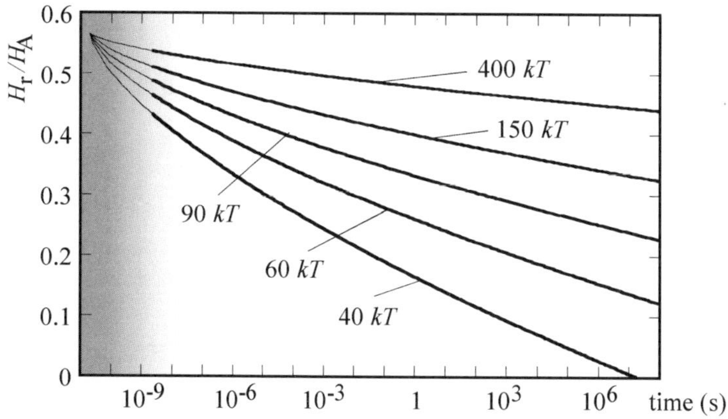

Fig. 6.2. Remanent coercivity as function of pulse width for various energy barriers for 2D-isotropic media (Eq. 6.12). At short times, (shaded area), gyromagnetic effects cause deviations from (6.12). Frequency factor is set to 30 GHz

2/3 and relates the coefficients back to the material parameters. Figure 6.2 illustrates the time dependence of the coercivity using equation (6.12) for various values K_1/k_BT.

If the term in square brackets approaches unity, the remanent coercivity (formally) becomes negative. This can occur only for strong thermal effects in the time scale of interest and indicates that the backward switch can no longer be neglected. Equation (6.12) was derived for magnetically non-interacting grains. From micromagnetic modeling, it is known that exchange coupling between the grains reduces remanent coercivity. A first estimate shows that weak exchange coupling primarily causes a reduction of the pre-factor (0.566 in (6.12) by about 16% [10].

It should be noted that (6.12) has been derived for isotropic media. Most of the commercial thin-film media have a slight orientation which somewhat increases the pre-factor (0.566) as well as the time dependence of coercivity. Therefore the values for K_1/k_BT derived from dynamic coercivity experiments have to be corrected for the orientation effect. It was also pointed out that the presence of a second order anisotropy constant, which causes an additional energy term $K_2 \sin^4(\vartheta - \vartheta_0)$ in (6.1), affects (6.12) [12]. For $K_2 > 0$, the short term switching field is hardly changed, but the energy barrier for small applied field is increased. A non-vanishing K_2 would therefore act as if a bigger volume were present. For pure Co, $K_2 \cong 0.2K_1$ [13] and one would therefore expect that K_2 has a non-vanishing contribution for commercial Co-containing thin-film alloys. Unfortunately, separate measurements of K_1 and K_2 in polycrystalline films are generally difficult to perform and measurements on single crystals [14] have been reported only for a few alloys. It is therefore difficult to estimate how much the second order anisotropy constant contributes to the total anisotropy for current thin-film media. It should be noted that most techniques to measure anisotropy do not report K_1 and K_2 separately; they rather report a combination of K_1 and K_2. For example, the "45° method" [15] – which is suitable for polycrystalline thin-film media – gives $K_1 + K_2$ for the total anisotropy constant.

6.2.3 Grain Size Distributions

For most magnetic materials, the time dependent magnetization follows the relation $\Delta m(t) = -\text{const} \cdot \log(t)$ rather than (6.8). In contrast to the case discussed earlier, real materials have a distribution of energy barriers rather than a single valued energy barrier. The logarithmic time dependence of the magnetization arises if the distribution of energy barriers that are thermally activated during the time range of observation is constant. For thin-film recording media, the grain size as well as the magnetic properties of the grains are not uniform. In this section the effect of a distribution in grain volumes is discussed. For simplicity, the grains are treated as cylinders with diameter D and height equal to the film thickness, δ. The grain size distribution of thin-film media typically follows a log-normal distribution with probability density, p, of the grain diameters:

$$p(D) = \frac{1}{\sqrt{2\pi} D \sqrt{\ln(1+\sigma^2)}} \exp\left\{ - \left[\frac{\ln\left(\frac{D}{\mu}\sqrt{1+\sigma^2}\right)}{\sqrt{2\ln(1+\sigma^2)}} \right]^2 \right\}, \tag{6.13}$$

where D is the grain diameter, σ the standard deviation, and μ is the mean. Equation (6.13) is defined such that the first moment, $\int D \cdot p(D)\, dD$, is identical to the mean, μ. The contribution of each grain to the magnetization (and the energy barrier) is then $\propto p(D) \cdot D^2$, which implies that the larger grains carry more weight. Further analysis can be simplified considerably using the *blocking criterion.* In superparamagnetism, it is customary to introduce a blocking temperature below which the magnetization is thermally stable and above which it is unstable [6]. Rather than defining a blocking temperature one can define a blocking volume, V_B, or diameter, D_B, respectively, such that all grains with $D < D_B$ have switched, but none with $D > D_B$. Mathematically, this assumption means that the double exponential term in (6.8) is replaced by a step function. Solving equation (6.8), $m_r(h,t) = 0$ together with (6.11), D_B can be calculated:

$$D_B = \frac{2}{\sqrt{\pi}} \sqrt{\frac{k_B T}{K_1 \delta \cdot f(\vartheta_0, h)} \ln \frac{f_0 t}{\ln 2}} \tag{6.14}$$

Then the magnetization is:

$$m_r(D_B) = \frac{1}{2}\left[1 - \text{erf}\left\{ \frac{2\ln(1+\sigma^2) - \ln\frac{D_B}{\mu}\sqrt{1+\sigma^2}}{\sqrt{2\ln(1+\sigma^2)}} \right\}\right] \tag{6.15}$$

From (6.15), it is evident that there is always a grain size D_B which corresponds to a given value of the remanent magnetization. This is also true for the point of zero remanent magnetization, i.e., remanent coercivity. Therefore equation (6.12) holds for a grain size distribution without any changes. If an anisotropy distribution is present in addition to that of the grain size, the analysis becomes more complicated [16]. In spite of its usefulness, it should be remembered that the blocking criterion is an approximation. Magnetization curves derived using the blocking criterion differ noticeably from those calculated directly from the Arrhenius–Néel equation.

6.2.4 Fast Switching

Magnetization reversal processes at very short times have to be analyzed using the Landau–Lifshitz equation, which is given here in the form suggested by Gilbert [17] (see also Chap. 3):

$$\frac{d\vec{M}}{dt} = -\mu_0 \left|\gamma\right| \left(\vec{M} \times \vec{H}\right) + \frac{\alpha}{M} \left[\vec{M} \times \frac{d\vec{M}}{dt}\right], \quad \gamma = -1.761 \times 10^{11} \frac{1}{T \cdot s} \tag{6.16}$$

Here α is a (phenomenological) damping constant. Equation (6.16) predicts a sharp increase of the switching field at very short times. The switching speed strongly depends on the damping constant. If the damping were zero, the precession would last eternally and the speed of magnetization reversal would be zero. Due to finite damping, the kinetic energy associated with the spin precession can be transported to the lattice and the magnetization can reverse. Therefore, the switching speed increases with increasing damping. Using (6.16), Kikuchi derived that the maximum switching speed occurs at critical damping $\alpha = 1$ [18]. On the other hand, measurements of the line width of ferromagnetic resonance experiments yielded damping constants of the order of 0.02 for materials similar to those used in thin-film media [19]. Equations (6.12) and (6.16) describe magnetization switching in either the viscosity regime or the gyromagnetic regime, respectively. There exists, however, a transition between the two time regimes, which contains elements of thermal as well as gyromagnetic switching.

An understanding of the magnetization reversal behavior in the transition regime requires an examination of the frequency factor f_0 that appears in (6.12). It has been shown that f_0 depends on the damping [5,12,20,21]:

$$f_0 = 2\sqrt{\frac{2}{\pi}} \frac{\alpha}{1+\alpha^2} \sqrt{\frac{K_1 V}{k_B T}} \mu_0 \left|\gamma\right| H_A \tag{6.17}$$

Since the damping constant describes the energy transport of the magnetization to the lattice and *vice versa*, the damping constant appears in both (6.16) and (6.17).

Using Néel's and his own expressions, Brown calculated the frequency factor f_0 for small iron particles and obtained $f_0 \approx 1$ GHz [22]. Since then, in almost every investigation of thermally activated magnetization processes, the frequency factor f_0 was set to the "universal" value of 1 GHz, independent of the material investigated. As can be seen from (6.17), f_0 depends on the anisotropy field and is therefore higher for modern hard disk media; for $H_A \approx$ 1000 kA/m and $\alpha \approx 0.04$, f_0 is of the order of 100 GHz.

The derivation of (6.6) was originally given by Brown [5] and assumes that the energy barrier to be surmounted is much larger than the thermal energy k_BT. Naturally, this assumption breaks down for fields that are close to the Stoner–Wohlfarth switching fields. In this case, the bond of the magnetization to its equilibrium orientation is weak such that the magnetization processes become very slow. Therefore, the magnetization needs "an extra kick" to finally switch and a slight increase of the switching field compared to that given by (6.12) is found (note that this leads to a field dependence of f_0). The transition regime between pure thermal switching and pure precession dynamics ranges down to times as long as $\approx 100/f_0 \cdots 1000/f_0$. Safonov and Bertram discuss this matter in detail in Chap. 3 of this book.

6.3 Measurement of Time Dependent Coercivity

6.3.1 Introduction

The most straightforward way to determine the time dependence of coercivity is to perform hysteresis loop measurements at different field sweep rates. The point of zero magnetization in a hysteresis loop is the coercivity rather than the remanent coercivity that has been discussed so far. Equation (6.12) should therefore be evaluated using remanence curves for which the field exposure time is varied. It should also be noted that the derivation of (6.12) assumes that the reversing field is switched on and off instantaneously, which is difficult to achieve experimentally.

Since conventional magnetometry is limited to the time range of roughly 1 second and above, several alternative techniques to measure the time dependence of coercivity have been suggested recently. Using a micro-strip to create field pulses of the order of several nanoseconds, Doyle and co-workers have measured the short-term coercivity of various tape media as well as thin-film media [23,24]. For tape media with a very small anisotropy, a steep rise of the coercivity due to precession effects could be observed [23]. It was also found that the coercivity at times near the precession regime tends to increase when compared to the prediction of (6.12). Rizzo and Silva [25], who investigated the remanent magnetization of a thin-film medium as function of time, have performed a similar experiment. Their magnetization data for a low coercivity thin-film medium as a function of time show a distinct kink at 10 ns, indicating a shift from an exponential to a logarithmic time dependence of the magnetization. Rubin et al. [26] and Moser et al. [27] record di-bits with pulsed fields from a recording head. The coercivity is identified with the current needed to achieve 50% of the maximum di-bit signal. Moser and Weller discuss this technique in detail in Chap. 5 of this book.

6.3.2 Dynamic Coercivity Measurements Using the "Reptation" Technique

Another technique that is simple to implement on a spin-stand tester uses the idea that the state of zero magnetization, that is, remanent coercivity, is associated with a maximum in noise power [28–30]. Consider a head that is energized by a dc current flying over a disk. As sketched in Fig. 6.3, every portion of the recording medium "sees" a field pulse with an approximate duration [31]:

$$t_F \approx \frac{g}{v} , \tag{6.18}$$

where g is the gap length of the writer and v is the linear velocity. Equation (6.18) has to be considered as a crude estimate; there are indications that the effective field duration is primarily determined by the duration for which the field is close to its maximum [32]. Since the head is simply excited by dc, magnetization dynamics in the head or rise time effects of the electronics play no role. Only the head to medium velocity and the field distribution of the head control the field rise time. Depending on the gap length of the (flying) head employed and the linear velocity, the time g/v can be varied between 2 and 50 nanoseconds. Longer field pulses can be realized with stationary heads; results with gap length as large as 650 μm have been reported [33].

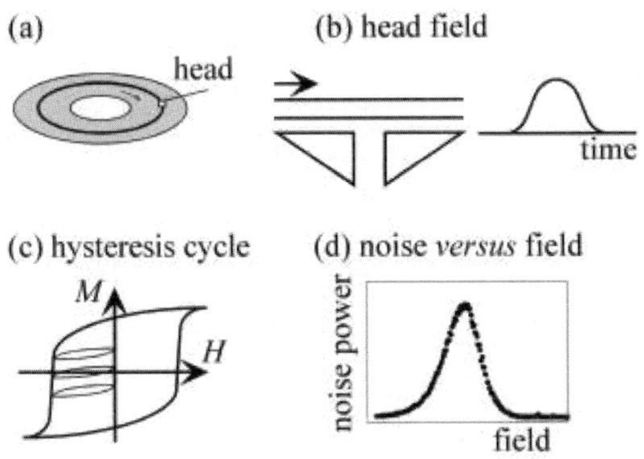

Fig. 6.3. Measurement of time dependence of coercivity on a spin-stand. While a dc excited head is flown over the disk (**a**), it presents field pulses (**b**) to the medium. **c** The time is changed by repeatedly erasing the medium and leads to a "reptation" process. **d** The magnetization state can be observed using the reverse dc erase noise power

A reverse dc erase noise measurement is performed as follows:

i. the medium is band-erased in, say, positive field direction;
ii. the medium is erased in negative field direction with current I_{E1};
iii. the noise power is measured.

Plotting the erase current as an independent variable *versus* the measured noise power yields a curve that peaks at an intermediate current, see Fig. 6.3d (see also Sect. 6.4.2). The peak current creates a head field that is identical to the remanent coercivity.

To measure the time dependence of the coercivity, the field exposure time has to be varied. Neither varying the head to medium velocity nor changing the gap length of the head is a useful approach, because the flying height – and therefore the head field acting on the medium - changes as well and the peak current location does not reflect the true coercivity change. A better method to vary the field exposure time is to change the number of revolutions for which the medium is exposed to the reverse field [31]. Thus step ii) has to be performed for a varying number of revolutions. As illustrated in Fig. 6.3c, multiple head passes cause the magnetization to repeatedly cycle through minor loops (*reptation*). In the viscosity regime, the field pulses accumulate and n short field pulses of duration t_F can be treated as if one long pulse with $t'_F = nt_F$ would have been applied.

Fig. 6.4 shows an example for a dynamic coercivity measurement over a wide time range. The data were taken with three different heads with gaps of 200 nm, 500 nm and 650 µm, respectively. As suggested by (6.12), the data are plotted versus $\left(\ln \frac{f_0 t}{ln2}\right)^{2/3}$ with f_0 = 30 GHz and follow the the-

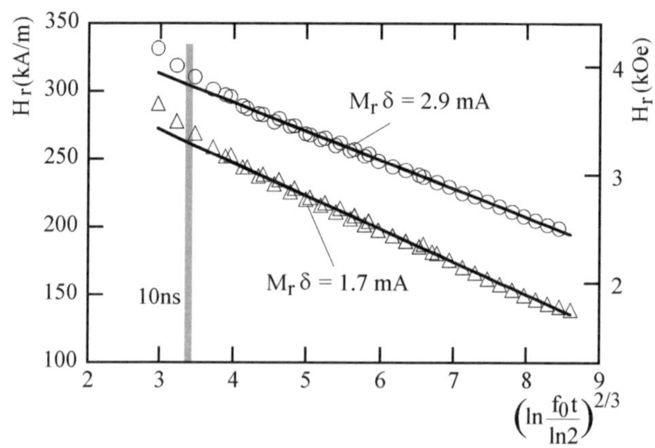

Fig. 6.4. Dynamic coercivity data over a wide range of times measured with the reptation technique. The lines give fits to the data using (6.12). The frequency f_0 was set to 30 GHz. The gray line roughly indicates at which time the viscosity regime ends and deviations from (6.12) are expected, after [33]

oretical formula well. A slight deviation from the theoretical curve occurs at very short times (the first four points were not included in the fits). As outlined above, this behavior is expected theoretically for times shorter than $100/f_0 \ldots 1000/f_0$, which roughly amounts to 10 ns for the present case. The upturn generally occurs for every head with a reasonable gap length of 100–500 nm [34]. Tarnopolsky et al. [30] pointed out that demagnetizing fields exist during the reverse dc erasure process, and a simple estimate using average demagnetizing fields showed that its effect should be negligible after the first head pass [31]. Igarashi et al. simulated the dynamic coercivity experiment micromagnetically using a Langevin approach and found that the effect of the demagnetizing field is more subtle than previously assumed [35]. For a medium with a somewhat greater demagnetization ratio $M_r\delta/H_A$ than those investigated here, it takes as many as 10 head passes for the effect of the demagnetizing field to become negligible. This explains why the dynamic coercivity data taken by the reptation technique data generally show an upturn at very short times [35]. Since the deviation between the data and the straight line fit to (6.12) occurs only for the first three points, it is likely that the effect of demagnetization is observed rather than a true deviation from (6.12). Using a micro-loop experiment, Mallary et al. reported dynamic coercivity data for an even wider range than that shown in Fig. 6.4 and again confirmed (6.12) [36].

Wachenschwanz and Alex [37,38] have proposed a variant of the reptation technique. Rather than monitoring the medium noise power of the remanent state, they record a square-wave with a very low density. For a sufficiently long wavelength, the rms value of the fundamental is proportional to $M_r\delta$ and can easily be measured by a spectrum analyzer. The erasure process is performed as described before. This approach has practical advantages, because the data evaluation is easier to perform. On the other hand, care must be taken that the read-back non-linearities of the MR transducers do not influence the results.

Figure 6.5 shows another example for a dynamic coercivity measurement as function of temperature. Increasing the temperature does not only increase the slope of the dynamic coercivity curve; it also introduces a shift. If (6.12) is valid, the shift must be due to a temperature dependence of the anisotropy field, H_A. Since the main constituent of the alloys used for recording media, Co, has a strong temperature dependence of the magnetocrystalline anisotropy [13], this result is not surprising. The data indicate that the temperature dependence of H_A is about 0.1–0.2% per °C for typical alloys. An important implication of this result is that the temperature dependence of the standard coercivity measured in a magnetometer is stronger than that of the material parameter H_A due to thermal activation. At short time scale, i.e., at the recording process, this effect is much less pronounced. For typical alloys for thin-film media, the temperature dependence of H_A is twice as strong as that of the saturation magnetization M_s [39].

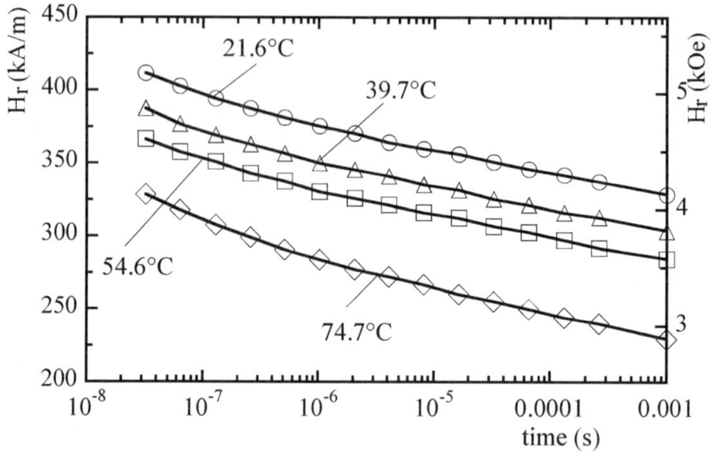

Fig. 6.5. Dynamic coercivity measurements using the reptation technique at various temperatures. The medium investigated is isotropic in plane with $M_r\delta = 4$ mA, $f_0 = 30$ GHz

6.3.3 Measurement of Energy Barrier Distributions

In the previous sections of this chapter, recording media were treated as an ensemble of non-interacting single-domain particles. They are manufactured using Co–Cr alloys, which are deposited by sputtering techniques. The sputtering process is controlled in such a way that the Cr primarily migrates to the boundaries of the grains. Inaba et al. have provided experimental evidence for the existence of these compositional fluctuations [39]. The Cr-rich grain boundaries suppress exchange coupling between grains, which is required to decrease medium noise. Therefore the manufacturing process inevitably leads to distributions of the saturation magnetization and the anisotropy of the grains as well as to exchange coupling between them. The simplified theory does not consider any distributions other than that of the grain size and it is desirable to measure the energy barrier distribution rather than to infer it from grain size measurements and macroscopic magnetic data.

The idea of the blocking criterion can be used to experimentally determine an energy barrier distribution. As discussed before, the dynamic coercivity measurement determines the time dependent coercivity for that particular grain volume which separates the magnetization into two equal fractions (switched and not switched, respectively). If the dynamic coercivity is evaluated for a different ratio between switched, m_{sw} and not switched magnetization, $1 - m_{sw}$, the energy barrier at the magnetization level m_{sw} is obtained. Varying m_{sw} thus yields the energy barrier distribution of the sample.

Energy barrier distributions can therefore be extracted from dc demagnetization curves measured for different field exposure times. Higgins et al. [40] performed these measurements using a magnetometer and showed that

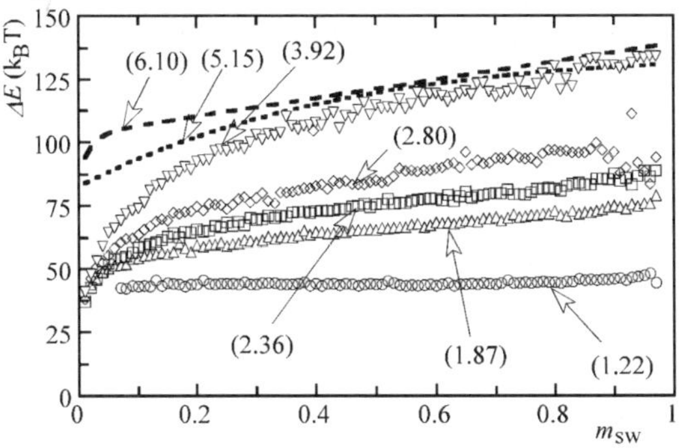

Fig. 6.6. Measured energy barrier distribution for the series of disks with properties given in Table 6.1. The *numbers in brackets* give the areal moments in mA. The curves shown for the disks with highest areal moment (*dashed lines*) are fits to the measured data

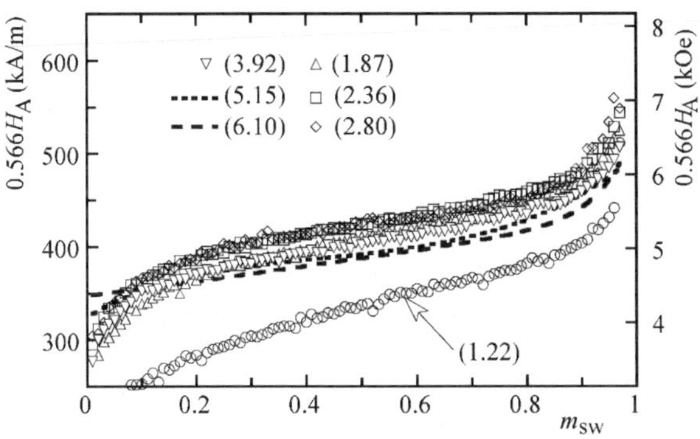

Fig. 6.7. Measured effective anisotropy field distribution for the series of disks with properties given in Table 6.1. The *numbers in brackets* give the areal moments in mA. The curves shown for the disks with highest areal moment (*dashed lines*) are fits to the measured data

the obtained energy barrier distribution is considerably flatter than that expected from grain size distribution data. Alternatively, the energy barrier distributions can also be measured using the reptation technique discussed before. The variant of the reptation technique that uses square-wave recording is particularly attractive, because the magnetization level m_{sw} can be identified much more easily than from the noise data. The data shown in Figs. 6.6 and 6.7 were taken on a special spin-stand tester [41] using inductive read-back to avoid head non-linearities associated with magnetoresistive transducers. Since the erasure is performed with a magnet that has a very large gap length of 2 mm, demagnetizing effects can be safely neglected. The dynamic coercivity data were taken by continuously monitoring the remanent

Table 6.1. Magnetic properties of the media investigated. Common grain size is 15 nm

disk #	$M_r\delta$ (mA)	H_r (kA/m)	H_A (kA/m)	ΔE (k_BT)
1	1.22	153	584	43
2	1.87	242	735	63
3	2.36	267	754	74
4	2.80	279	744	87
5	3.92	290	712	114
6	5.15	282	680	122
7	6.10	270	652	122

magnetization, $m_r(h,t)$ as function of field and field exposure time. The intersections of the $m_r(h,t)$ curves with the targeted magnetization levels, m_{sw}, yield pairs of fields and times that lie on the dynamic coercivity curve (see [42,43]). Figure 6.6 shows energy barrier measurements for a set of disks, the magnetic properties of which are given in Table 6.1. All disks have grains with easy axes at random in the film plane. The mean grain size is around 15 nm, but the grains tend to be bigger for the thicker films with higher areal moment (the areal moment is the product of remanent magnetization and film thickness, $M_r\delta$). Since thin-film media are always magnetized fully to their saturation value, the areal moment is generally quoted as a design parameter. The data confirm that the energy barrier distribution is much narrower than that expected from the grain size distribution data. As suggested in [33], exchange coupling can have a stabilizing effect, because it can tie the magnetization of the small grains to the more thermally stable magnetization of the larger ones. The general flattening of the energy barrier distribution has also been simulated theoretically [44]. Fitting the data to (6.12) not only yields data for the energy barrier, but also provides values for the pre-factor 0.566 H_A in (6.12). The existence of weak intergranular exchange coupling reduces the pre-factor somewhat. A value of $0.45\ldots0.5H_A$ is a reasonable estimate that yields good agreement with independent measurements of the anisotropy field.

The interpretation of the measurement of the energy barrier distribution is not quite straightforward. On the one hand, a direct application of the blocking criterion – as it has been assumed so far – introduces noticeable errors as compared to the exact solution for the time dependent magnetization. Moreover, (6.12) is only valid for non-interacting or weakly interacting grain ensembles. In a strongly interacting system, the energy barriers of each grain do not only depend on the applied field, but also on the magnetization state of its neighbors. Then the interpretation of the energy barrier measurement becomes very complicated.

6.4 Recording Physics

In current hard disk drives, the area allocated to one bit ranges from 0.02 to 0.15 μm^2. The size of the individual grains is much smaller than that of a bit, so that many grains "fit" into a bit. The number of grains per bit, or more precisely, the number of grains per transition, determines the signal-to-noise ratio, SNR. Due to the onset of superparamagnetic effects, the grain size can no longer be reduced arbitrarily and the traditional recording models need to be re-evaluated. The following sections discuss the effect of finite grain size on transition width and on SNR.

6.4.1 Magnetization Transitions

High-density recording requires the written transitions between the bits to be as narrow as possible. Williams and Comstock [45] suggested a simple slope model that has been very successful in predicting the transition width in longitudinal recording. The Williams–Comstock model treats the medium as a continuum; therefore the model lacks grain size information and information about the easy axis dispersion of the grains. Clearly a magnetization transition from $-M_r$ to $+M_r$ cannot be smaller than the grain size. In tape recording, this effect has been named "length loss", see e.g. [46]. It is straightforward to show that, for a magnetic medium with identical and perfectly aligned grains positioned at random, the transition parameter, a, is limited by [46]:

$$a = \frac{D}{\pi} , \tag{6.19}$$

where D is the grain size. Since the easy axes of the grains in thin-film media are not aligned, the interplay of the recording head field vector and the angle dependence of the switching field of the grains leads to a broadening of the transitions. Mathematically, the transition broadening is described by replacing the factor π in Eq. (6.19) with R ($R < \pi$). The numerical value of R is typically between 2 and 3 and depends primarily on the ratio of writing gap length and grain size. Since the slope model already considers the simultaneous occurrence of the effect of the demagnetizing fields and the switching field distribution, the limiting effect of the grain size on the transition width can be modeled as a fictitious switching field distribution, $SFD_{eq,D}$ [33]:

$$SFD_{eq,D} = \frac{\pi D}{2RH_r} \frac{dH_{tot,x}}{dx}, \quad R \approx 2 \ldots 3 . \tag{6.20}$$

The SFD term to be used in the slope model is $\sqrt{SFD_0^2 + SFD_{eq,D}^2}$, where SFD_0 is the conventional switching field distribution. As expected, taking into account a finite grain size therefore increases the transition width. It should also be remembered that H_r is time dependent; as a first estimate, the "writing coercivity" is $H_r \approx \frac{H_A}{2}$, see also (6.12).

6.4.2 Medium Noise

It is well known that achieving a high SNR requires the recording media to have very small particles [47–49]. Magnetic transitions cannot be written at any location of the medium, because, microscopically, the transitions need to follow the grain structure (because the grains are single-domains). Therefore, the exact position of a written transition will always deviate from the intended position, and this deviation appears as timing jitter in a recorder. Conceptually, transition jitter should not be termed "noise", because it does not simply add to the signal as, e.g., Gaussian white noise associated with a resistor. As Minhuin points out [50], the noise – while being an *additive* noise to the medium magnetization – is *multiplicative* in nature at read-back, and therefore appears as modulation noise. A microstructure together with the mean magnetization and its variance is sketched in Fig. 6.8.

A very pictorial approach to understand noise is a micro-track model [47,51,52]. Let us suppose that the volumetric packing fraction of the medium is 100%. Assuming square-shaped grains, one readily arrives at the micro-track model as shown in Fig. 6.9. The randomness of the media is coded in the down-track positions of the grains. Each micro-track thus "starts" at a different location. Consider first a general state of these micro-tracks *without* a transition being written. If all grains have the same probability to switch, and are assumed not to interact with one another, the variance of the *magnetization*, σ_M^2, is [30,53,49]:

$$\sigma_M^2 \propto \left(1 - pm^2\right) \tag{6.21}$$

Here, m is the reduced magnetization and p is the volumetric packing fraction of the grains. In the micro-track model as introduced above, $p = 1$, and the magnetization independent contribution vanishes. Equation (6.21) means that the media are quietest "in the bit". Hence, the dc-saturated state

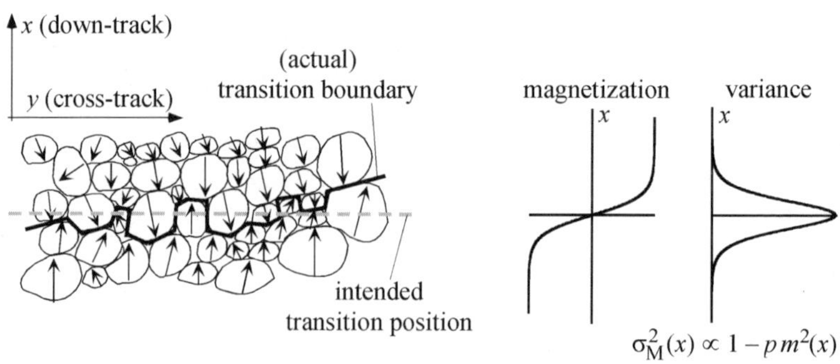

Fig. 6.8. Sketch of a transition written on an idealized, granular medium. Microscopically, the transition has to follow the grain structure of the medium so that the actual transition boundary fluctuates around the intended writing location. At the transition center, the magnetization is zero with a maximum variance

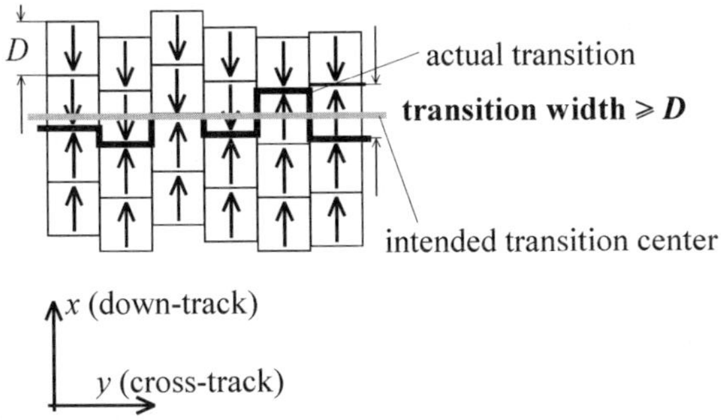

Fig. 6.9. Micro-track model. The minimum transition width is equal to the grain size. Due to finite grain size, the transition locations deviate from the intended positions

is that with the lowest noise. If transitions are recorded, the magnetization m is smaller than 1 and the noise increases with increasing linear recording density. This is tantamount to stating that the noise is concentrated at the transitions.

Caroselli and Wulf [52] applied the micro-track formalism to magnetization transitions and interpreted the magnetization derivative, dM/dx, as a probability density function which is used to calculate the variance of the magnetization. Then the following expression is obtained for the standard deviation of the zero-crossings of the transitions; i.e., the transition jitter:

$$\sigma_x = C_T a \sqrt{\frac{s}{R_W}} \,, \tag{6.22}$$

where R_W is the reader width, s is the cross-track correlation length and C_T is a constant depending on the shape of the magnetization transition, namely $C_T = \sqrt{\frac{\pi}{2}}$ for a transition shape according to an error function and $C_T = \frac{\pi^2}{4\sqrt{3}}$ for a transition shaped like a hyperbolic tangent. For the simple example given above, the width of the micro-tracks is identical to the grain size, D.

Transition jitter can (formally) be converted into a signal-to-noise ratio [54]:

$$SNR = \frac{\sqrt{\pi}}{2} U^{1.5} \frac{B}{\sigma_x} \exp\left(-\tfrac{U\pi}{4}\right) \,, \tag{6.23}$$

where B = transition spacing, $U = PW_{50}/B$ with PW_{50} = pulse width. The reader is cautioned that several SNR definitions can be found in the literature. Equation (6.23) is the ratio of the rms values of a square-wave recorded at a density $1/(4B)$ and the integrated noise voltage excited by a square-wave of density $1/(2B)$.

6.5 The "Superparamagnetic Limit"

This section discusses the limit of conventional longitudinal recording as it is presently understood. Prior to discussing the limiting factors, a review about thermal decay measurements along with the implications of thermal effects on the recording process are given.

6.5.1 Thermal Decay Measurements

So far, dynamic coercivity and energy barrier distribution measurements have been discussed. Although these measurements are related to thermal stability and are very useful, the ultimate test to properly assess the thermal stability of recording media is a decay measurement of a recording. These measurements have been extensively be reported in the literature [26,33,38,55–68]. Most of these measurements were done on experimental media with very small areal moment at room temperature. Since hard disk drives operate at elevated temperature, the thermal decay test should be performed at elevated temperature.

Experimental. Media in commercial hard disk drives do not show as much decay as experimental media and a decay measurement becomes a much more challenging task. As an example, consider a recorded signal decaying at a rate of 1% per order of magnitude (decade) in time. Then a signal change of only 5% has to be detected during a typical measurement ranging from 0.1 to 10000 seconds. It is not easy to perform spin-stand measurements to such a high accuracy, because the heads tend to drift across the track and the read sensitivity of the MR transducers fluctuate. The drift problem can be overcome using track scans or servo schemes. A common approach to compensate for head sensitivity fluctuations is to use an aged control track [58]. Since recorded signals decay approximately linearly with $\log(t)$, the signal does not reduce very much after a suitable aging period and can be regarded as approximately constant. A measurement scheme using a separate, aged control track yields much cleaner data, but the data quality remains relatively poor. The best results are obtained if the control track (CT) is not recorded onto a separate track, but recorded together with the track of interest (TOI) on physically the same track [71]. The common track is therefore split up into a CT portion and a TOI portion. Since the CT and TOI data are acquired within one revolution, head sensitivity fluctuations are caught in situ and excellent measurement quality is obtained. As an example, Fig. 6.10 shows how effectively a split-track measurement removes head sensitivity fluctuations [71].

Although the signal amplitude of the CT does not age significantly during the time period in which the TOI data are taken, it still continues to age and produces a small systematic error. An iterative data processing

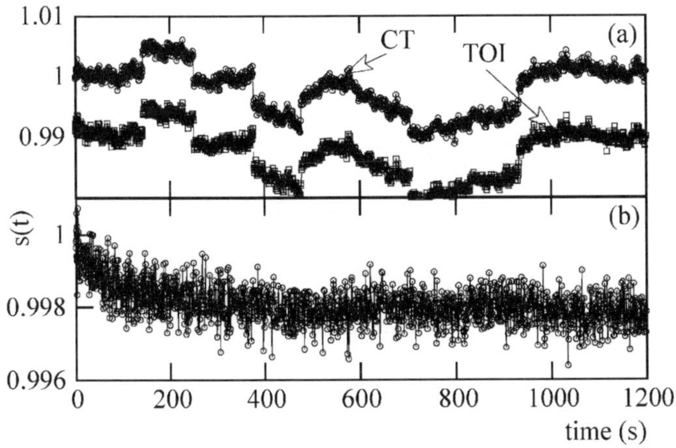

Fig. 6.10. Example for split-track testing. An aged control track, CT serves as a reference signal to take out head sensitivity fluctuations. **a** shows the raw data for the CT and the TOI (track of interest); **b** shows the normalized data, $s = \text{TOI}/\text{CT}$. © IEEE Transactions on Magnetics [71]

scheme can successfully remove the remaining error [71]. The thermal decay has to be measured at elevated temperature to properly assess stability. On a spin-stand, it is most appropriate to locally heat the disk, using, for example, a halogen bulb [59,71]. The most elegant way to measure the disk temperature is to use a four-wire resistance measurement of a separate head that is flown on the disk [71]. It is desirable to span a wide time range for a thermal decay measurement. Economic considerations limit measurement times to about 10 to 20 hours. On the other hand, using a standard set-up, the first datum can be taken after one revolution, which corresponds to roughly 10 milli-seconds. In most cases, the data acquisition takes some time and the first datum can be obtained 0.1 to 1 second after recording. This limits the accessible time range to about 5 to 6 orders of magnitudes in time. Dhagat et al. reported decay measurements starting at times as short as 200 μs [68]. These very short times have been achieved using separate record and read heads, which have been placed closely together on the same track.

Since the signal decay is approximately linear in $\log(t)$, it became customary to report thermal decay as percent signal change per decade (in time). It should be remembered, however, that all decay curves must have a negative curvature at short times and a positive curvature at long times. Since most experiments cannot identify the initial "100%" of the recorded signal and therefore assign the 100%-value to an arbitrary time, the reported decay rate is greater than the "true" decay rate [68].

Signal Decay: Dependence on Linear Density. Most of the thermal decay measurements reported have been performed on media that were designed to show significant thermal decay [55–68]. For recording applications, these media are of no importance and the discussion here explicitly includes slowly decaying media. Thermal decay measurements of the series of disks with properties given in Table 6.1 were performed at ambient as well as elevated temperature (75°C) and are given in Figs. 6.11 and 6.12, respectively.

Clearly, the expectation is that the amount of thermal decay should increase with increasing temperature and increasing linear recorded density

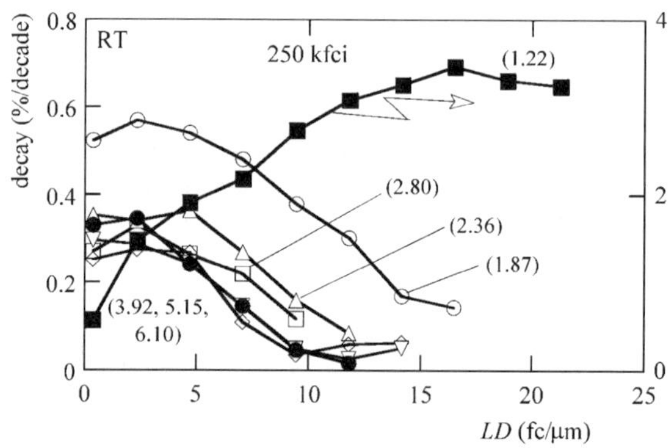

Fig. 6.11. Linear density dependence of thermal decay for the series of disks with properties given in Table 6.1. The numbers in parenthesis are the areal moments in mA. Note that numbers for the disk with areal moment of 1.1 mA report to the right hand axis. Test time 1 hour, room temperature. © IEEE Transactions on Magnetics [71]

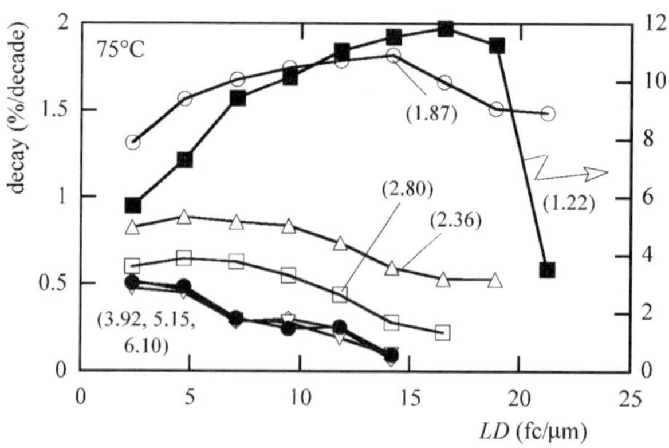

Fig. 6.12. Same as Fig. 6.11, but for 75°C temperature. © IEEE Transactions on Magnetics [71]

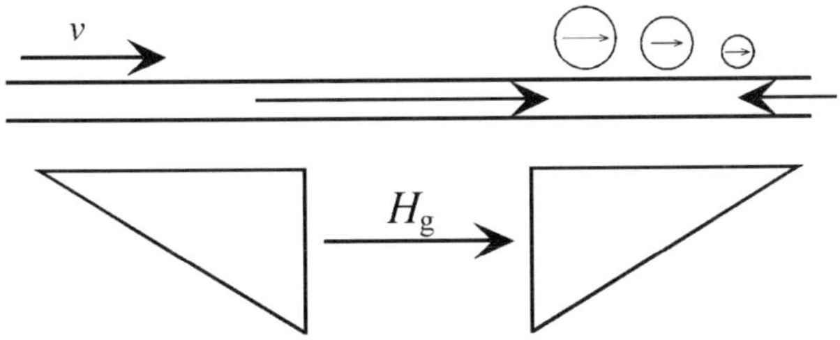

Fig. 6.13. Illustration of the recording process. Due to thermal activation, the (time dependent) switching field of the small grains is reduced and writing occurs further downstream. This is equivalent to the effect of a switching field distribution

(see (6.6) and (6.11)), see also [69,70]. As Figs. 6.11 and 6.12 illustrate, for relatively stable media, the measured thermal decay peaks at surprisingly low linear density and essentially vanishes at high recording density. Only for media that have a very low areal moment – and thus show much thermal decay – does the decay increase with linear density as expected. On the other hand, the measured thermal decay follows the theoretical expectation and increases with increasing temperature.

The remainder of this chapter discusses the linear density dependence of thermal decay. Firstly, it is relatively straightforward to explain why there is any thermal decay at low linear density. Inspection of the energy barrier distributions (Fig. 6.6) shows that the initial switching of the magnetization is associated with very small energy barriers and small effective anisotropy fields[1]. (Note that the tail and not the mean of the energy barrier distributions controls thermal decay.) For low linear density, the rms signal as measured with the spectrum analyzer is proportional to $M_r\delta$ and therefore the observed decay is virtually identical to that occurring at zero applied field. If the linear density is increased, the average demagnetizing field on the media increases (because there are more transitions per unit length) and more grains become susceptible to thermal agitation. Therefore the mere presence of grains with small energy barriers is not sufficient to explain the linear density dependence of thermal decay.

A more detailed analysis [72] shows that the data can be explained only if the energy barrier and the effective anisotropy field distributions are taken into account. Fig. 6.13 illustrates the physical processes at the instant of writing a transition. Grains with small switching field are recorded further downstream, while those with greater switching fields are recorded closer to the edge of the gap of the writing head. These different switching fields can be due to different anisotropy fields, but they can also be due to the time dependence of the coercivity that is caused by different grain volumes. (Al-

[1] As evident from Figs. 6.6 and 6.7, this low energy barrier tail cannot be detected for media with high areal moment (disks 6 and 7).

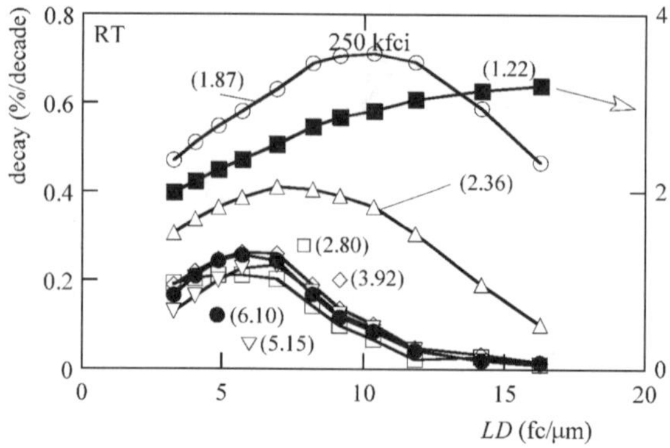

Fig. 6.14. Calculated thermal decay for the series of disks with properties given in Table 6.1. Temperature 293 K

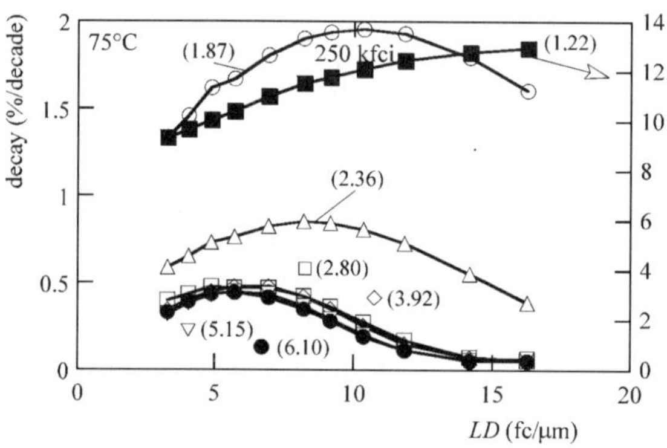

Fig. 6.15. Calculated thermal decay for the series of disks with properties given in Table 6.1. Temperature 348 K

though the time scales at recording are very short, thermal activation effects can not be neglected completely!) Therefore the recording process *inherently* causes phase shifts of the recording locations between grain populations. Due to these phase shifts, different grains experience the *different* demagnetizing fields – a break with the assumption implied in simple model considerations. This mechanism explains why there is less decay than anticipated for high recording densities. Conversely, a monotonic increase of the thermal decay with increasing linear density occurs only if the switching field distributions are very narrow. Then the less thermally stable grains are subjected to the same demagnetizing fields as the more stable ones and the decay increases with linear density. Figures 6.14 and 6.15 show theoretical results corresponding to the experimental ones shown in Figs. 6.11 and 6.12.

6.5.2 The "Dilemma": the Superparamagnetic Limit

An increase of areal density can be achieved only if the medium SNR is increased. As can be seen from (6.22), this requires smaller grains as well as a smaller transition width (transition parameter a). In longitudinal recording the transition parameter is mainly determined by the demagnetization ratio, $M_r\delta/H_r$, (note that H_r is the short-term coercivity) and, to a lesser extent, by the switching field distribution. As already discussed, the grain size itself also contributes to the transition width. Even for advanced media, the simple slope model from Williams–Comstock [45] gives reasonable estimates for the transition parameter. Therefore, in view of (6.22), the most effective way to decrease jitter, or conversely, to increase SNR, is to decrease the ratio $M_r\delta/H_r$. In a similar fashion, a smaller cross-track correlation length (or grain size), reduces transition jitter. Thus, all measures to increase SNR inevitably lead to a reduction of the thermal stability – with the exception of increasing anisotropy (field), which of course increases the writing coercivity. To first order, maintaining thermal stability means that the energy barrier, ΔE, has to be sufficiently large. The required energy barrier that ensures that the state of magnetization is stable for (say) ten years is roughly 40 k_BT, see (6.6). This applies to the most critical case, namely the highest temperature to which the media are exposed in a drive and the strongest possible demagnetizing field. Assuming a typical temperature dependence of the anisotropy of $\partial K_1/\partial T \approx 0.2\%/°\mathrm{C}$, a value of $K_1V/k_BT \approx 40$ at 75°C leads to $K_1V/k_BT \approx 53$ at room temperature. Since the demagnetizing fields further reduce the energy barriers, the K_1V values have to be increased even more. Depending on the demagnetization ratio, M_s/H_A, reasonably stable media require energy barriers of about $60\ldots70$ k_BT. Thus the product K_1V cannot be arbitrarily reduced and any volume reductions have to be countered by an increase in anisotropy. Since writing fields cannot substantially increase due to material limits, the possible anisotropy increase remains moderate and the grain dimensions (cube root of the volume!) remain essentially fixed.

Within these boundary conditions, further media optimization can be achieved only by narrowing the energy barrier and switching field distributions. In addition, lowering the head-to-medium separation offers a gain in SNR. On the other hand, head-to-medium separations as low as 10 nm are already considered, which, apart from the tribological problems associated with such a low spacing, offer a limited potential for improvement. In summary, the traditional means of increasing medium SNR (reduction of intergranular coupling, decreasing $M_r\delta/H_r$) fail once the energy barriers approach $60\ldots70$ k_BT. Consequently, future medium developments are expected to yield only small improvements of medium SNR.

The data indicate that thermal decay can either appear as uniform signal decay, as pulse broadening (increasing decay with linear density), or as a mixture thereof. For a pure pulse broadening mechanism, the harmonics of a low-density square-wave recording decay at exactly the same rate as the fundamental recorded at the correspondingly higher density [68]. In general, the amount of thermal decay depends on the pattern that is recorded, because the demagnetizing field is pattern dependent. This induces non-linearities [60,65,67], which add to the effect of signal loss and lead to a further degradation of the recording performance. Error rate degradation as function of time has been reported by Taratorin et al. [60,65]. It should also be mentioned that the signal decay is accompanied by a change in noise; in particular, the dc noise increases during the decay process [63].

It should not be forgotten that the areal density "limit" of a recording system does not depend only on the recording media, but also on the other components in a recording system, such as the head and the channel. The channel requires a minimum SNR at the input to recover the data. Although channels have been continuously improved, i.e., less SNR is required for a correct detection, channel gains have always been moderate when compared to the gains obtained from heads and media. In principle, medium SNR can be traded off against head sensitivity within limits, but the total SNR available to the channel can at best approach the medium SNR. The complexity and the different requirements of recording systems make it very difficult to give reliable estimates for the maximum achievable recording density.

6.6 Alternatives to Conventional Longitudinal Recording

6.6.1 Anti-ferromagnetically Coupled Media

In 2000, both Fujitsu and IBM proposed an alternative type of medium for longitudinal recording [73,74]. The main idea is sketched in Fig. 6.16. In the simplest case, such an anti-ferromagnetically coupled (AFC) medium consists of two layers coupled through a very thin spacer layer, typically made of 8 Å Ru. The coupling is through a RKKY-type interaction and is anti-ferromagnetic in nature; that is, the anti-parallel state of the two magnetic layers is favored. A hysteresis loop for a single "particle" of such a medium is sketched on the right in Fig. 6.16. A high applied field forces

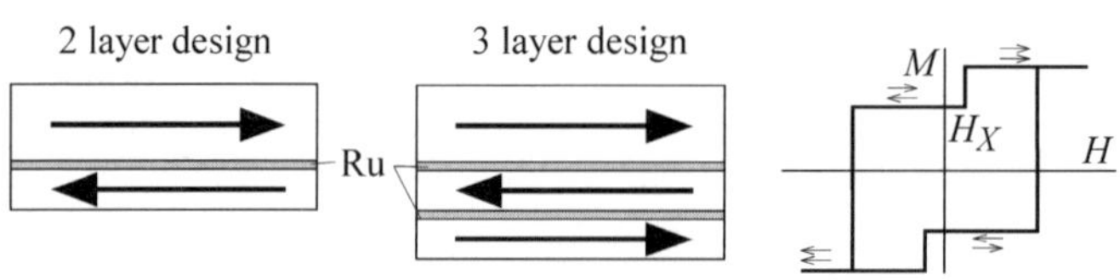

Fig. 6.16. AFC media: to increase thermal stability, additional magnetic volume is coupled anti-ferromagnetically to the main recording layer. Two and three-layer designs have been proposed. On the right hand side, a hysteresis loop for a two-layer medium is sketched

the two magnetizations to be parallel; the field energy is thus stronger than the energy associated with the anti-ferromagnetic coupling. At sufficiently small field, H_X, the magnetizations switch to a scissored state. Since, in the scissored state, thermal agitation has to reverse both sub-grains, it was argued that AFC media have effectively higher grain volumes that they therefore show enhanced thermal stability [73].

It is easy to show that H_X depends on the relative thicknesses of the two layers and the coupling strength, J. The thicker the bottom layer, the greater the magnetization change and the smaller the field H_X. Since the bottom layer is highly thermally active, the transition from the parallel to the anti-parallel state should occur relatively fast, which means that H_X should be smaller than zero. It should be noted that the grain area does not enter into this consideration, because the (exchange) coupling energy is proportional to the grain area, $E_X \propto J \cdot \mathit{area}$, and all other energies are proportional to the volume, $E \propto HM \cdot \mathit{area} \cdot \delta$.

It is still under discussion, whether areal density can be gained with AFC media as compared to conventional media. For one AFC grain, it has been proven that the energy barrier at zero field is indeed $(K_1V)_1 + (K_1V)_2$ for sufficiently strong coupling, but the same calculation yields that the switching field is increased by almost the same amount [75]. Therefore, to first order, almost the same stabilizing effect can be achieved by using a conventional material with a higher anisotropy field [75]. It is important to realize that the stabilizing effect as such does not automatically mean that AFC media have a higher recording potential; only if the required recording field is not increased by the same amount, can AFC media be optimized to yield higher SNR and thus higher recording density. It should also be mentioned that AFC media experience smaller demagnetizing fields at storage, which results in less thermal decay. It was estimated that two-layer AFC media could gain between 1 and 2 dB in medium SNR over conventional media [75]. The opposing magnetizations of the top and the bottom layer do not only reduce the demagnetizing fields, but they also reduce the flux that the read-back head senses. This leads to higher demands for the read head sensitivity, especially for the case of a two-layer AFC medium.

6.6.2 Perpendicular Recording

In the nineteen seventies and eighties, many researchers considered perpendicular recording superior to longitudinal recording. Although considerable efforts were undertaken, perpendicular recording never held its promise and then it fell a little bit into oblivion. In view of the superparamagnetic limit, there is renewed interest in perpendicular recording. The following paragraphs discuss potential advantages of perpendicular recording (see also Chap. 8 of this book).

The simplest way to realize perpendicular recording is to use a perpendicularly oriented medium instead of a longitudinally oriented medium and

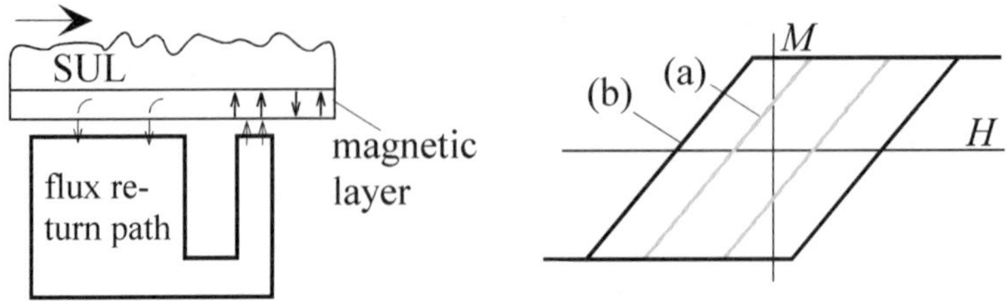

Fig. 6.17. Perpendicular recording. An efficient perpendicular recording system employs a pole head in conjunction with a magnetically soft underlayer underneath the medium (left). The medium needs to have sufficient coercivity to withstand the film demagnetizing field

keep the recording head the same. There is general agreement that such a system has at most marginal advantages over longitudinal recording. The most promising system for perpendicular recording is sketched in Fig. 6.17. In this case, recording is accomplished with a "pole head" in conjunction with a magnetically soft underlayer, SUL, which is placed underneath the medium. A second, auxiliary pole has a much wider area and provides a flux return path. The magnetically soft underlayer belongs physically to the medium and magnetically to the head. The recording takes place at the trailing edge of the main pole. The read-back is accomplished magnetoresistively and is not further discussed here.

The key advantage of this type of record process is that the medium is effectively subjected to the deep gap field of the head as opposed to the fringe field as in a conventional ring head. In fact, the simplest way to visualize this is to consider the surface of the pole to be a charge sheet which has a magnetic image in the SUL. The two charge sheets form the same arrangement as a conventional head; the only difference is that it is turned sideways such that the medium travels through the deep gap region. Therefore, the effective write field is increased by up to a factor of two, implying that perpendicular media can have a higher anisotropy field.

The medium design deserves further considerations. For the following simple discussion, the grains are assumed perfectly oriented in the perpendicular direction as well as exchange de-coupled; that is, they only interact magnetostatically. Early medium designs for perpendicular recording had relatively small coercivities and loop shapes (in the perpendicular direction) similar to (a) as indicated in Fig. 6.17. In this case, the demagnetizing field caused by the full saturation magnetization is higher than the coercivity and the medium is partly demagnetized at zero applied field. (It should be noted that "long bits" experience almost the full film demagnetizing field.) The internal field, H_i, is $H_i \cong H_a - M_\perp$, and automatically adjusts itself to the coercivity if the external field is removed. Consequently the demagnetizing field has reduced the energy barriers of the individual grains to the point at which they are very close to switching. Media of this type are therefore highly susceptible to thermal agitation, which has also been reported [55,76]. In addition to the thermal stability aspect, any deviation from saturation creates magnetization noise, $\sigma_M^2 \propto \left(1 - pm^2\right)$, see (6.21). Therefore, media with loop shapes according to (a) create unnecessary noise "in the bits". Clearly perpendicular media should have a hysteresis loop shape as indicated by curve (b) in Fig. 6.17.

For the simplified assumptions made here (perfect orientation, no correction for the internal demagnetization, no exchange interaction between grains), it is easy to derive the magnetic parameters that optimize thermal stability and SNR. The most critical situation occurs "in the bit", where the demagnetizing field is greatest. Using (6.10):

$$|H_r\left(\vartheta_0 = 0, t\right)| = H_A\left(1 - \sqrt{\frac{k_B T}{K_1 V} \ln \frac{f_0 t}{\ln 2}}\right) = M_s \ ,$$

it is straightforward to derive that the minimum volume is supported if the condition

$$H_A = 3M_s \tag{6.24}$$

is fulfilled. If (6.24) is fulfilled and the volume is sufficiently large, the hysteresis loop shape automatically follows curve (b). Bertram and Williams [77] have given a more in depth discussion of the recording potential of perpendicular recording and estimated that the potential areal density of perpendicular recording is roughly five times that of longitudinal recording. I personally believe that the difference between longitudinal recording and perpendicular recording has been overestimated, because the recent high areal density demonstrations have used significantly different media designs for longitudinal recording than those assumed for the comparison.

6.7 Summary and Outlook

Traditionally, there have always been many debates about the future of magnetic recording. Most of these debates have been rather negative and have "predicted" that the magnetic recording technology could no longer be extended and would be replaced by alternative, more viable technologies. In hindsight, it is easy to say that these predictions have been plain wrong. As a consequence, any "prediction" that magnetic recording will reach its limit lacks credibility.

In contrast to all previous "limits", the "superparamagnetic limit" represents a *physical barrier* rather than an *engineering barrier*. Many analysts have based their predictions on engineering rather than on physical barriers; and up to now, the recording industry has never failed to overcome these engineering barriers. The superparamagnetic effect represents a physical barrier for magnetic recording, such as the wavelength of light does for optical recording. In my opinion, approaching this physical barrier means that the areal density growth will have to slow down significantly.

Slowing down the areal density growth makes it possible for alternative technologies to intercept conventional longitudinal recording. Apart from perpendicular recording that has been briefly discussed in Sect. 6.6 (and more comprehensively in Chap. 8), thermally assisted recording as well as recording on patterned media (Chap. 7) are currently under discussion. All of these technologies (even perpendicular recording) add significantly to the complexity and cost of a recording system, and it remains to be seen whether any of these technologies will replace longitudinal recording.

Acknowledgments. I am greatly indebted to many colleagues who have contributed to aspects of this work. Special thanks are due to R. M. Brockie and G. C. Rauch for careful proofreading of the manuscript.

References

1. S. H. Charap, P. L. Lu, and Y. He, IEEE Trans. Magn. **33**, 978–983 (1997)
2. compare press releases from IBM, 20.3 Gbit/in^2 (1999), Fujitsu 20.9 Gbit/in^2, Seagate 23.8 Gbit/in^2 (1999), Read-Rite/HMT/Komag 26.5 Gbit/in^2, (1999), IBM 35.3 Gbit/in^2, Seagate 44.2 Gbit/in^2 (2000), Read-Rite/HMT/Komag 50 Gbit/in^2, (1999), Hitachi 52 Gbit/in^2 (perpendicular), Fujitsu 56 Gbit/in^2, Read-Rite/Samsung 60 Gbit/in^2 (perpendicular), Read-Rite/HMT/Komag 63 Gbit/in^2.
3. E. C. Stoner and E. P. Wohlfarth, Phil. Trans. Roy. Soc. A **240**, 599–642 (1948)
4. L. Néel, Compt. Rend., Acad. Sci, Paris **228**, 664–666, (1949)
5. W. F. Brown Jr., Phys. Rev. **130**, No. 5, 1677–1686 (1963)
6. C. P. Bean and J. D. Livingston, J. Appl. Phys. Suppl. **30**, 120S–129S (1959)
7. E. F. Kneller and F. E. Luborsky, J. Appl. Phys. **34**, 656–658 (1963)
8. M. P. Sharrock and J. T. McKinney, IEEE Trans. Magn. **17**, 3020–3022 (1981)
9. R. H. Victora, Phys. Rev. Lett. **63**, 457–460 (1989)
10. H. N. Bertram and H. J. Richter, J. Appl. Phys. **85**, 4991–4993 (1999)
11. H. J. Richter and R. Y. Ranjan, J. Magn. Magn. Mat. **193**, 213–219 (1999)
12. H. N. Bertram, X. Wang, and V. L. Safonov, IEEE Trans. Magn. **37**, (2001), to be published
13. D. M. Paige, B. Szpunar, and B. K. Tanner, J. Magn. Magn. Mat. **44**, 239–248 (1984)
14. N. Inaba, Y. Uesaka, and M. Futamoto, IEEE Trans. Magn. **36**, 54–60 (2000)
15. H. Miyajima, K. Sato, and T. Mizoguchi, J. Appl. Phys. **47**, 4669–4771 (1976)
16. S. M. Stinnett, J. W. Harrell, A. F. Khapikov, and W. D. Doyle, IEEE Trans. Magn. **36**, 148–153 (2000)
17. T. L. Gilbert, Phys. Rev. **100**, 1243 (1955)
18. R. Kikuchi, J. Appl. Phys. **27**, 1352–1357 (1956)
19. N. Inaba, Y. Uesaka, A. Nakamura, M. Futamoto, and Y. Sugita, IEEE Trans. Magn. **33**, 2989–2991 (1997)

20. V.L. Safonov, J. Magn. Magn. Mat. **195**, 523–529 (1999)
21. V. L. Safonov and H. N. Bertram, J. Appl. Phys. **87**, 5681–5683 (2000)
22. W. F. Brown, J. Appl. Phys. Suppl. **30**, 130S–132S (1959)
23. L. He, W. D. Doyle, L. Varga, H. Fujiwara, and P. J. Flanders, J. Magn. Magn. Mat. **155**, 6–12 (1996)
24. S. M. Stinnett, W. D. Doyle, P. J. Flanders, and C. Dawson, IEEE Trans. Magn. **34**, 1828–1830 (1998)
25. N. Rizzo, T. J. Silva, and A. B. Kos, IEEE Trans. Magn. **36**, 159–165 (2000)
26. K. Rubin, J. S. Goldberg, H. Rosen, E. Marinero, M. Doerner, and M. Schabes, Mat. Res. Soc. Symp. **517**, 261–266 (1998)
27. A. Moser, D. Weller, M. E. Best, and M. F. Doerner, J. Appl. Phys. **85**, 5018–5020 (1999)
28. H. Aoi, M. Saitoh, N. Nishiyama, R. Tsuchiya, and T. Tamura, IEEE Trans. Magn. **22**, 895–897 (1986)
29. H. N. Bertram, K. Hallamasek, and M. Madrid, IEEE Trans. Magn. **22**, 247–252 (1986)
30. G. J. Tarnopolsky, L. T. Tran, A. M. Barany, H. N. Bertram, and D. R. Bloomquist, IEEE Trans. Magn. **25**, 3160–3165 (1989)
31. H. J. Richter, S. Z. Wu, and R. K. Malmhäll, IEEE Trans. Magn. **34**, 1540–1542 (1998)
32. M. Igarashi, F. Akagi, A. Nakamura, H. Ikekame, H. Takano, and K. Yoshida, IEEE Trans. Magn. **36**, 154–158 (2000)
33. H. J. Richter, IEEE Trans. Magn. **35**, 2790–2795 (1999)
34. H. J. Richter, T. Thomson, and E. T. Yen, IEEE Trans. Magn. **36**, 2465–2467 (2000)
35. M. Igarashi, F. Akagi, A. Nakamura, H. Ikekame, K. Yoshida, and Y. Sugita, IEEE Trans. Magn. **36**, 2148–2150 (2000)
36. M. L. Mallary, A. F. Torabi, and R. Beauregard, IEEE Trans. Magn. **36**, 2477–2479 (2000)
37. D. Wachenschwanz and M. Alex, J. Appl. Phys. **85**, 5312–5314 (1999)
38. M. Alex and D. Wachenschwanz, IEEE Trans. Magn. **35**, 2796–2801 (1999)
39. N. Inaba, Y. Uesaka, and M. Futamoto, IEEE Trans. Magn. **36**, 54–60 (2000)
40. B. E. Higgins, A. F. Torabi, M. L. Mallary, IEEE Trans. Magn. **37**, (2001), to be published
41. R. D. Fisher and J. L. Pressesky, IEEE Trans. Magn. **25**, 3414–3416 (1989)
42. Y. Estrin, P. G. McCormick, and R. Street, J. Phys. Condens. Matter **29**, 4845–4851 (1989)
43. R. Street and D. C. Crew, IEEE Trans. Magn. **35**, 4407–4413 (1999)
44. H. Zhou, H. N. Bertram, A. Torabi, and M. Mallary, IEEE Trans. Magn. **37**, to be published (2001)
45. M. L. Williams and R. L. Comstock, *17th Ann. AIP Conf. Proc.*, vol. 5, 738–742 (1972)
46. H. J. Richter and R. J. Veitch, J. Magn. Magn. Mat. **155**, 335–337 (1996)
47. H. N. Bertram, *Theory of Magnetic Recording* (Cambridge Univ. Press, Cambridge, UK, 1994)
48. J. C. Mallinson, IEEE Trans. Magn. **5**, 182–186 (1969)
49. J. C. Mallinson, IEEE Trans. Magn. **27**, 3519–3531 (1991)
50. V. B. Minhuin, IEEE Trans. Magn. **36**, 2077–2090 (2000)

51. T. C. Arnoldussen: "Theoretical considerations of media noise", in: *Noise in Digital Magnetic Recording*, T. C. Arnoldussen and L. L. Nunnelley (eds.), (World Scientific, Singapore, 1992), pp. 101–139.
52. J. Caroselli and J. K. Wolf, SPIE Proc., Coding and Information Storage **2605**, 29–38 (1995)
53. T. J. Silva and H. N. Bertram, IEEE Trans. Mag. **26**, 3129–3139 (1990)
54. G. J. Tarnopolsky and P. R. Pitts, J. Appl. Phys. **81**, 4837–4839 (1997)
55. B. M. Lairson, W. Liu, and A. P. Payne, J. Appl. Phys. **79**, 7920–7925 (1996)
56. Y. Hosoe, I. Tamia, K. Tanahashi, Y. Takahashi, T. Yamamoto, T. Kanbe, and Y. Yajima, IEEE Trans. Mag. **33**, 3028–3030 (1997)
57. Y. Hosoe, T. Kanbe, K. Tanahashi, I. Tamai, S. Matsunuma, and Y. Takahashi, IEEE Trans. Magn. **34**, 1528–1533 (1998)
58. Y. Zhang and H. N. Bertram, IEEE Trans. Magn. **34**, 3786–3793 (1998)
59. P. Dhagat, R. S. Indeck, M. W. Muller, J. Appl. Phys. **85**, 4994–4996 (1999)
60. A. Taratorin, D. Cheng, and E. Marinero, J. Appl. Phys. **85**, 5309–5311 (1999)
61. M. Futamoto, Y. Hirayama, N. Inaba, Y. Honda, K. Ito, A. Kikugawa, and T. Takeuchi, IEEE Trans. Magn. **35**, 2803–2807 (1999)
62. A. Moser and D. Weller, "Thermal processes and stability of longitudinal magnetic recording media", IEEE Trans. Magn. **35**, 2808–2813 (1999)
63. Y. Zhang and H. N. Bertram, IEEE Trans. Magn. **35**, 4326–4338 (1999)
64. D. Weller and A. Moser, IEEE Trans. Magn. **35**, 4423–4439 (1999)
65. A. Taratorin, D. Cheng, and E. Marinero, IEEE Trans. Magn. **36**, 80–85 (2000)
66. E. N. Abarra, I. Okamoto, K. Sato, Y. Yoshida, H. Akimoto, and Y. Uematsu, IEEE Trans. Magn. **36**, 86–91 (2000)
67. D. Wachenschwanz, M. Alex, L. He, and M. Nichols, IEEE Trans. Magn. **36**, 98–103 (2000)
68. P. Dhagat, A. Jander, R. S. Indeck, M. W. Muller, and H. J. Richter, IEEE Trans. Magn. **36**, 528–531 (2000)
69. R. S. Indeck, P. Dhagat, A. Jander, and M. W. Muller, J. Magn. Soc. Japan **21 (S2)**, 205–210 (1997)
70. A. Moser, D. Weller, N. D. Rizzo, and T. J. Silva, J. Appl. Phys. Lett. **77**, 1505–1507 (2000)
71. H. J. Richter and R. M. Brockie, IEEE Trans. Magn. **37**, (2001), to be published
72. H. J. Richter, unpublished
73. E. N. Abarra, A. Inomata, H. Sato, I. Okamoto, and Y. Mizoshita, J. Appl. Phys. Lett. **77**, 2581–2583 (2000)
74. E. E. Fullerton, D. T. Margulies, M. E. Schabes, M. Carey, B. Gurney, A. Moser, M. Best, G. Zeltzer, K. Rubin, H. Rosen, and M. Doerner, J. Appl. Phys. Lett. **77**, 3806–3808 (2000)
75. H. J. Richter and E. Girt, IEEE Trans. Magn. **37**, (2001), to be published
76. W. Cain, A. Payne, M. Baldwinson, and R. Hempstead, IEEE Trans. Magn. **32**, 97–102 (1996)
77. H. N. Bertram and M. L. Williams, IEEE Trans. Magn. **36**, 4–9 (2000)

7 Patterned Media

Gordon F. Hughes

Patterned media is a potential new recording disk technology for future high density disk drives. The 60–130% yearly growth of disk drive areal bit density ("AD") is driving magnetic recording technology towards four apparent physics limits: stored bit thermal decay [1], signal-to-noise ratio loss ("SNR") and corresponding bit error rate increase [2], maximum switching speeds and magnetization limits of magnetic materials [3]. These apparent limits will be discussed further in Sect. 7.3.

Figure 7.1 schematically illustrates a patterned magnetic film on a conventional disk substrate, similar to todays continuous magnetic film disks, but with the film patterned into individual bits. This chapter will discuss the potential of such patterned media to "detour around" the physics limits above, necessary technology developments, drive engineering issues, and finally product introduction and extensibility (Sect. 7.14).

Disk drive technologists have been remarkably successful for nearly half a century in designing around AD limit predictions [4]. An example was a 1973 prediction [5] that digital magnetic recording would "limit out" in 1984 at 20 kbpi. This was a correct prediction of the commercial bit density in 1984 but was not a limit. Hard physics were not involved in this prediction, merely a truism that capacity, access time, and cost "cannot indefinitely improve" and that technology limiting effects would be felt "an order of magnitude before physics limits were reached." Disk drive bit densities are now an order of magnitude higher than that 1984 prediction, in part because the rotating disk data storage industry developed new technology to design around limits (including thin film disks, magnetoresistive heads, and new channel codes).

New technology alternatives may again be needed to design around current apparent limits. Today, these appear to be hard physics limits, and they

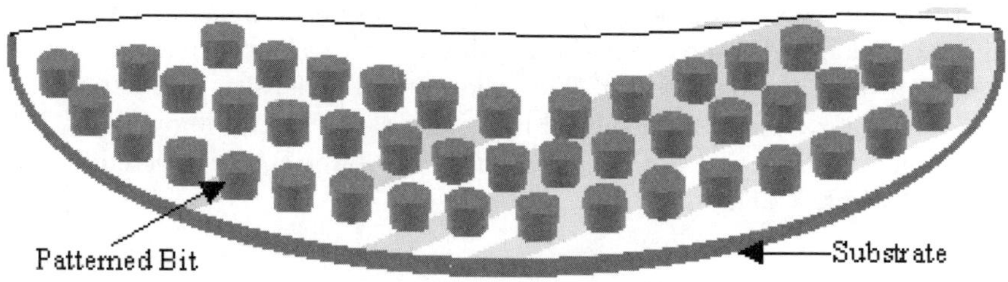

Fig. 7.1. Patterned hard disk

should be seriously addressed. Disk drives may encounter one or more of these physics limits by the year 2006 at an AD of 0.1-1.0 Terabits/in^2. Developing alternative technologies may require several years, and integrating them into products may require several more. Appropriate alternative technologies will allow maximum use of existing rotating disk magnetic technology, and also allow future extensibility to higher AD. Alternative technologies include patterned media, thermal writing using atomic force microscope ("AFM") probe current [6] and probe head arrays [7].

Patterned media addresses SNR, switching speed and magnetic material issues as well as thermal decay. It offers several practical advantages. Magnetic near field recording can still be used (with no wavelength limit). Rotating disk storage technology and manufacturing investments can continue to be used.

7.1 Conventional Thin Film Disk Media

Today's hard disk drive storage media are continuous thin magnetic films of polycrystalline cobalt alloys deposited by sputtering onto nickel-plated aluminum or glass disk substrates [8]. See Chap. 2. These storage media have been called 'featureless' to distinguish them from bit memories such as semiconductor random access memory ("RAM") chips. In the latter, individual bit positions are lithographically patterned into the silicon chip. It may be noted that RAM cost per bit is several orders of magnitude higher than disk memory. This is related to their orders-of-magnitude lower density limitations due to connections for data access, power, and heat removal. Patterned media at high density need not be high cost.

The polycrystal grain boundaries of film media provide magnetic exchange isolation, the required high film coercivities of 3–5 kOe, and media noise control. Drive mechanical tolerances are larger than grain sizes so there is random granularity noise in the playback signal. The grains are kept sufficiently small that recorded bits have 50–1000 grains, thus controlling this noise by averaging many grains per bit.

Recorded bit positions on film disks are magnetically defined by track servo and bit synchronization signals, pre-recorded between the user data sectors. These "sector servos" are used to synchronize both the data sector recording and playback processes. Tracking signals occur once per data sector, and a bit clock (derived from the readback data waveform) maintains bit timing through the data sector. This allows the bit density to be about ten times the track density for a typical bit aspect ratio "BAR" about 10:1. This servo configuration may not be sufficient at the lower BAR expected at very high AD (Sect. 7.11).

7.2 Patterned Media

Patterned recording media differs from film media in that the disk bit positions are discrete magnetic "islands" separated by a nonmagnetic matrix or by media material with altered magnetics. Individual patterned magnetic bits replace featureless films (Fig. 7.2). Each bit has only two stable magnetic states to allow binary recording. Since only a single bit 'grain' is needed it is much larger for a given areal density than the 50–1000 grains per bit of film media.

Work on patterned magnetic 'domains' dates to 1973 [10–14]. Discrete *track* patterning predates discrete bits, dating from 1972 [15,16]. This early work discusses the elimination of track edge noise by patterning. Discrete track patterning could be an initial entry point for commercial products using patterned disk technology (Sect. 7.14). Lithographic patterning of permalloy rectangles and their magnetic properties was reported in 1991 [17].

Much of featureless disk technology can be used in patterned disks. New requirements arise in finding magnetic materials having bit orienting aniso - tropy and single domain binary switching (Sect. 7.3.3) [18,19]. Vertical recording film disks already have orienting H_k and can be patterned (Sect. 7.6.3). Vertical disk underlayers appear to improve recording performance but may not be mandatory (Sect. 7.10.4).

Patterned media will share many of the normal technology issues of high-density disks. These include flyability, tribology, cost/yield and corrosion.

Initially, the need is for research media to allow head, disk, channel, flyability and tribology development. Work is also needed on patterned low-cost

Fig. 7.2. 2 × 2 mm patterned region; AFM (A) and MFM images (B); higher magnification images (C) and (D), respectively. Pattern period is about 100 nm, made with a 60 s exposure to a 1 pA 30 keV beam (courtesy IBM Almaden Research Center) [9]

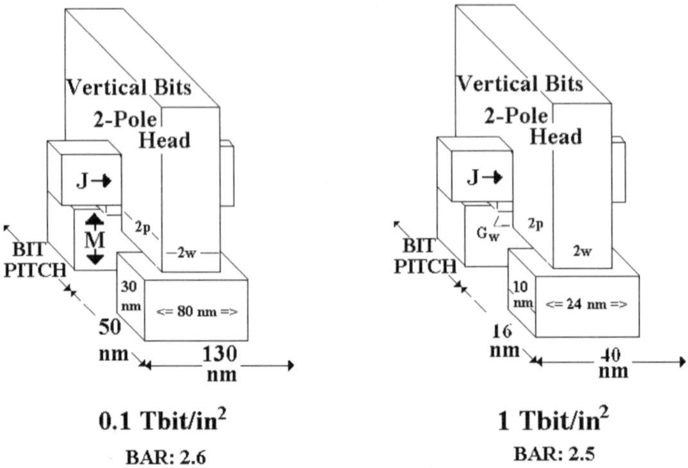

Fig. 7.3. Example two-pole write head and bit geometries, 0.1 and 1 Tbit/in^2

fast disk manufacturing processes (less than 1 minute/disk). Fast parallel deposition/patterning is required, with low defect density. Candidate processes are discussed in Sect. 7.5.

Introducing patterned media into disk drives requires solving engineering issues such as write synchronization and disk centering eccentricity servoing (Sect. 7.13). Physics limits are being traded for engineering issues. The latter can be studied and probably resolved. The former cannot be directly violated.

Figure 7.3 shows two example patterned head/disk recording geometries [3]. Miniaturized conventional hybrid write and MR read heads are assumed (Sect. 7.9). The parameters are shown in Table 7.1. The write head shown has a single turn in the gap for fast write field risetime (because write current density $\mathbf{J}_w$ applies drive MMF directly to the poletips [3]). The bit velocity V is in the downtrack direction along the bit pitch direction, as is the pole length 2p and the writer gap length G_w. The writer width 2w is less than the bit width to allow writing with offtrack tolerances.

7.3 Physics Limits: Conventional vs. Patterned Media

Magnetic recording scaling rules are useful in predicting the consequences of increasing areal density. For the same magnetic materials (e.g., constant magnetization and coercivity), the magnetostatic fields remain constant if all dimensions are reduced by a scale factor $S \,\alpha \sqrt{AD}$.

7.3.1 Thermal Decay

A stored binary bit represents a volume of ordered magnetic spins whose magnetic energy competes with the Boltzman thermal energy (which is attempt-

Table 7.1. Glossary of symbols, and 1 $Tbit/in^2$ example parameters, for vertical recording with two-pole writer and MR reader

Term	Definition	1 $Tbit/in^2$ value/Units
AD	Areal bit density, 1/BT	$Tbits/in^2$
2a	Patterned bit down-track length	10 nm
2b	Patterned bit cross-track width	26 nm
B	Bit pitch distance (>2a)	16 nm
BAR	Bit aspect ratio, T/B	2.5
2c	Patterned bit thickness	10 nm
d	Head-media tops physical spacing	6 nm
Film media	"Continuous film disk media"	
G_w	2-pole vertical writer gap length	6 nm
Gbps	Bit transfer rate, Gbits/sec	1.3-2
H^d	Media demag field, bit center	kOe
H^w	Write head field, bit center	kOe
H_k	Bit or head anisotropy field	kOe
J_w	Head write current density	$G\text{-amp/cm}^2$
J_{ws}	Head saturation J_w for M_s^p	$G\text{-amp/cm}^2$
M_s^b	Bit saturation magnetization	400 emu/cc
M_s^p	Head pole saturation	800–1200 emu/cc
2p	Two pole head pole length (down-track)	10 nm
P2L	Write head trailing pole length	nm
PES	Servo position error signal, bit and track	nm
PF_x	Downtrack patterning fraction	2a/B
PF_y	Crosstrack patterning fraction	2b/T
RRO	Repeatable runout disk eccentricity	µm
S	Size scaling factor	$\alpha\sqrt{AD}$
SNR	Playback peak amplitude: noise	peak:rms
T	Track pitch distance (>2b)	40 nm
2w	Head pole width (cross-track)	18 nm
x	Down-track (along-bit) axis	nm
y	Cross-track axis	nm
z	Media-normal (vertical) axis	nm
V	Head-media velocity	32 m/s

ing to disorder the spins into random noise). In the simplest micromagnetic models [1] the expected time scale for this disordering is roughly

$$\tau \approx e^{(H_c M_s V_g/kT)}\ \text{nsec}$$

This problem exponentially grows with areal density (see Chaps. 5 and 6). Experimental evidence is shown in [20]. Thermal decay theory and experiment for playback signal amplitude decay is shown in [21], and for SNR decay in [22]

This problem is near-term because film media has 50–1000 media grains per bit to average out asynchronous granularity noise [23,24]. Patterned media has only one grain per bit, giving a potential areal density extensibility of $50^{2/3}$-$1000^{2/3}$, or 13–100 times. (The magnetic recording scaling rules imply $V_{bit}\,\alpha\, AD^{3/2}$.)

7.3.2 Signal to Noise Ratio

Magnetic recording scaling rules make bit readback SNR drop as areal density increases. The magnetoresistive (MR) readback signal scales with the size scaling factor S [25]. Electronic noise power scales up as 1/S (with the read channel bandwidth), if disk size and rpm are constant. So SNR drops as $SNR\,\alpha\, S^{3/2}\,\alpha\, AD^{3/4}$ Noise is also increased by thermal decay, overwrite, and nonlinear distortion [2,26].

Patterned media may improve SNR because single-domain bits need not have film media transition noise and nonlinear bit shift [19]. Discrete bits need not have track edge noise [27]. However, high bit packing 2a/B or 2b/T (Table 7.1) will cause magnetostatic interaction field "crosstalk" noise in patterned media by rotating the bit magnetization away from the vertical or horizontal orientation direction.

7.3.3 Switching Modes and Speed

For sub-nsec switching, the gyromagnetic ratio requires increased drive fields. High-speed magnetization dynamics data clearly shows this effect [28–30]. In these experiments, the driving fields are applied uniformly to the entire magnetic film under study. This situation differs from write heads, where the write current square wave switches writer yoke magnetization first, and nsec-scale switching has to propagate 5–10 µm to the pole tips in order to write media bits. Landau-Lifshitz calculations suggest that this may be slow [3,31]. If magnetization propagation from writer drive coil to the pole tips turns out to be a speed limitation, it appears that heads and bits smaller than 100 nm in size can switch in fractional nsec using practical drive currents. There are experimental indications in single pulse coplanar waveguide experiments that flux propagation can occur at 10^5 m/s through 1/4 µm thick permalloy films, suggesting dynamical flux propagation by modes resembling surface magnetostatic spin waves [32]. Picosecond magnetic switching has been demonstrated, using extremely high currents [33,34].

Spin valve MR readers may also be nearing gyromagnetic ratio switching speeds [35]. However, the media bit read flux is close to the active MR element area (within tenths of a micron) so field-driven fast switching should occur.

Media orientation (horizontal or vertical) can be unpredictable if there are several polycrystals per patterned bit [45]. Only two single domain static states should exist. (M can be allowed to "flower" away from the average magnetization direction at bit edges [36].) Bits larger than about 100 nm (greater than the exchange length) are often multidomain [18,37,45].

Bit aspect ratios need to be greater than unity (Sect. 7.10.1), so shape anisotropy energy tends to make the stable magnetization directions cross-track. For vertical or horizontal recording, orienting anisotropy is needed to keep the stable magnetization directions vertical or horizontal. System modeling suggests that 2–6 kOe H_k is sufficient [3].

7.3.4 Magnetic Material Limits

Recording physics requires that *film* disk coercivity increase with bit density in longitudinal recording [8]. This correspondingly requires the writer head pole saturation $4\pi M_s^p$ to increase, as high as 20 kG in the future. This is close to known material limits (e.g., 24 kG Co-Fe).

Patterned media requirements theoretically appear to be more modest. At 1 Tbit/in^2 (Sect. 7.10, Table 7.1) expected requirements are 800-1200 emu/cc head M_s^p (1015 kG), 300–400 emu/cc bit M_s^b and 2.0–3.5 kOe bit orienting anisotropy (2-3.5 kOe disk coercivity) [3].

7.4 Media Magnetic Materials

Patterned media magnetic materials have included iron [38], cobalt [39,53], amorphous cobalt [40], soft nickel [18,19,41,52,53], permalloy [42,60], Co/NiO [43], Co-Pt [57] and NiFe/Cu multilayers [42]. Post-patterning of film disks (Sect. 7.6.3) allows use of present cobalt alloys for horizontal media, and Co–Cr and Co/Pt multilayers [44] for vertical media

7.5 Media Fabrication for R&D and Manufacturing

Conventional continuous polycrystal media can be sputtered and then post-patterned (Sect. 7.6.3). Starting with a disk having known flyability, tribology, and homogeneous magnetics may be advantageous in producing patterned disks with these necessary features. This may also be an early way of producing test disks for research. Serial, maskless patterning methods can be used such as steered electron beam or X-ray beam lithography [10,42]. These techniques are slow, but can pattern mm size "patches" for test disks [37,39,45]. Reported feature sizes have been generally ≈100 nm or larger. The smallest feature sizes reported with electron beam lithography are arrays of 12 nm dots with 25 nm period [46], which suggests a potential patterned bit density of 1 Tbit/in^2.

Mass production electron beam methods are under development for semiconductor wafer fabrication. Electron beam projection mask lithography is one such promising technology [47,48]. Disk substrates could be thereby patterned before magnetic media deposition, or conventional film media can be post-patterned. It offers parallel mask lithography over 8-inch substrates, and

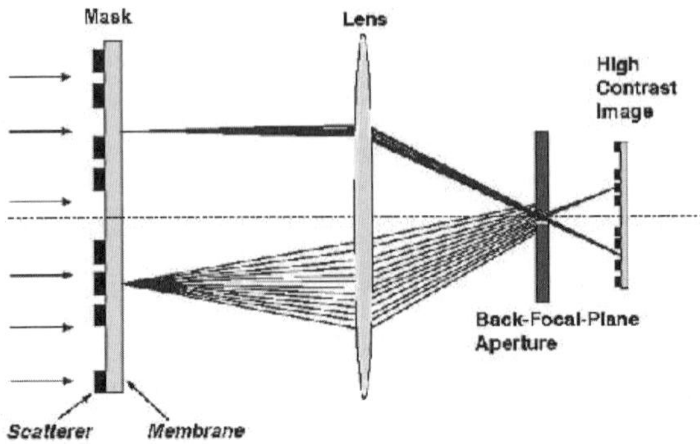

Fig. 7.4. Projection electron beam lithography uses high atomic number electron scatterer on low atomic number membrane (SCAPEL®, courtesy Bell Labs, Lucent Technology [47])

the large-scale uniformity required to produce synchronous bits on concentric tracks for rotating disk storage. Rates of thirty 8-inch wafers/hour are projected with 80-nm feature size. Producing 2.5-inch disks instead of 8-inch wafers projects to a rate of $30(8/2.5)^2 \approx 300$ disks/hour.

Figure 7.4 shows one such system. It is an experimental proof-of-concept projection electron-beam lithography system based upon the Bell Labs SCALPEL® concept [47] ("SCattering with Angular Limitation Projection Electron-beam Lithography"). PREVAIL is another projection system, from IBM and Nikon [48].

Its initial design demonstrates feasibility of the electron optics, scattering mask, and resist platform. Preliminary results indicate lithographic potential for the production of sub-0.18 µm features.

Interferometric lithography is an alternative parallel process, capable of quickly patterning 50 cm by 50 cm glass substrates with ≈100 nm features [49–51]. Reported work is on rectangular arrays, not circular tracks of patterned bits. It also does not appear to be capable of customized subpatterning for bit and track servo (Sect. 7.11).

Another alternative is evaporation of cobalt films onto pre-patterned substrates (Sect. 7.6.1) [43,51]. Another is nickel electrodeposition into prepatterned holes [41,42,52,53].

7.6 Media Patterning Technologies

Three classes of patterning technologies have been reported: pre-pattern the substrate before media deposition, produce the patterning during media deposition, or post-patterned bits after media deposition. These will be discussed in turn.

7.6.1 Pre-Patterning Before Media Deposition

Substrates have been directly patterned by electron beam lithography [52,53], followed by media deposition. Most commonly the substrate material is silicon and direct or mask electron beam lithography is used (Sect. 7.5).

7.6.2 Patterning During Media Deposition

Self-organizing systems (see Chap. 9) produce small-scale uniform patterns by controlled precipitation, growth, or annealing of particles [54,55]. Magnetic media can also be electrochemically deposited in the pores of anodic alumina [56]. Arrays have been fabricated by a combination of chemical vapor deposition and scanning tunneling microscopy [38]. These systems have potential to narrow the present wide grain size distribution of conventional film media and alleviate the thermal decay problem. Smaller grains decay first, reducing the playback SNR. Reducing grain size variability may therefore reduce thermal decay.

However, self-organizing systems do not appear to have the large-scale uniformity required to produce synchronous bits on concentric tracks for rotating disk storage, nor to produce servo sub-patterns (Sect. 7.11.2).

7.6.3 Post-Patterning After Media Deposition

Flyability, tribology, and corrosion issues may be simplified by post-patterning conventional continuous polycrystal film media. Post-patterning may offer early R&D disks for testing patterned media issues. Recording systems considerations appear to favor vertical or horizontal recording on patterned media, both of which require magnetic orienting anisotropy in the patterned bits (Sect. 7.10.1).

Vertical film media offers this needed orienting H_k, which stems from the vertical process direction of media deposition. Soft underlayers are commonly used with vertical film media and offer recording advantages for patterned disks as well [3]. The soft underlayer can be a continuous film or patterned for servo purposes [14,16].

Conventional horizontal film media normally has random crystal anisotropy axes primarily in the media plane. Producing sufficiently strong circumferential orientation for horizontal patterned media appears problematic.

If polycrystal film media is post-patterned, multidomain bit states may arise [45] (Sect. 7.3.3). Increasing the inter-grainular exchange of the film media might allow the bits to remain single domain.

Physical patterning may be advantageous for bit magnetic isolation but might damage the media properties [42]. Ion beam patterning of *magnetic properties* is an alternative to physical patterning [44,57,58]. The coercive force of irradiated regions can be made three times lower than mask protected areas. A potential drawback is that this may cause magnetostatic coupling between bits, a potential source of bit signal variation and readback noise.

Ion beam mixing of metallic multilayers is another alternative [59]. Destroying the magnetic properties of patterned bits by ion beam exposure (rather than their surrounds) can produce "anti-dot" arrays [60].

7.7 Media Substrates

Patterned media substrates have included Al_2O_3 [18,19], silicon wafers up to 5-cm diameter [49,53], and SiO_2 [39]. Glass up to 50 cm × 50 cm [49] has been used as substrates for nonmagnetic chrome dots. Continuous magnetic films with coercivity "patterned" by beam heat are reported in [58,59].

7.8 Head-Disk Interface: Head Flight and Tribology

After patterning by material removal (media or substrate), head flight hydrodynamics is no longer over a plane media surface. Patterning is on a scale far smaller than the mm-size of the air bearing surfaces and might be treated similarly to disk surface roughness [61,62]. The effective disk surface lies between the peaks and valleys of the roughness, and fly height d (above the top plane of the patterning) will decrease accordingly.

Flight over patterned disks has been included in air bearing numerical computer models [63] and characterized in terms of downtrack and crosstrack patterning packing fractions $PF_x = 2a/B$ and $Pf_y = 2b/T$ (Table 7.1). If the areal packing PF_x*PF_y is near unity (patterned bits occupy most of the disk surface area), d will be nearly that of the unpatterned disk, and if PF_x*PF_y is small, d will decrease by nearly the patterning depth. The flight dynamics and the static fly height vary with PF_x and PF_y [63].

If disk drive embedded servo wedges (Sect. 7.11.1) are patterned with packing fractions differing from the patterned data bit areas there may be flight dynamics issues when the air bearing surfaces encounter the servo wedges.

Patterned disk tribology appears to be unknown territory. Post-patterned film disks may retain their tribological properties if physical patterning does not unduly modify the tops of the bits (or if magnetic patterning is used instead of physical), assuming that the head air bearing dimensions are adjusted to maintain the same fly height as on the film disk.

It seems reasonable that patterning may best be done before disk finishing processes; i.e., before carbon deposition, organic lubing and the final burnish/glide to remove asperities.

7.9 Patterned Media Heads

Conventional hybrid recording heads appears to be theoretically suitable for vertical or horizontal patterned media recording, perhaps made smaller in scale. The writer-to-reader separation and offset tolerancing must also scale. These ring head two-pole writers have good field gradients for horizontal media. (The arguments and parameters in this section are from [3].) Short pole writers (trailing pole length P2L comparable to write gap length G_w) also have good field gradients for *vertical* media. Single-pole probe writers also appear satisfactory (G_w large compared to P2L). Writer pole magnetization needs appears modest: 10–16 kG for 100 Gb/in^2.

Short pole writers can avoid the geometric problem of overwriting adjacent tracks due to disk drive rotary actuator head skew. This skew can put a writing pole with large P2L over two or more tracks simultaneously. Typically, there is a 20–30° head-to-track skew variation across a 3.5-inch disk radius. If P2L is larger than the track pitch the trailing edge of the pole will not stay within 1/2 track of centerline over this skew range. Two-pole short-gap writers improve the situation because their high writing fields are concentrated near the gap edges, for the same magnetic reasons that allow P2L to exceed track pitch in horizontal film disks today.

Short pole writers made with Ni–Fe pole materials (e.g., by FIB) may act as *hard* magnetic materials due to shape anisotropy, and continue writing bits after the write current is turned off. Increasing the pole material anisotropy H_k to several thousand Oe theoretically removes this problem [3]. This anisotropy is oriented cross-track as it is today. Raising H_k will lower the writer ring head permeability $\mu = 1 + 4\pi M_s^p / H_k$ which will lower the writer efficiency and raise the required write current density. Over 10^9 amp/cm^2 current densities are required due to this factor in [3], where a thermal scaling argument is made that these densities roughly compare to 10^7 amps/cm^2 today in writer temperature rise. Lower permeability may be compensated for by small writer size and single turn write coil in the gap (Fig. 7.3). However, the question of electromigration failure must be addressed [64].

Scaled versions of current magnetoresistive readers also appear workable. It may be necessary to increase the MR ΔR/R, perhaps to 50% or higher to get sufficient signal. This is to offset the SNR penalty caused by the increased electronic noise due to GHz read channel bandwidths. These bandwidths are simply caused by high bit density, coupled with the high rpm required of fast disk drives (see Sects. 7.10.4 and 7.14 for smaller disk alternatives).

7.10 Patterned Media Recording Systems

Based on [3], recording theory will be discussed for:

- Horizontal and vertical patterned media recording modes and switching;
- Playback pulse calculations from recording physics;

- Channel partial response targets and equalizers suitable for these playback pulses;
- BER simulations of these recording systems.

7.10.1 Recording Modes

Patterned bits must have only two stable states for binary recording. Uniaxial anisotropy appears to be necessary to produce this situation. Without uniaxial anisotropy, circular vortex magnetic states can be stable in patterned dots such as permalloy, as small as 60-nm diameter [65]. (While binary data could be stored using the two circular directions of such vortices, suitable writer and reader heads are not known.)

Vertical, horizontal, and transverse recording are all theoretically possible with patterned media. All three modes would produce a playback pulse from each patterned bit. This differs from film disk recording where playback pulses are only produced by transitions between opposite successive NRZ channel bits.

Patterned media playback waveforms have DC content, since there is no polarity alternation physics constraint. Bits can be written all positive or negative (see Sect. 7.12 for experimental proof). A RLL (d,k) code k-constraint is not necessarily required for bit timing since $d = 0$ in all current codes and k is unnecessary for clocking since each bit produces a read pulse (but a code spectral null at the Nyquist all-1's frequency *will* require a k-constraint). However, DC-free or DC charge constrained modulation codes will likely be desirable to control low frequency and DC content, allowing continued use of present ac-coupled channel electronics.

Due to rotating disk drive mechanics considerations, the bit aspect ratio will have to remain above 1:1. Orienting magnetic anisotropy will be needed for both vertical and horizontal media. Otherwise, the stable bit states will lie in the crosstrack direction due to shape anisotropy ($b > a$, Table 7.1). Crosstrack magnetization is a transverse recording mode with unfavorable writer and reader field gradients compared to vertical or horizontal modes. This is because writer and reader gap lengths need to be comparable to the bit dimension in the recording direction. Since the bit aspect ratio BAR > 1, the head gap length will be comparable to the bit track width, which is larger than the bit length. This will allow large downtrack fringe field extent, potentially overwriting the bit just written and escaping downtrack.

Recording and playback processes for patterned media can be simulated using standard magnetostatic theory from recording physics. Single domain bits have uniform magnetic charge on simple geometric surfaces, e.g. rectangular bits. Analytic formulas can be derived for the magnetic fields from these bits. Fields from small head poles can also be approximated by uniform charges on the pole surfaces. Thereby, total three-dimensional vectorial writing fields can be calculated as a writing head pole moves over a bit to be written, including the write current rise and fall time. These vectorial recording fields start write bit switching with the field at an angle to the vertical or horizontal bit magnetization. This causes faster switching with lower write fields [3,66]. Using normal 2D recording physics instead of 3D can be misleading with single-domain switching and BAR near unity.

Stoner-Wohlfarth coherent switching formulas can be used to determine when the write bit switches and whether a neighboring bit becomes sufficiently disturbed to switch. Using worst-case patterns for the write bit (giving the highest write field), the minimum required writing pole magnetization M^p can be calculated. Bit writing margins can be found by finding the higher head pole M^p that switches an adjacent bit, using another worst-case bit pattern (smallest possible switching field). Using these tools, recording designs can be tested and optimized.

The resultant designs can be tested for magnetization dynamics with Landau-Lifshitz calculations to verify single domain switching, and to calculate the head and bit switching speeds. These calculations show three-dimensional bit switching. Bit magnetization moves through the crosstrack direction between the two stable states. Faster switching may result from this effect (apparently caused by the crosstrack shape anisotropy). These calculations are shown in [3] at 100 Gbits/in^2. The 1 Tbit/in^2 results shown below are from similar calculations.

7.10.2 Playback Process

The isolated playback pulse from a single patterned bit may be calculated from recording physics using the Potter mathematical model for a magnetoresistive head [25] and uniform magnetic charge on the bit surfaces orthogonal to their orienting directions [67]. Referring to Fig. 7.5, a vertical bit has plus charge on its top surface, spaced at the fly height d from the MR reader, producing a large sharp unipulse. An equal charge on its bottom surface produces an opposite polarity long pulse with low amplitude, since the bottom

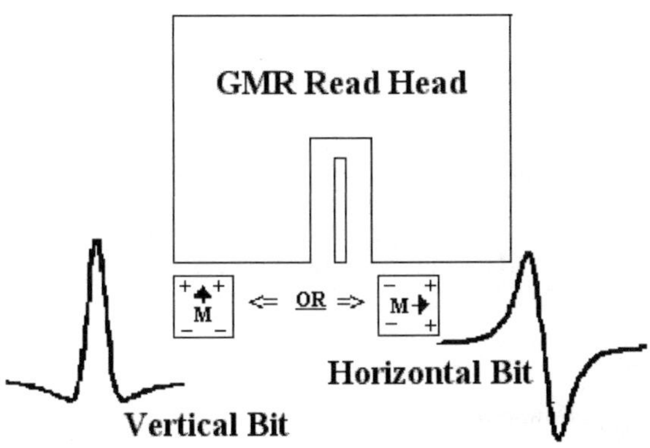

Fig. 7.5. Patterned bit recording modes and playback pulses: [3,67] (© 1999 IEEE)

surface is spaced nearly four times further (at d + 2c if no soft underlayer is present (Table 7.1)).

A horizontal patterned bit produces two opposite polarity pulses from its leading and trailing surface charges, spaced apart by the bit length 2a. The result is the dipulse shown in Fig. 7.5 which is wider than the vertical pulse because of the effective magnetic spacing d + c, i.e., nearly three times the magnetic spacing d of the vertical (top) charge.

Patterned media playback pulses can have the full amplitude of a film media pulse even though the media magnetization change is $\pm M_r$ instead of $\pm 2M_r$. There only needs to be sufficient playback flux to swing the MR reader equally close to saturation [67].

7.10.3 Recording Channels

Patterned media can use conventional partial response and Viterbi bit detection PRML channels, which are simple variants of today's disk drive channels. A vertical channel bit pulse is shown in Fig. 7.6, with a modified EPR target of similar shape and spectral response. Suitable targets for 100 Gbits/in^2 are shown in [67]. At 1 Tbit/in^2 the best performing channel has the simplest pulse – the unipulse from vertical patterned media with a soft underlayer (so the bit bottom charge is gone, along with the consequent undershoots).

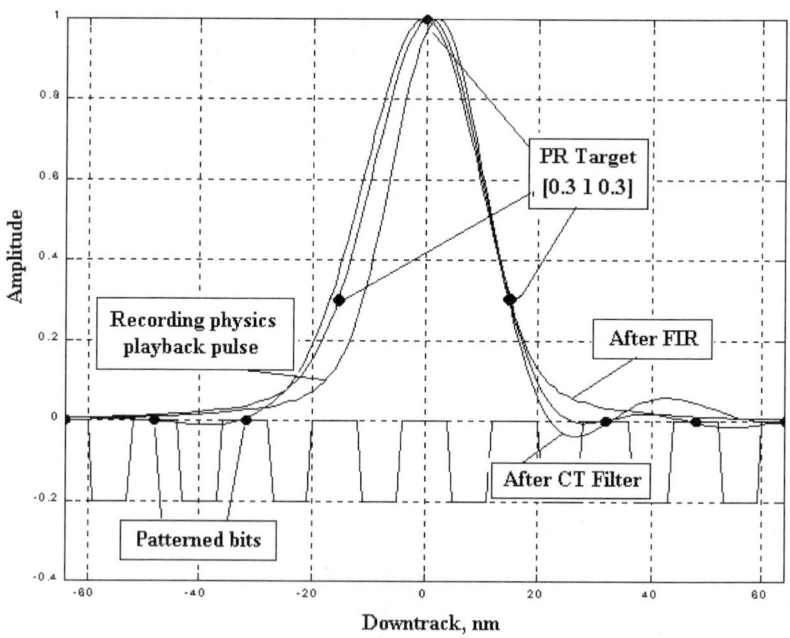

Fig. 7.6. Read pulse; after analog CTF filter; after FIR filter [67] (© 1999 IEEE)

Table 7.2. Recording candidates at 0.1 and 1 Tbit/in^2

Patterned Media:	**0.1 Tbit/in^2**	**1 Tbit/in^2**	
Bit orientation	Vertical	Vertical	
Writer	1 pole	2 poles	
Disk dia,Gbits/s, krpm	**3.5"**, 1, 11.5	**2.5"**, 2, 11.5	**1.0"**, 1.33, 10
Bit size LxWxH, nm	30x80x30	10x26x10	
Decay Exponent	500	68	
FCI (disk outer zone)	508,000	1,588,000 (16 nm bit pitch)	
TPI	200,000	635,000 (40 nm track pitch)	
Bit Aspect Ratio	2.6	2.5	
Fly Height	10 nm	3.5 nm	6.0 nm
Write Disturb Margin	350%	45%	-
MR Reader Gap	33 nm	26 nm	
MR ΔR/R	16%	75%	50%
SNR (Elec, Pk:rms)	18 dB	9.8 dB	11.3 dB
PR Channel Bit	[+1 +1]	[0.3 1 0.3]	
CTF	3rd Order Cheby-I, BP 0.8Fnyq		BP 0.5Fnyq
FIR	15-tap least-square		
Bit litho error σ*fci	7%	3%	3%
BER	10^{-9}	$5 \cdot 10^{-4}$	$6 \cdot 10^{-5}$

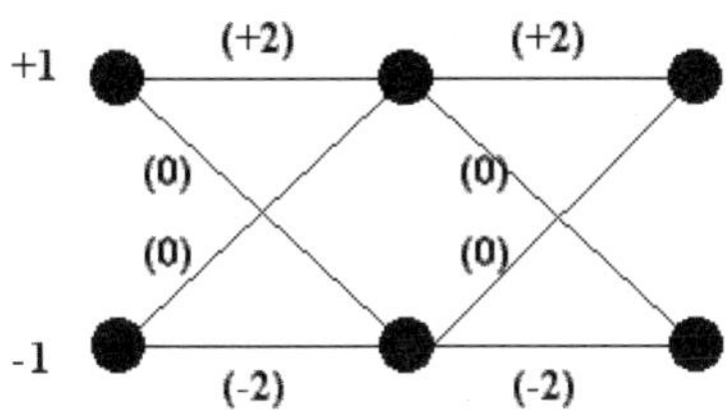

Fig. 7.7. PR4 trellis for patterned media. Channel ±1 state is bit magnetization

Figure 7.6 shows a vertical playback pulse calculated from 1 Tbit/in^2 parameters (Tables 7.1 and 7.2). Also shown is the playback pulse after passing through a Chebyshev-I continuous time filter ("CT Filter") with two-symmetric-zeros boost; and then after a 5-tap least-square FIR equalizer [67] with a modified EPR4 target of [0.3 1 0.3].

An example patterned media channel trellis for PR4 is shown in Fig. 7.7. The branch amplitudes represent the channel (single) bit impulse response. (In conventional film media channels the branch amplitudes represent the media magnetization step response.) This trellis resembles a rearranged conventional film media PR4 trellis with ternary amplitudes of $+2, 0, -2$, and an open eye pattern with no transitions from $+2$ to -2 or the reverse. A sequence of $+1$ patterned bit magnetization states produces a sequential positive $(+2)$ DC response. Zero amplitude sequences remain, but with different channel bit sequences (e.g., $-1, +1, -1, +1, \dots$). The minimum error event in the trellis is a single bit flip [67] with a minimum distance $d^2 = 2^2 + 2^2 = 8$.

7.10.4 BER Simulations

A Matlab channel simulation was used to find 1 Tbit/in^2 parameters for best BER. Figure 7.8 and Table 7.2 shows the results. With a one-inch disk diameter at 10 krpm a $6 \cdot 10^{-5}$ BER is predicted with 3% patterning lithography errors (i.e., the bit center positioning 3σ error is 9% of a bit cell B). This BER is limited by the read channel bandwidth of 325 MHz (50% of Nyquist) for 1.33 Gbits/sec transfer rate at the outer zone of the 1-inch disk. This gives an electronic SNR of only 11.3 dB (MR resistive and preamp noise). With the 2.5-inch, 11-krpm disk shown, the read channel bandwidth is 750 MHz for 2 Gbits/sec and the SNR and resultant BER significantly degrade.

Media transition noise, soft underlayer noise, and bit magnetostatic coupling noise are absent in the simulation so that the increased MR $\Delta R/R$ shown was able to combat the high electronic noise. The 100 Gbit/in^2 predictions are from [67].

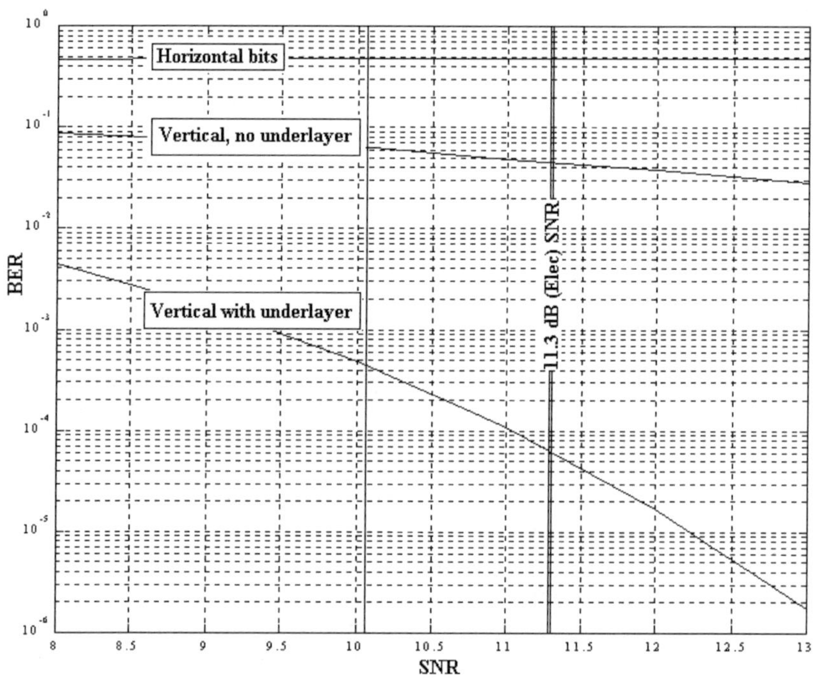

Fig. 7.8. Bit error rate vs. signal to noise for 1 Tbit/in^2 at outer zone ID of one-inch disk, horizontal and vertical bit orientation, and vertical with underlayer

7.11 Servo Systems

At ultra-high areal densities, today's sampled data sector servos may be inadequate to insure bit and track positioning during write and read. This may well be true for any rotating disk recording scheme at Tbit/in^2 densities.

Using the patterned media candidate design numbers in Table 7.2, bit and track servo recording systems consequences can be analyzed.

7.11.1 Patterned Media Sector Servo Systems

Conventional embedded sector-servo track and bit synchronization can be used with patterned media. In fact, this offers the manufacturing advantage of eliminating servowriters. A permanent magnet could uniformly magnetize the servo pattern in a single pass. Or servo schemes could be used which ignore bit polarity.

There will be repeatable runout "RRO" due to centering and clamping errors (as with optical disk drives). A feed-forward servo algorithm can eliminate such centering RRO. For a disk spinning at 10 krpm with a 30 µm centering error, the feed-forward servo would have to develop a zero-to-peak acceleration of $\ddot{r}_{max} = 30 \cdot 10^{-6}(2\pi 10000/60)^2/9.8 = 3.5\ g$. An RRO eccentricity error signal can be derived from the time intervals between servo sector timing marks, and RRO parameters stored in the drive reserved data area, on a per-head basis. They can be periodically updated by the residual eccentricity error signal.

Such servo techniques might be a step toward removable disk drives at high areal density (but tribology and cleanliness issues need to be addressed).

For rough servo sizing calculations, a 4 Kbyte sector data size means a sector length of 500 µm, or 25 µsec under Table 7.2 conditions (1,588,000 bpi at the outer zone radius of a 1-inch disk at 10,000 rpm). Mechanical disturbance frequencies should remain below 10 kHz, while sector times become shorter due to high bit density. The sector servo rate is 40 ksamples/sec so the servo bandwidth could be as high as the Nyquist rate of 20 kHz. (The microactuator response may be the bandwidth limiting factor.)

Bit tracking during writing is not as critical as during reading (where data clocking can be used). Write timing errors up to ±1/4 bit position and ±1/4 track position may still insure writing on the correct bit. (This can be analyzed by the micromagnetic methods in [3].) If writing is done by the present method of synchronizing at sector start, and open loop during sector data writing, the head and disk disturbance accelerations must be kept below roughly 1.3 g on average. (If G_{avg} is the average acceleration, 1/4 of a 16-nm bit pitch in a 25 µsec sector time gives $G_{avg} = \left[2 \cdot 4 \cdot 10^{-9}/\left(25^{-6}\right)^2\right]/9.8 = 1.3\ g$.) This includes non-repeatable spindle bearing runout (which must scale for any rotating disk storage system). Air bearing spindle NRR has been reported in the 2–5 range [68].

7.11.2 Servo During Writing

At ultra-high areal densities, conventional sampled data sector servos may be inadequate to insure bit and track positioning during write and read. This will be an issue with *any* electromechanical recording scheme at Tbit/in^2 densities. Bit position errors are currently corrected by clock signals derived while reading, but track position errors are not. Current bit aspect ratios of 10–15:1 allow this, but low bit aspect ratios may require bit *and* track servo error signals during *both* write and read.

Servoing by reading during bit writing brings a serious electrical interference problem that might be termed the "120 dB crosstalk problem." With readback signals of about 100 µv p-p [67], the roughly 5 volts p-p on the writer coils must cause no more than roughly 5 µv crosstalk to be coupled into the read servo signal, which is an attenuation requirement of $20\log(10^6)$ or 120 dB. Out of band techniques such as buried servo have been proposed to accomplish this [69], as well as hybrid recording head schemes [70,71].

A systems approach to attain this 120 dB total might be to break it up into several serially applied isolation steps:

- First, use the inherent shielding between hybrid write and read heads.
- Second, use write pulse echo cancellation.
- Third, pattern bit and track servo dither or pilot tones into the data bits.

Inherent shielding between hybrid write and read heads would provide 26 dB if the writer-reader coupling was $<5\%$. This shielding includes all crosstalk paths from the read/write preamp chip, the head suspension interconnect wires, and between the writer and reader heads. Chip-on-suspension appears to be necessary for Ghz data rates, and will likely increase the overall shielding by reducing the interconnect distance from today's 50 mm to at little as 4 mm [72].

Write pulse echo cancellation might employ test sectors with no data bits (only a few adjacent tracks on each disk surface, to avoid causing flight dynamics problems over low density patterned areas). The write head might be energized with a low frequency squarewave, producing isolated crosstalk echo pulses in the read channel, of alternating polarity. These echo pulses could be processed through the bit and track PES algorithms (below). The resultant cancellation amplitude would be stored in reserved data areas on the drive on a per-head basis, for each pulse polarity. During subsequent data writing, the write pulse polarities are known from the data being written, the writer-to-reader downtrack offset is known, and so the appropriately signed echo pulse amplitude can be subtracted in the bit and track servo channels.

Servo marks can be sub-patterned into the disk data bits [14,39,73]. Patterning bit and track servo pilot tones into the data bits could be done at frequencies well below the data band (analogous to optical disk tracking off data). The data signal can be attenuated by averaging (low pass filtering) the continuous servo signals.

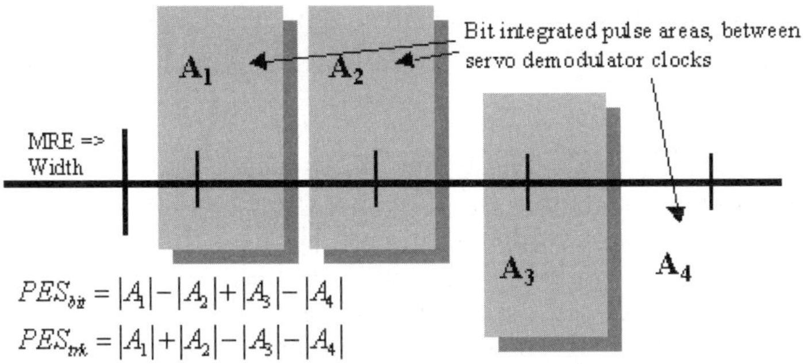

Fig. 7.9. Tracking on data: PES averages of hundreds of quad-bit servo patterns

The details of such data-tracking servo designs are complex and may include dealing with interactions between the two servo signals. Modifications in the read/write preamp and channel chips may be required. Only a simple conceptual scheme will be described here to illustrate the ideas (Fig. 7.9). Identically pattern each successive group of four data bits ("quadbit") in a sector. The first two bits are each offset by an amount Δx in the *downtrack* direction, closer to each other than the nominal bit pitch B. Both bits are offset *crosstrack* by the same amount Δy. The second pair of bits in each quadbit pattern are again offset closer to each other downtrack and both offset crosstrack, but in the opposite direction to the first pair. The relative offsets Δx/2a and Δy/2b would be comparable to the total offtrack budget, commonly ≈15% today (the offsets are exaggerated in the figure).

Bit and track servo position error signals PES_{bit} and PES_{trk} can be derived by integrating the playback pulses to get their *absolute-value* areas between bit clock times, similar to today's method of integrating sector servo playback pulses between servo clock times. The algebraic sum shown of the four absolute pulse areas is the bit clock PES_{bit}. If the writing clock drifts late from the nominal bit centers (vertical hash lines in Fig. 7.9) the read pulses will no longer be centered on the clock intervals. The first bit pulse area A_1 will increase and the second bit pulse area A_2 will decrease. Similarly for the third and fourth bit pair. Adding four pulses as shown cancels out the crosstrack servo error. An offtrack PES_{trk} is similarly derived with the bit servo error cancelled.

Averaging these quad PES signals using two phase lock oscillators with hundred-bit time constants should provide further attenuation of the write head crosstalk. Note that the bit clock derived from the data during writing must be synchronous for any such scheme to work, just as today's bit clock derived from the read data must be synchronous with no bit slips.

Using data-based servo schemes places requirements on the channel modulation code. The playback pulse amplitude area (absolute value) from each bit should have the partial response intersymbol interference from adjacent bits removed, for example by a bit decision directed timing channel. Or a soft-output Viterbi algorithm detector (SOVA) could be employed, to minimize servo loop delay time [74].

Additionally, at these areal densities the writer-reader downtrack and crosstrack offsets of each individual head must be measured much more accurately than today. (These head offset variations are caused by wafer fab mask alignment errors, and writer-to-reader shield layer thickness tolerances.) Today, these offsets can be measured for each head during drive servowriting, and stored in the drive reserved data area. But new methods may be required with patterned media. The downtrack offset might be measured by 1) prewriting a sector with negative bit polarities, 2) turning on positive polarity writing halfway down the sector on the next disk revolution, and 3) measuring the position of the first bit that switched, on the third revolution. If this was done on a set of successive tracks instead of a single one, the crosstrack offsets might be measured also. Fine track positioning might be required to find the precise crosstrack offsets where successive tracks were first written. Fine bit positioning may also be necessary.

The formidable precision requirements implied by Tbit/in^2 areal densities (for any mechanical recording technology) may be seen by noting that 15% of the bit pitch B = 16 nm is only 2.4 nm! Of course, the bit and track servos are averaging hundreds of bits, so the servo lithography errors can be larger than 2.4 nm.

An alternative servo scheme proposal uses the "thermal cooling" signal from the MR sensor to sense proximity to patterned bits (or patterned tracks, Sect. 7.14). Experimental data indicates that this can allow servo during writing as well as during reading [75].

7.12 Experimental Demonstrations

Many high areal density patterning demonstrations have been made, beginning in 1991 [17]. Features down to 50–100 nm have been achieved [13,18,37,76]. Demonstrations of binary-state stable bits, writable with disk drive heads have been reported [19]. Playback pulse signals from patterned bits have been demonstrated, using disk drive heads mounted in AFM mechanics, [41,77]. Read pulse shapes follow recording physics predictions [66].

Figure 7.10 shows a read pulse waveform taken in this manner with a commercial disk drive MR reader head [66]. The bits are vertically oriented 170 nm Ni columns, spaced about 250 nm apart in order to get isolated pulses with the lower density head used. Note that each bit produces a playback pulse, and that pulse polarity need not alternate (unlike conventional film media).

Digital recording and servoing remains to be demonstrated under disk drive conditions of flying heads on high rpm disks, with high frequency data channels showing stable uniform playback pulses and practical bit error rates.

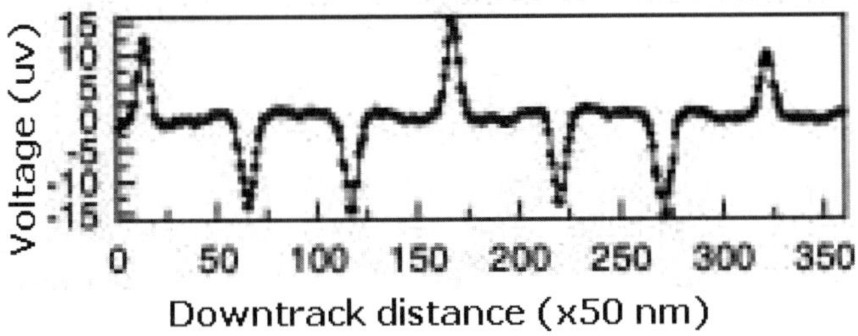

Fig. 7.10. Experimental playback waveform [66]

7.13 Technology, Systems, and Engineering Issues

A number of technology and systems issues must be addressed, to make patterned media disk recording systems practical. These issues include:

• Recording system design candidates (Sect. 7.10):
Should the disk bit orientation be horizontal, vertical, or vertical with soft underlayer?
Single pole or dual pole write heads? (Sects. 7.9, 7.10.1)
• No published experimental disk drive recording data exists:
Only low velocity write and read of sparsely patterned bits (Sect. 7.12)
• Fabricating experimental R&D and low-cost production heads and disks:
Head design candidates are discussed in Sect. 7.9
Tbit/in^2 electron beam lithography has been demonstrated [46]
Potential fast 8-inch semiconductor wafer processes exist (Sect. 7.5)
Synchronous disk drive channels favor lithographic patterning (Sect. 7.5).
• Recording channel design and silicon implementation:
Theory BER of $6{\cdot}10^{-5}$ possible using "rearranged" silicon (Sect. 7.10.4)
Lithography patterning errors are a new "noise" source, replacing film disk transition noise [78]
• Servo systems needed for bit and track following [3,23,66], Sect. 7.11: Disk centering runout must be tracked (Sect. 7.11.1)
Conventional sampled bit and track servos may work (Sect. 7.11.1)
Write synchronization: a new problem? (Sect. 7.11.2)
• Patterning lithography and hard bit errors are unknown:
Theoretical work has been reported [67,78]
• Head flight dynamics:
Fly design computer models are appearing [63]
• Patterned disk tribology:
May have similar difficulty as flying on patterned track plastic optical disks
• Hidden problems lurk in new technology:
Need to start R&D with sufficient lead-time margin

7.14 Product Introduction and Extensibility

Patterned media may first appear in commercial drives as disks with patterned tracks and servo marks (but not patterned bits) [79]. This may eliminate track edge noise [15,27] and thereby allow the bit aspect ratio to be lowered to increase areal density. Raising areal density with TPI more than with BPI will allow retaining the popular 3.5-inch disk size, retaining high RPM for low data latency, and reducing increases in data rate. The latter alleviates pressure on magnetic switching speeds and read channel silicon clocking speed.

Elimination of track edge effects by patterning tracks also has beneficial effects on write and read head width tolerances. For example, it eases the requirement that track edges of old overwritten data not interefere with reading new data.

Introducing patterned media by patterning tracks but not bits also has the considerable advantage of minimizing new technology risks. For example, conventional servos and PRML recording channels may be used. The MR reader head thermal spacing signal can be used for patterned track following during writing as well as reading [75].

Servo writers could be eliminated, which is an increasingly difficult and expensive process. Servo write time increases proportionally with TPI for a fixed disk size and rpm so this practical issue could itself limit AD increases by TPI.

Disk defect mapping in drive manufacturing final test may be eliminated by optical or magneto-optical defect mapping during disk fabrication. This is not currently possible with continuous film disks because their fine granularity makes actual bit positions indefinite in drives.

Smaller disks may be needed at very high AD even though this appears counterproductive to AD increases (GBytes per disk is the economically significant factor, not AD). However, such disk size reductions have occurred repeatedly since the IBM Ramac was introduced in 1957 with 24-inch disks. Note the 1-inch disk performance advantages in Table 7.2.

Removable media drives might again become possible at the high AD of fixed disk drives (issues such as head/disk contamination do need to be addressed).

Patterned media has excellent theoretical extensibility. Thermal decay limits are manageable at areal densities of Tbits/in^2. Switching speeds can achieve fractional-nsec Landau-Lifshitz limits in write heads and disk bits [3], and in read heads [35], permitting transfer rates of many Gbits/sec.

Acknowledgments. The author thanks Mladen Todorovic for suggestions on this chapter, as well as for valuable discussion on this technology along with Sheldon Schultz, Jack Wolf, Eric Boerner, and Bruce Terris.

References

1. S. Charap, P. Lu; Y. He, IEEE Trans. Mag. **33**, no. 1, 978–83 (1997)
2. Y. Zhang, H. N. Bertram, IEEE Trans. Mag. **35**, no. 5 (pt. 3), 4326–38 (1999)
3. G. Hughes, IEEE Trans. Mag. **36**, no 2, (2000)
4. E. Daniel, C. Denis Mee, and M. Clark, *Magnetic Recording: The First Hundred Years* (IEEE Press, New York, 1999)
5. D. M. Bowers, Modern Data, May 1973
6. C. Haginoya, et al., Appl. Phys, Lett. **75**, no. 20, 3159–61 (1999)
7. B. D. Terris et al., J. Vac. Sci. & Tech. B **15**, no. 4, 1584–7 (1997)
8. H. N. Bertram, *Theory of Magnetic Recording* (Cambridge University Press, 1994)
9. B. D. Terris, C. T. Rettner, and M. E. Best, 2001 Intermag Conf. Paper FA–03
10. R. Barr and D. Bazzarre, J. Electrochem. Soc. **121**, no. 3, 110C (1974)
11. H. Heitmann, B. Hill, J.-P. Krumme, and K. Witter, Philips Technical Review **37**, no. 8, 197–206 (1977)
12. W. Wernsdorfer et al., J. Mag. and Magnetic Materials **151**, pt.1–2, 38–44 (1995)
13. R. New, R. Pease, and R. White, J. Vac. Sci. & Tech. B **12**, no. 6, 3196–201 (1994)
14. J. Cuomo, T. DiStefano, R. Hollis, and R. Srinivasan, IBM Tech. Discl. Bull. **27**, pt. 1B, 505–7 (1984)
15. L. Shew, IBM Tech. Discl. Bull. **15**, no. 3, 752–3 (1972)
16. A. Hoagland, IBM Tech. Discl. Bull. **20**, no. 7, (1977)
17. J. F. Smyth, S. Schultz, D. R. Fredkin, D. P. Kern, S. A. Rishton, H. Schmid, M. Cali, and T. R. Koehler, J. Appl. Physics **69**, no. 8, pt. 2A, 5262–6 (1991)
18. R O'Barr and S. Schultz, J. Appl. Phys. **81**, no. 8, 5458–60 (1997)
19. M. Todorovic, S. Schultz, J. Wong, and A. Scherer, Appl. Phys. Lett. **74**, no. 17, 251618 (1999)
20. A. Moser, D. Weller, and M. F. Doerner, Appl Phys. Lett. **75**, no. 11, 1604–6 (1999)
21. Y. Zhang and H. N. Bertram, IEEE Trans. Mag. **34**, no. 5, pt. 2, 3786–93 (1998)
22. Y. Zhang and H. N. Bertram, IEEE Trans. Mag. **35**, no. 5, pt. 3, 4326–38 (1999)
23. R. White, R. Newt, and R. Pease, IEEE Trans. Mag. **33**, no. 1, 990–5 (1997)
24. D. Weller and A. Moser, IEEE Trans. Mag. **35**, no. 6, 4423–39 (1999)
25. R. I Potter, IEEE Trans. Mag. **10**, no. 3, 502–8 (1974)
26. A. Taratorin, D. Cheng, and E. Marinero, IEEE Trans. Mag. **36**, no. 1, pt. 1, 80–5 (2000)
27. J. M. Chue, D. Baral, and G. Dunbar, J. Appl. Phys. **85**, no. 8, 5306–8 (1999)
28. T. J. Silva and T. M. Crawford, IEEE Trans. Mag. **35**, no. 2, pt. 1, 671–6 (1999)
29. Y. Acremann et al., Science **290**, no. 5491, 492–495 (2000)
30. C. Alexander Jr., J. Rantschler, T. J. Silva, and P. Kabos, J. Appl. Phys. **87**, no. 9, pt. 1–3), 6633–5 (2000)
31. E. D. Boerner, H. N. Bertram, and G. F. Hughes, J. Appl. Phys. **85**, no. 8, 5318–20 (1999)

32. T. J. Silva, M. R. Pufall, P. Kabas, "Nonlinear magneto-optic measurement of flux propagation dynamics in thin permalloy films,", submitted to J. Appl. Phys.
33. H. C. Siegmann et al., J. Mag. and Magnetic Materials **151**, no. 1–2, L8–12 (1995)
34. C. H. Back et al., Science **285**, no. 5429, 864–7 (1999)
35. L. Varga, A. Tanaka, K. Nagasaka, and R. Kondo, J Appl. Phys. **85**, no. 8, 5852–4 (1999)
36. M. E. Schabes and H. N. Bertram, J. Appl, Phys. **64**, no. 3, 1347–57 (1988)
37. R. New, R. Pease, and R. White, J. Vac. Sci. & Tech. B **13**, no. 3, 1089–94 (1995)
38. S. Wirth, S. Von Molnar, M. Field, and D. D. Awschalom, J. Appl. Phys. **85**, no. 8, 5249–54 (1999)
39. S. Y. Chou, R. C. Shi, and L. Kong, J. Appl. Phys. **81**, no. 8, 4673 (1997)
40. W. Wernsdorfer et al., Phys. Rev. B **53**, no. 6, 3341–7 (1996)
41. S. Y. Yamamoto, R. O'Barr, S. Schultz, and A. Scherer, IEEE Trans. Mag. **33**, no. 5, 3016–18 (1997)
42. J. L. Duvail et al., J. Appl. Phys. **84**, no. 11, 6359–65 (1998)
43. S. Landis et al., Appl Phys. Lett. **75**, no. 16, 2473–5 (1999)
44. B. D. Terris et al., Appl. Phys. Lett. **75**, no. 3, 403–5 (1999)
45. R. New, R. Pease, and R. White, J. Mag. and Mag. Matls. **155**, no. 1–3, 140–5 (1996)
46. O. Dial, C. C. Cheng, and A. Scherer, J. Vac. Sci. & Tech. B **16**, no. 6, 3887–90 (1998)
47. W. K. Waskiewicz et al., SPIE-Int. Soc. Opt. Eng., 255–63 (1997)
48. H. C. Pfeiffer, Microelectronic Engineering **53**, no. 1–4, 61–6 (2000)
49. J. P. Spallas et al., J. Vac. Sci. & Tech. B, **14**, no. 3, 2005-7 (1996)
50. A. Fernandez, P. J. Bedrossian, S. L. Baker, S. P. Vernon, and D. R. Kania, IEEE Trans. Mag. **32**, no. 5, .4472–4 (1996)
51. T. A. Savas, M. Farhoud, M. Hwang, H. I. Smith, and C. A. Ross: J. Appl. Phys. **85**, no. 8, 6160–2 (1999)
52. J. Wong, A. Scherer, M. Todorovic, and S. Schultz, J. Appl. Phys. **85**, no. 8, 5489–91 (1999)
53. M. Farhoud et al., IEEE Trans. Mag. **34**, no. 4, 1087–9 (1998)
54. A. Sugawara, G. G. Hembree, and M. R. Scheinfein, J. Appl. Phys. **82**, no. 11, 5662–9 (1997)
55. S. Sun, C. B. Murray, D. Weller, L. Folks, and A. Moser, Science **287**, no. 5460, 1989–91 (2000)
56. H. R. Khan, O. Loebich, and G. Rauscher, Thin Solid Films **275**, no. 1–2, 207–9 (1996)
57. C. Chappert et al., Science **280**, no. 5371, American Assoc. Adv. Sci. 1919–22 (1998)
58. T. Devolder et al., J. Vac. Sci. & Tech. B **17**, no. 6, 3177–81 (1999)
59. H. Bernas et al., Nuclear Ins. & Meth. Phys. Res. B **148**, no. 1–4 (1999)
60. A. Yu. Toporov, R. M. Langford, and A. K. Petford-Long, Appl Phys. Lett. **77**, no. 19 (2000)
61. W. Hua, B. Liu, and G. Sheng, IEEE Trans. Mag. **35**, no. 5, pt.1, 2460–2, (1999)
62. M. J. Donovan, D. B. Bogy, IEEE Trans. Mag. **31**, no. 6, 2994–6 (1995)

63. N. Tagawa, International Joint Tribology Conference, Seattle, October 1–4, 2000
64. E. M Williams, *Design and Analysis of Magnetoresistive Recording Heads* (Wiley, New York, 2001)
65. L. Buda, IEEE 2001 Intermag Conference paper CE-03
66. M. Barbic, PhD dissertation, Univ. Calif. San Diego, 2000
67. G. F. Hughes, IEEE Trans. Mag. **35**, No. 5, 2310–12, (1999)
68. A. C. Swan, J. C. Harrison, F. E. Talke, IEEE Trans. Mag. **32**, no. 3, 1727–32 (1996)
69. M. K. Haynes, IEEE Trans. Mag. **17**, no. 6, 2730–4 (Nov. 1981)
70. D. H. Martin, IBM Technical Disclosure Bulletin, 3228–29 (Jan. 1977)
71. J. Schneider and G. Trippel, IBM Technical Disclosure Bulletin, 4426–27 (April 1977)
72. B. McInerney and M. Sheperek, Data Storage, 34–38 (December 2000)
73. X. Lin, J. G. Zhu, and W. Messner, J. Appl. Phys. **87**, no. 9, pt. 1–3, 5117–19 (2000)
74. J. Li, K. Narayanan, and C. N. Georghiades, 2001 Intermag Conf. paper CA–12
75. G. J. Smith, H. H. Ottesen, EEE Trans. Mag. **36**, no. 5, pt. 1, 2232–4 (2000)
76. K. J. Kirk, J. N. Chapman, S. McVitie, P. R. Aitchison, and C. D. W. Wilkinsonn, Appl Phys. Lett. **75**, no. 23, 3683–5 (1999)
77. S.Y. Yamamoto and S. Schultz, Appl. Phys. Lett. **69**, no. 21, 3263–5 (1996)
78. S. K. Nair and R. M. H. New, IEEE Trans. Mag. **34**, no. 4, pt. 1, 1916–18 (1998)
79. Y. Kaneda, IEEE Trans. Industrial Electronics **42**, no. 3, 257–62 (1995)

8 Perpendicular Recording Media

Randall Victora

Perpendicular recording has been frequently proposed as a technique for surpassing the limits placed on longitudinal recording by thermal fluctuations. Basically, superparamagnetism will be a problem when the product of the anisotropy K and the granular volume V is no longer sufficiently greater than the product of the Boltzman constant k_B and the temperature T. At this point thermal fluctuations will erase stored data, the granular volume can no longer be decreased, and the rapid density growth of the recording industry will end.

Perpendicular recording is thought to offer several advantages relative to longitudinal recording. Several of these advantages stem from the nature of the destabilizing field generated by neighboring bits. In longitudinal recording, a bit boundary forms a charged domain wall. This stored charge generates strong destabilizing fields that work to widen the transition and switch the magnetization of grains into unintended directions. The strength of these fields increases with media thickness. As density is increased, the number of charged domain walls and the requirement for sharp transitions increase, thus aggravating the problem. In perpendicular recording, a bit boundary places plus and negative magnetostatic charges of grains on opposite sides of the boundary in close alignment. This is magnetostatically favorable. In other words, magnetostatics favors transitions and prefers high density. Furthermore, thicker media exhibit ever more favorable magnetostatics. Thus, the granular volume can be increased without affecting the areal density.

These two effects account for most of the perceived benefit of single layer perpendicular media. However, it is generally argued that addition of a magnetically soft underlayer generates the largest benefits for perpenicular recording. This soft underlayer is placed between the recording layer and the substrate and serves to image the magnetic recording head. This effectively doubles the field generated , thus allowing the coercivity and anisotropy field to be doubled. This, of course, stabilizes the grain.

There are also several secondary advantages to perpendicular recording. Longitudinal recording for disk drives employs 2-d isotropic media meaning that the anisotropy axes of individual grains are not aligned. This produces noise that would be eliminated by the highly aligned perpendicular materials. This also implies that track edges are sharper in perpendicular recording. This is a consequence of the range of coercivities found in the disoriented longitudinal grains that may be written by the stray side fields generated by the recording head. In perpendicular recording, the coercivities will be more uniform owing to the better orientation and the track edge will be sharper.

The promise of perpendicular recording can only be met if the media meets some stringent requirements. An obvious example is coercivity. The advantage offered by a large write field is the ability to record on high coercivity materials that will be able to resist thermal fluctuations. Simulations suggest [1] that approximately 65% of the saturation magnetization of the head will be available as the write field for the track widths and flying heights necessary at future densities. For new head materials such as FeTaN that offer saturation magnetizations of 20 kGauss, this translates to 13 kOe write field. For coherently rotating magnetizations within a grain, the coercivity will be substantially smaller [2] at a slightly oblique applied field direction. For example, an applied field 15 degrees off the perpendicular axis will reduce the coercivity by 40%. Therefore, the maximum writable perpendicular coercivity becomes 22 kOe. Leaving a margin for opposing demagnetization fields suggests an ideal perpendicular coercivity of 15–20 kOe depending on the media saturation magnetization. If a proposed material cannot reach this, then the material cannot fully exploit perpendicular recording and should probably be discarded.

The next requirement is perfect remanence, i.e., squareness of one. Unlike longitudinal recording, the position of maximum destabilizing field is found at the middle of the perpendicular bit where the demagnetization field tends to reverse the magnetization. This corresponds to the point of zero applied field, but full demagnetization field on a hysteresis loop, i.e., the remanence. Note that hysteresis loops are usually measured on time scales of minutes, but recording media must resist thermal fluctuations for periods of years. Thus the constraint on a normal hysteresis loop is actually more severe than perfect remanence. It must, in fact, exhibit the saturation magnetization for substantial reversing fields. The distance to the point where the magnetization first decreases is called the nucleation field. The proper constraint is, therefore, a high nucleation field. This ensures that the magnetization in the center of the bit will not switch even over extended time scales.

Small grain size is necessary for ensuring adequate Signal-to-Noise (SNR) and thus good Bit Error Rate (BER). Although the ratio of bit area to grain area is continuously declining owing to improved signal processing, it is clear that grain size must decrease to accommodate ever higher areal densities. For example, if perpendicular recording is to reach 1 Terrabit/inch^2, then even with aggressive signal processing able to interpret 20 grains/bit, the grain diameter must be reduced to 6 nm where an allowance of 0.4 nm has been made for a nonmagnetic grain boundary. Proposed perpendicular recording materials should therefore have paths for reaching such tiny dimensions.

In recent years, a primary focus of longitudinal media fabrication has been the reduction of exchange coupling between grains. Exchange coupling causes jitter in the placement of the boundaries between bits which leads to increased noise and decreased BER. Excess exchange coupling can also adversely affect the performance of perpendicular recording media. However, the adverse effect may be weaker because, unlike longitudinal recording, magnetostatic effects counteract the exchange coupling. In particular, at the bit boundaries, magnetostatic effects tighten the transition. Therefore a small amount of exchange interaction, perhaps 80% of the demagnetization interaction, will not be able to broaden the transition or induce jitter. On the other hand, deep inside the bit, where demagnetization effects are most detrimental, the exchange interaction will help stabilize substandard grains and reduce the DC noise. Thus a small amount of exchange interaction, e.g., less than 10% of the bulk exchange interaction, may actually be helpful. Emphasis must be placed on the word small; most perpendicular media discussed in the literature suffers from excessive exchange and this will be the primary focus of researchers in the near future.

The final requirement is a small spacing between the soft underlayer and the recording layer. Longitudinal recording requires the distance between the recording head and the recording layer of the media to be small in order to reduce spacing loss. Perpendicular recording suffers this same constraint, but also benefits from a small distance between the head imaged in the soft underlayer and the recording layer. Reduction of this "effective spacing loss" requires the seed layer typically used to start the proper growth of the recording layer to be small, perhaps a few nm. The search for such seed layers is an active topic among media researchers.

Five requirements on perpendicular recording layers have been listed. There are other constraints dealing with obvious issues such as durability and corrosion resistance. The rest of this chapter will examine three basic material choices (compounds, alloys, and superlattices) and evaluate their properties against these requirements. The chapter will end with a discussion of soft underlayers.

8.1 Compounds

8.1.1 Barium Ferrite

Barium ferrite ($BaFe_{12}O_{19}$) forms [3] a hexagonal unit cell consisting of 64 atoms: It is a ferrimagnet in which three of the Fe^{3+} sites generate spin in a direction opposed by the other two sites. The crystal exhibits a resulting magnetization of 360 emu/cm^3 and uniaxial magnetocrytalline anisotropy of 3.4×10^6 ergs/cm^3, both evaluated at room temperature. The superexchange interaction generates a Curie temperature of 750 K.

Barium ferrite has several strengths and weaknesses for use as a perpendicular recording material. Probably, its primary attributes are durability and corrosion resistance. In particular, the material is fully oxidized. Thus, exposure to atmosphere is unlikely to harm it. This has produced proposals for its use in contact recording, i.e., head and media in intimate contact without an intervening protective C overcoat. Its major weakness stems from this strength: the processing used to fully oxidize and properly arrange the Fe and Ba atoms usually requires high temperatures with subsequent formation of large grains. It is also worth noting that the highest potential coercive force $H_c = 2\ K/M$ is 19 kOe, implying that nearly perfect grains must be formed in order to be a high density recording material. In practice, coercivities are rarely, if ever, found to exceed 10 kOe.

The potential for contact recording was demonstrated by Morisako et al. [4], who deliberately recorded at low velocities so that the head would rub the disk. No scratches were visible. This sample was deposited in a facing targets sputtering system and then annealed in air at 900 °C for 100 sec. Film thickness was 100 nm and coercivity was approximately 3 kOe. This paper, like several others finds that coercivity and magnetization decrease with film thickness below a cutoff value of, in this case, 100 nm. The recording performance was plagued by high (apparently media) noise and low $D_{50} =$ 180 kfrpi given the small spacing loss afforded by the contact recording.

Improvement is frequently sought by use of a seed layer. For example, a thin amorphous Ba–Fe–O underlayer enabled the same group [5] to successfully deposit at 525 °C without a subsequent anneal. K. Noma et al. [6] used a ZnO seed to obtain a mean crystallite size of 25 nm at a deposition temperature of 600 °C. The associated magnetization $4\pi M_s$ was 5 kG and the perpendicular anisotropy was 3.3×10^6 ergs/cm^3. The facing targets sputtering was performed at a substrate temperature of 600 °C using a partial Xe pressure of 0.1 Pa, a partial oxygen pressure of 0.02 Pa, and a partial argon pressure of 0.08 Pa. Thick AlN seed layers have also been found [7] to induce perpendicular anisotropy.

Z. Zhuang et al. [8] found that the combination of a Pt underlayer and a Barium-rich deposition was beneficial for the promotion of small grains and good perpendicular orientation. The films were deposited using rf diode sputtering in a pure Ar gas and then rapidly annealed at 800 °C for about 60 sec. Figure 8.1 shows that good orientation could be achieved if the oxygen partial pressure was minimized. The plateletlike grains had a size of 30 nm. Perpendicular coercivities greater than 5 kOe could be obtained for thicknesses exceeding 80 nm; however, thinner materials were limited to 3 kOe.

In an attempt to surmount these difficulties, several novel structures have also been deposited. Yoshi et al. [9] found that alternate deposition of Fe_3O_4 and the barium ferrite layer did not decrease the need for a high deposition temperature. A. Morisako et al. examined strontium ferrite both with [10] and without [11] Cr additives. In neither case, was the grain size reduced to acceptable levels.

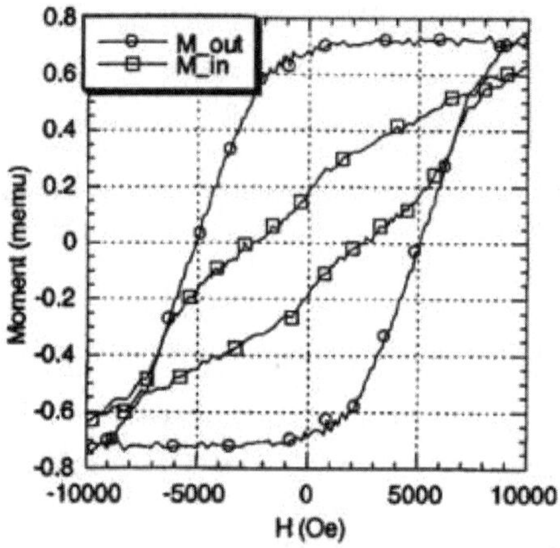

Fig. 8.1. In-plane and perpendicular MH loops of 900-Å-thick barium-rich barium ferrite thin films with a 150-Å-thick Pt interlayer. (Courtesy Laughlin, [8])

8.1.2 Cubic Compounds

Several potential compounds based on the inverse spinel crystal structure have also been proposed. These cubic compounds are ferrimagnets, in which the magnetic ordering is induced by the superexchange interaction. Like barium ferrite, their primary attributes are durability and corrosion resistance. They also have been proposed for contact recording and they also require high processing temperatures that promote the formation of large grains. Unlike, barium ferrite they do not naturally possess a uniaxial crystalline anisotropy. Thus, a seed layer that compresses a crystalline axis and thus induces a uniaxial anisotropy is a key feature of research in this area. Presumably, a lack of such layer has led to the reported difficulties in obtaining an adequate perpendicular coercivity.

Perhaps the strongest argument for this media type is provided in Fig. 8.2. Yamamoto et al. [12] subjected three types of media, a Co–Cr thin film with no overcoat, a commercialized Co–Ni hard disk, and a Co–γFe_2O_3/NiO film without protective layer to a recording test at high scratch load. It can be seen that the Fe_2O_3 based material showed the best resistance. In fact, this disk survived 18,000,000 passes under a more normal head loading of 900 g without loss of output. This media uses the NiO layer to provide a tensile stress at the interface with the γ-Fe_2O_3. The Co dopant serves to increase the anisotropy for a given compression. Anisotropies up to 3×10^6 ergs/cm^3 and coercivities up to 6 kOe were obtained. Recording tests were made at relatively low densities (up to 100 kfrpi) and showed good signal to noise ratio. However, referenced material [13] suggests that the media has a columnar structure with diameter 30–50 nm. In addition, the tested media was 90 nm thick. These values are incompatible with the high recording densities

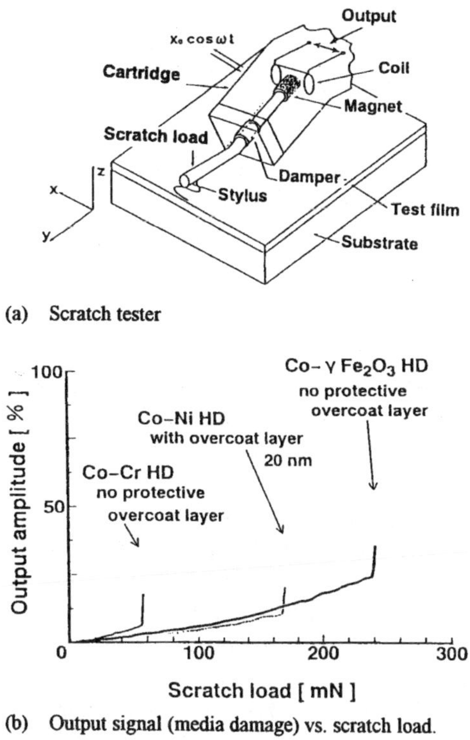

Fig. 8.2. Scratch test for Co–γFe_2O_3/NiO disk. (Courtesy Yamamoto, [12])

now being considered and it is unclear whether this material combination is compatible with more stringent requirements, e.g., average diameter of 6 nm and thickness of 25 nm.

A method for plating Co ferrite based compounds at low temperatures has been proposed. [14] Use of a 110 nm Zn ferrite seed layer allows the Co ferrite thickness to be reduced to 50 nm while retaining a coercivity of 3.8 kOe. However, further reduction of the Co-ferrite layer to 25 nm reduces the coercivity to 2 kOe, apparently owing to poor initial layer growth. The films can exhibit high surface roughness; doping with Ni reduces this problem. [15] This Ni doping also reduced the grain size to 40–50 nm. Kitamoto et al. has also proposed [16] the use of a CoCr/NiZn ferrite doublelayered film for perpendicular tape.

N. Matsushita et al. [17] find that Co–Zn ferrite can be deposited at 250 °C using facing target sputtering. A (111) orientation was detected; this places the (100) easy axis at an angle of 35 degrees to the film normal. Both the in-plane and perpendicular coercivities had a modest value of 1.5 kOe. Their best recording performance was reported in a later paper [18], using a 50 nm thick film.

8.2 Alloys

8.2.1 CoCr Based Alloys

In 2000, H. Takano and collaborators [19] reported a laboratory demonstration of 52.5 Gbit/inch^2 using CoCrPt atop a soft underlayer (see Fig. 8.3). This demonstration illustrated the high signal-to-noise ratio afforded by Co based alloys. In particular, such materials can be produced with small grain size. Weak intergranular exchange coupling can be induced by annealing Cr to the grain boundaries. Both effects are essential for low noise. The demonstration also illustrated the weakness of these materials: low coercivities that make the media susceptible to thermal decay. Although these materials could, in principle, attain the same coercivities as the longitudinally oriented Co-based alloys, it is thought that the lack of a seed layer along the c-axis allows excess stacking faults to appear and thus reduces the observed anisotropy and coercivity.

Details [20] of this low noise media are instructive. The material was deposited on a 5 nm seed layer. The recording layer thickness was 20 nm and the magnetic cluster size was 13 nm. The hysteresis loop revealed a coercivity of 2.6 kOe and a squareness of 0.98. The saturation magnetization of 250 emu/cm^3 implies a maximum demagnetization field of 3.1 kGauss. The relatively large value of the demagnetization field compared to the coercivity has prompted speculation that the material may exhibit a weak intergranular exchange coupling. A BER of 10^{-6} was achieved at a linear density of 590 kbpi and a track density of 89 ktpi.

Other interesting results reported by the Hitachi group concern the role of dopants and sputtering deposition pressures. T. P. Nolan et al. [21] show that grain nucleation is best accomplished at low argon pressure, even though the majority of the deposition is undertaken at high pressure, at least for the specific case of CoCrPt atop TiCr seeds. In particular, they advocate depositing the template layer (top of the seed) and the nucleation layer (bottom of the recording layer) at low pressure in order to preserve good orientation. It has been demonstrated by other workers [22] that well oriented CoCrPt

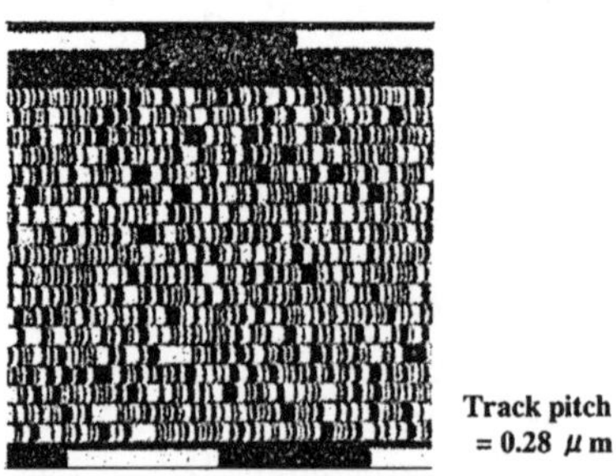

Fig. 8.3. MFM image of the recorded pattern. (Courtesy Takano, [19])

with squareness up to 0.96 can be obtained if grown on a 50 nm Ti seed atop hydrofluoric etched Si(111) substrates. If 12 nm of Ag is placed between the Ti and the Si, then a coercivity of 4.6 kOe is obtained, although the squareness is decreased. These results imply that high squareness is possible and motivate the search for alternate seeds and processes. Y. Hirayama et al. [23] present data illustrating superior SNR for CoCrTa compared to CoCrPt or CoCrPtTa. However, they are unable to deposit CoCrTa with unity squareness. It is unclear, however, if these results are intrinsic to the materials or could be remedied with alternate seeds or processing.

Other groups have also reported attempts to increase the squareness of CoCrTa. T. Shimatsu [24] et al. found that a 5 nm Ti seed layer yielded a squareness near 0.6. X. Sing [25] et al. tried various combinations of a dual seed layer: Co_3O_4/Pt. The maximum squareness obtained was less than 0.4. C. Yang et al. [26] used a 5 nm C seed to obtain a squareness of 0.65. In all cases, the reported squareness is associated with inadequate coercivity. K. Ishikawa et al. [27] reported the deposition of thick (150–200 nm) CoCrTa on tape substrates. The measured coercivity was less than 1500 Oe in all cases. The addition of 3.5% Mo to CoCrTa was shown [28] to decrease grain size and increase SNR. However, the SNR appears to be low (12 dB at $\lambda = 0.34$ μm) and details of the recording measurement that might explain the cause of this performance are not provided. In any case, coercivity remains small.

N. Honda and collaborators [29] fabricated a series of CoCrNb alloys atop a 70 nm Ti seed layer. They found that the dc erased noise was proportional to $M_s^{3/2}\delta(1\text{-SQ})^{1/2}$ where M_s is the saturation magnetization of the media, δ is the recording layer thickness, and SQ is the squareness (ratio of remanent magnetization to saturation magnetization) taken from the hysteresis loop (see Fig. 8.4). The dependence on squareness is particularly instructive: if the squareness drops only slightly to 0.9, the noise voltage will increase to more than 30% of the erased (remanence equals zero) value. One of their better materials was a 50 nm thick $Co_{74}Cr_{22}Nb_4$ alloy film. It had SQ = 0.74, M_s =

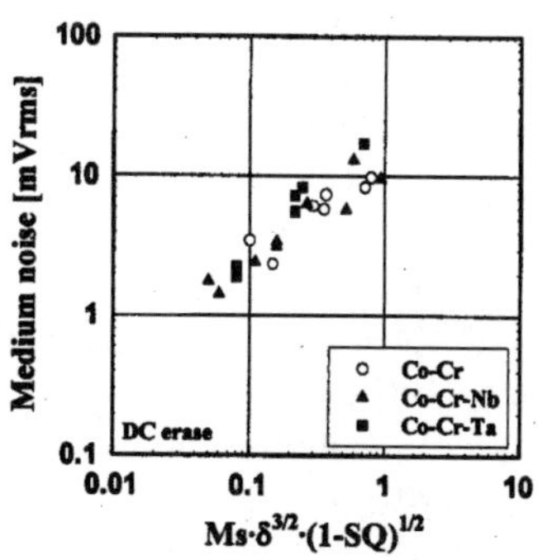

Fig. 8.4. Dependence of DC erased noise on $M_s^{3/2}\delta(1\text{-SQ})^{1/2}$ for various Co–Cr alloy films. (Courtesy Honda, [29]) © 1998 IEEE

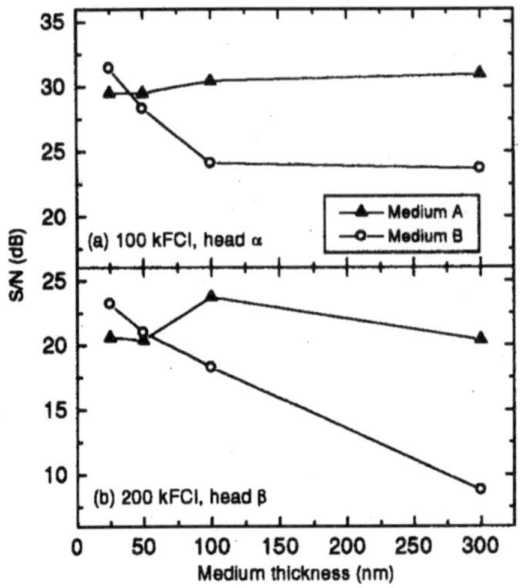

Fig. 8.5. S/N dependence on the medium thickness for two different heads α (**a**) and β (**b**). Medium A is $Co_{72}Cr_{28}$ and medium B is $(Co_{72}Cr_{28})_{78}Pt_{22}$. (Courtesy Ikeda, [32]) © 1998 IEEE

280 emu/cm^3, and perpendicular coercivity of 3175 Oe. Recording results were extrapolated from a track pitch of 2.7 µm to a track pitch of 0.85 µm and suggested that the media was compatible with 5 Gbit/in^2 recording. These results were obtained without a soft underlayer; previous work [30] suggested that a soft underlayer could be used to solve overwrite problems. In very recent work, the authors argue that [31] the addition of Pt enhances their materials sufficiently to make them compatible with 50 Gbit/inch^2 recording.

Undoped CoCr is the most frequently studied alloy for perpendicular recording Unfortunately, most workers have been unable to obtain coercivities beyond 3 kOe. Ikeda and collaborators argue [32] that a solution to this problem is to use very thick CoCr media. Since thermal stability derives from a product of the anisotropy and the volume, it may be possible to have thermally stable grains that have a small areal cross-section and a small coercivity, all compensated by a large thickness. Clearly this argument depends on the thermal activation volume correlating to the grain thickness, which needs to be shown. It is found that CoCr, unlike CoCrPt, reaches a maximum SNR at a medium thickness of 100 nm when recorded at 200 kfci (see Fig. 8.5). It is further argued that additional optimization is achieved by breaking the single thick layer into two parts, with the upper part having less Cr content and more anisotropy. For example Y. Sonobe finds advantages for top layers of composition [33] $(Co_{72}Cr_{28})_{78}Pt_{22}$ and [34] $Co_{78}Cr_{22}$. It is worth noting that media of this thickness greatly reduce the positive effects of a soft underlayer which, however, would not be needed unless the coercivities can be greatly increased.

8.2.2 Ordered Alloys

The Co alloys described in the previous section rely on structural, rather than chemical, order to produce their uniaxial anisotropy. Thus the purpose of elevated substrate temperatures or annealing steps is to move some of the Cr atoms to the grain boundary. This segregation need not be complete and, consequently, the temperatures involved need not be large. In particular, small average grain diameter can be retained.

In contrast, most chemically ordered materials require high processing temperatures. This typically encourages the formation of large grains. However, chemical ordering often has the important advantage of producing large uniaxial anisotropies that may translate into high coercivities. For example, Leonhardt et al. [35] reports the growth of $CrPt_3$ in the $L1_2$ structure. Deposition at 650 °C followed by an 850 °C anneal was used to produce films with coercivity 7.9 kOe and magnetization of 200 emu/cm^3. Squareness was 0.97. However, the average grain size was 35 nm, even in a film produced at lower temperatures.

Suzuki and collaborators [36] deposited ordered FePt in the face-centered tetragonal structure (fct) at temperatures of 300–375 °C. This structure is a natural superlattice alternating (001) monolayers of Fe and Pt. (001) orientation was obtained using a 1 nm MgO seed atop an (100) oriented FeSi soft underlayer. The combination of 5 kOe coercivity and squareness of 1.0 was obtained, although the loop shear was sharper than generated by the demagnetization field. This suggests that some intergranular exchange persisted despite the use of Ar pressures elevated to promote grain segregation. Recording measurements were made using a single pole write head (main pole CoZrNb) and MR read head with track width 1.2 µm and shield to shield spacing 0.2 µm. Figure 8.6 reveals good SNR for a 12.5 nm thick film at the relatively low densities recorded. Excellent thermal stability for linear den-

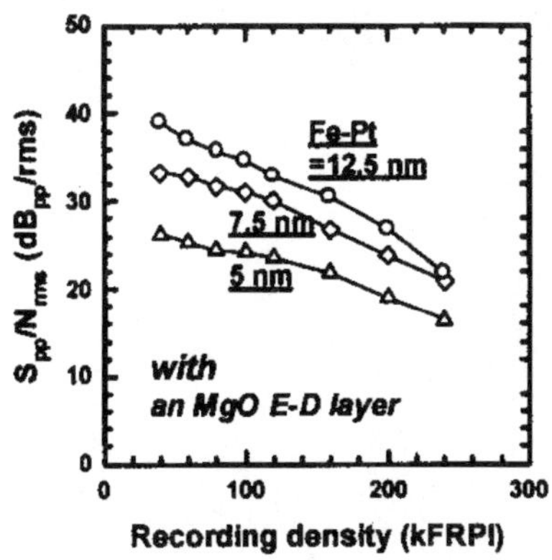

Fig. 8.6. SNR for composite media with various Fe–Pt thicknesses. (Courtesy Honda, [36]) © IEEE 2000

sity 5 kFRPI was also measured over 4 decades in time. These results suggest that fct FePt warrants additional research and development.

8.3 Superlattices

Superlattice media offer performance very different than that provided by alloy media. Superlattices are formed by repeating a basic structure, e.g., a bilayer of Co and Pd (see Fig. 8.7). This creates a secondary, or super, lattice. Many superlattices generate large perpendicular anisotropies from their numerous interfaces. For example, Co/Pt and Co/Pd superlattices employing monolayers of Co have been shown experimentally [37,38] and theoretically [39,40] to exhibit uniaxial anisotropies of $6.5–8.5 \times 10^7$ ergs/(cm^3 of Co) when forming lattices of high structural and chemical uniformity. The corresponding magnetizations range from 2000–2500 emu/(cm^3 of Co) [41]. The resulting anisotropy fields of 60–80 kOe offer the potential of enormous coercivities. In reality, the same experimental conditions that provide high crystal quality, e.g., Molecular Beam Epitaxy, also generate excessively large grains that are exchange coupled. This reduces the coercivity. Nonetheless, several groups have obtained coercivities beyond 10 kOe. Furthermore, it has been found both experimentally [42] and theoretically [43] that the anisotropy can be approximately tripled by employing a repeating quadrilayer of Co/Pt/Pd/Pt where the Pt layers are approximately one monolayer thick. Essentially the extra interfaces introduced between the spin polarized Pt and Pd make additional large contributions to the anisotropy owing to the large spin orbit energies of the Pt and Pd. In short, it is not expected that the obtaining of adequate coercivity will be a problem in these materials.

In contrast, the hysteresis loops typically measured in these materials exhibit a very sharp drop of magnetization versus applied field as the loop crosses through the coercivity. This is indicative of strong intergranular exchange coupling [44]. It is also associated with excessive transition noise that

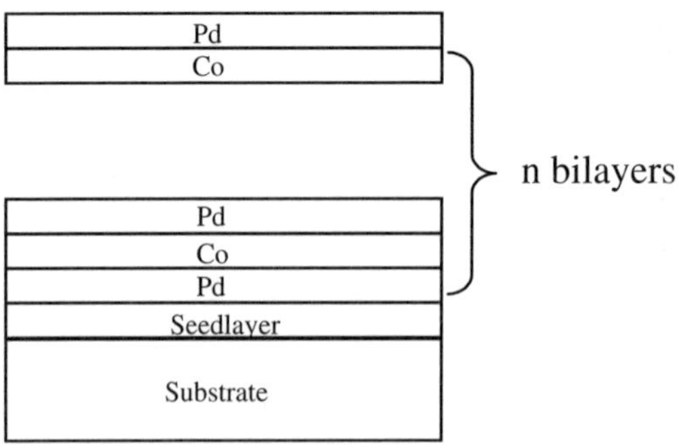

Fig. 8.7. Schematic diagram of a Co/Pd superlattice

is, of course, undesirable. To date, it is unclear whether this exchange coupling is a consequence of the history of Co/Pt and Co/Pd (they were studied extensively for use as magneto-optical media where this loop shape is considered to be desirable) or whether it is intrinsic. Possible solutions include the addition of dopants such as boron that migrate to the grain boundaries or the deposition of physically separated grains

Superlattices also offer some useful secondary advantages. Proper choice of the nonmagnetic Pt or Pd layer makes the magnetization easily adjustable to meet system needs. The high Pt and Pd content makes the materials highly corrosion resistant. This is an important consideration in view of the very thin protective carbon layers necessary for future recording densities. They can be deposited at relatively poor vacuum pressure [45]. A negative feature is the requirement of many layers. Typically, this is accomplished by rotating a substrate beneath two distinct sources, e.g., Co and Pt. This technique would require modification of most current media manufacturing equipment.

In 1998, Lairson and coworkers [46] reported perpendicular superlattice media that outperformed a leading longitudinal medium at densities beyond 260 kfci. In particular, the superlattice media offered a superior BER to the longitudinal medium that was a prototype for a 2.5 Gbit/inch^2 product with a coercivity of 2.2 kOe. The key innovation was the use of a $Co_{0.73}B_{0.27}/Pt$ multilayer in which elevated temperature was used to move the B to the grain boundaries. It was argued that this broke the exchange coupling and reduced the transition noise to the extent that an additional 3 dB of integrated SNR relative to the longitudinal disk was achieved at high densities. Low Non-Linear Transition Shift and good partial erasure were also demonstrated. The media was formed by sputter depositing 20 (0.5 nm Pt/1.0 nm CoB) bilayers on a 27 nm seed atop NiFe substrates. It is unlikely, however, that small grains of this material had good resistance to thermal fluctuations based on the low coercivity of 2–3 kOe and less than perfect squareness.

In a series of papers [47–49] beginning in 1999, Peng et al. addressed the problem of the excessively thick seed layer used by Lairson. Based on earlier results for superlattices designed for magneto-optic recording, it was shown that a 2 nm indium tin oxide (ITO) layer provided an excellent seed for superlattice growth. In particular, high quality hysteresis loops with high squareness and perfect demagnetization limited shear (see Fig. 8.8) were produced. Initially, high pressure Ar sputtering of the Co/Pd was used to separate grains and reduce intergranular exchange coupling. A particular sample consisting of 13 (0.028 nm Co/1.0 nm Pd) was extensively analysed [50] using micromagnetics including a novel treatment of thermal fluctuations. The sample had mean grain diameter 12.1 nm and standard deviation 3.6 nm. This was represented by 625 grains taken to be randomly generated Voronoi cells with the correct physical dimensions (Voronoi cells are illustrated in Fig. 8.9). $M_s = 350$ emu/cm^3 and grain spacing of 2.9 nm were obtained from electronic structure calculation and the experimentally measured magnetization.

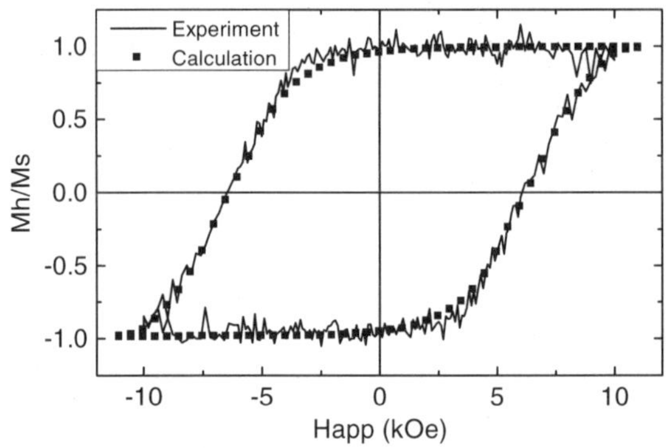

Fig. 8.8. Predicted hysteresis loop vs. experimental (VSM) curve for a Co/Pd superlattice thin film. (Courtesy Xue, [50])

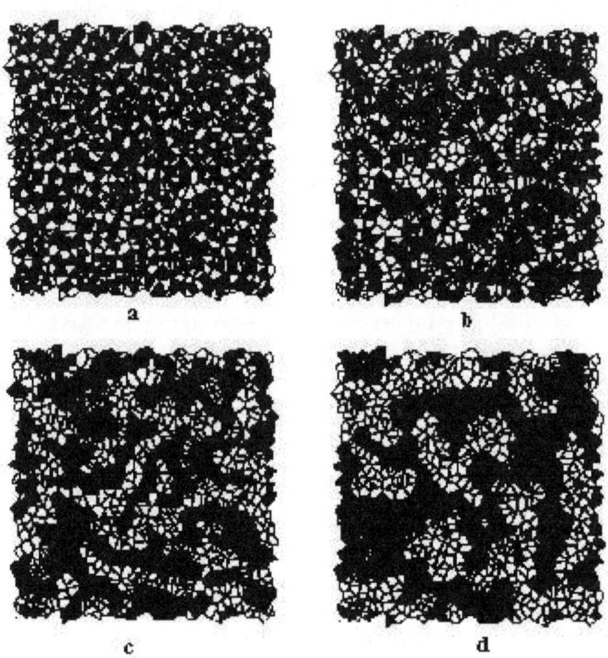

Fig. 8.9. AC demagnetization calculation of a Co/Pd film with varying intergranular exchange coupling constant (**a**) $J_{ex} = 0$, (**b**) $J_{ex} = 0.25$, (**c**) $J_{ex} = 0.5$, (**d**) $J_{ex} = 0.75$

The hard direction hysteresis loop was used to calibrate the anisotropy field and anisotropy dispersion angles. It was found that the uniaxial crystalline anisotropy $K_u = 2.0 \times 10^6$ ergs/cm^3 and that the anisotropy axes did not deviate more than 4° from the film perpendicular. Exchange coupling was taken to be zero based on the large intergranular spacing and studies of the domain size after AC demagnetization. (An example of domain size versus exchange is shown in Fig. 8.9. It can be seen that domains both enlarge and change shape with increasing exchange interaction. A detailed correlation of exchange versus correlation length can be found in [49].) Predictions for the hysteresis lop at a sweep rate of 50 Oe/sec are shown in Fig. 8.8. Note that the hysteresis loops and, in particular, the coercivities are in excellent agreement with a deviation of less than 1%. It was found that individual grains reversed by coherent rotation.

Later experimental work focused on CoB/Pd superlattices deposited at elevated substrate temperature. Recording measurements of a 0.31 nm $Co_{0.80}B_{0.20}$/1.0 nm Pd sample found a large signal-to-noise ratio at low densities for a read head width of 0.675 μm. Noise increased by 5 dB at a linear density of 400 kfci. However, the signal decreased very rapidly with density for this relatively high flying head. (Head to media spacing was a nominal 30 nm with additional spacing loss caused by the 10 nm C layer on the media and the protective layer on the head.) This is apparently caused by the performance of the MR head being adversely affected by the proximity of the soft underlayer [49]. In spite of this difficulty, a BER of 10^{-6} was found at a recording density of 400 kfci using a differentiated signal fed into an ordinary longitudinal channel. Other recording measurements show low write current for this 6 kOe media, suggesting that the soft underlayer is successfully aiding the write process. As expected, the non-linear transition shift (7% at 400 kfci) and overwrite (more than 50 dB) are excellent.

Wu and coworkers have studied the influence of hysteresis loop shape on noise properties. In one influential paper [51] they showed that use of high sputtering pressure increases the shear of the hysteresis loop and reduces medium noise. A later paper emphasized [52] the importance of large negative nucleation fields in reducing bulk dc noise. Takano et al. suggest [53] that elevated deposition temperature is helpful in reducing intergranular exchange. Segregation of the Cr in CoCr/Pt is also aided by temperatures above 150 °C. Chen studied [54] Co/Pd, CoCr/Pd, and CoCrTa media and found that $Co_{88}Cr_{12}$/Pd exhibited better SNR than pure Co/Pd.

8.4 Soft Underlayers

Reported results for soft underlayer development are far less numerous than those for the recording layer. This is somewhat surprising given that it is common for integrated soft underlayer noise to far exceed that produced by the recording layer owing at least partly to its large relative magnetization.

One major cause of noise from the soft underlayer is the presence of domain walls. It is difficult to produce a soft material covering the disk surface that does not exhibit domain walls. At best, the domain wall will be pinned at a few spots and remain stationary as the read head approaches. The detected noise will then be a single spike that might be removed by signal processing. Alternatively, multiple domain walls might be formed at, for example, midpoints of bits. It is possible for noise to dominate signal in this case.

An alternate noise source is funnelling of stray fields into the media and head by the soft underlayer. For example, the lowest reluctance path of an applied field perpendicular to the media will, for a sufficiently high permeability underlayer, be through the soft underlayer (parallel to the plane of the disk) and then through the media and into the head. Thus, a small stray field can be magnified by orders of magnitude owing to this collection of flux. In extreme cases, this can rewrite the media. Frequently, it will disrupt operation of the sense (GMR) head by shifting it from its bias point and thus distorting the signal.

An obvious solution is to reduce the permeability of the soft underlayer. Both micromagnetic simulation [1] and finite element simulation [55] show that perpendicular and in-track permeabilities as low as 50 do not significantly weaken the head fields or gradients. However, this corresponds to an anisotropy field exceeding 200 Oe that must be oriented crosstrack so as not to introduce coercivity in the perpendicular or in-track direction. This is a significant materials challenge.

In 1994, Ando and coworkers proposed a solution to this problem that employed a hard layer to bias the soft underlayer. In a series of papers [56–58] they showed that a CoSm layer beneath a soft CoZrNb soft underlayer successfully reduced the permeability and the number of domain walls in the soft underlayer. This stabilized the signal against stray fields and reduced underlayer noise. The enhanced anisotropy also increased the high frequency suceptibility. This is an important consideration in view of the very high data rates considered for the future.

Takahashi et al. exploited [59] this idea by pinning a soft NiFe layer with an antiferromagnetic FeMn layer that introduces an exchange bias. However, NiFe has a relative low magnetization and FeMn has a low blocking temperature of 150 °C and poor corrosion resistance. Jung and Doyle further exploited [60] this idea by constructing multilayers of FeTaN/IrMn. The multilayering insures that all of the FeTaN is near a biasing layer. This reduces the permeability to 70 over a frequency range of 0–2 GHz (see Fig. 8.10). Despite this low permeability, the hysteresis loops in the circumferential direction appear to offer little coercivity (see Fig. 8.11).

There are, of course, other considerations for soft underlayers. Gong and coworkers have shown [61] that Ti (0001) can be deposited atop soft fcc Co, Co alloy and NiFe/Co underlayers. This is important because Ti(0001) promotes the growth of well-oriented Co alloy perpendicular media. Kita-

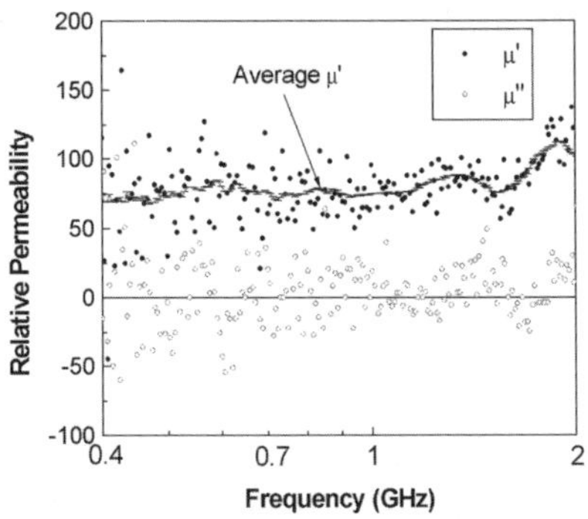

Fig. 8.10. Relative permeability of radially oriented FeTaN20/[IrMn10/FeTaN20]$_9$ multilayer films. (Courtesy Doyle, [60])

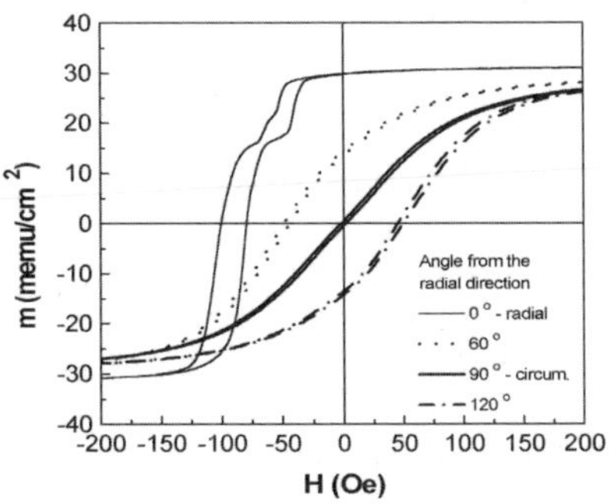

Fig. 8.11. Hysteresis loops of radially oriented FeTaN20/[IrMn10/FeTaN20]$_9$ multilayer films. (Courtesy Doyle, [60])

moto et al. have plated [62] Ni–Zn ferrite underlayers and then deposited a CoCrTa recording layer directly on top. They find correlations between the properties of the ferrite layer and the recording layer. Ichihara find [63] that NiFe/$Co_{67}Cr_{33}$ multilayers can be deposited with permeabilities of approximately 5000. Iwatsubo et al. sputter deposited [64] Fe films with saturation magnetization 21 kGauss and coercivities less than 5 Oe. Hokkyo and coworkers found [65] that a Cr seed produced FeSiAl soft underlayers with good surface smoothness. This measure reduces medium noise. Finally, it is worth noting that the demonstration of 52.5 Gbits/inch^2 was achieved [19] using an FeTaC soft underlayer. However, spike noise from the soft underlayer was observed.

Acknowledgements. The author wishes to thank J. Judy, B. Lairson, and H. Takano for useful private communications concerning the content reported here. He also wishes to thank J. Xue for scanning and inserting the figures.

This work was supported by NSF/DMR-9971573 and the National Storage Industry Consortium EHDR program.

References

1. M. Khan and R. H. Victora, IEEE Trans. Mag. **37** (to appear)
2. S. Chikazumi and S.H. Charap, *Physics of Magnetism* (Krieger, Florida, 1964), p. 284
3. H. Kojima, Fundamental Properties of Hexagonal Ferrites with Magneto-plumbite Structure, in: *Ferromagnetic Materials*, vol. 3 E. P. Wohlfarth (ed.) (North Holland, Amsterdam, 1982), pp. 305–392
4. A. Morisako, M. Matsumoto, and M. Naoe, IEEE Trans. Magn. **34**, 1630 (1998)
5. A. Morisako, X. Liu, M. Matsumoto, and M. Naoe, J. Appl. Phys. **81**, 4374 (1997)
6. K. Noma, N. Matsushita, S. Nakagawa, and M. Naoe, J. Appl. Phys. **81**, 4377 (1997)
7. N. Hiratsuka, E. Niizuma, and K. Kakizaki, J. Magn. Soc. Jpn. **21**, S2, 77 (1997)
8. Z. Zhuang, M. Rao, R. M. White, D. E. Laughlin, and M. H. Kryder, J. Appl. Phys. **87**, 6370 (2000)
9. Y. Hoshi, Y. Kubota, H. Onodera, and H. Shinozaki, J. Appl. Phys. **81**, 4690 (1997)
10. A. Morisako, M. Saitoh, and M. Matsumoto, J. Appl. Phys. **87**, 6872 (2000)
11. A. Morisako, M. Saitoh, and M. Matsumoto, J. Appl. Phys. **85**, 4732 (1999)
12. S. Yamamoto, T. Andou, N. Kanamaru, T. Nakamura, H. Kurisu, M. Matsuura, T. Doi, and K. Tamari, J. Magn. Soc. Japan **21**, Supplement S2, 51 (1997)
13. T. Doi and K. Tamari, J. Appl. Phys. **79**, 4887 (1996)
14. Y. Kitamoto, F. Zhang, M. Abe, and M. Naoe, J. Appl. Phys. **87**, 6878 (2000)
15. F. Zhang, Y. Kitamoto, M. Abe, and M. Naoe, J. Appl. Phys. **87**, 6881 (2000)
16. Y. Kitamoto, M. Abe, and M. Naoe, IEEE Trans. Mag. **32**, 3824 (1996)
17. N. Matsushita, K. Noma, A. Morisako, S. Nakagawa, and M. Naoe, IEEE Trans. Magn. **32**, 3578 (1996)
18. N. Matsushita, M. Ichinose, S. Nakagawa, and M. Naoe, IEEE Trans. Magn. **34**, 1639 (1998)
19. H. Takano, Y. Nishida, M. Futamoto,, H. Aoi, and Y. Nakamura, J. Magn. Magn. Mat., to appear
20. A. Kikukawa, Y. Honda, Y. Hirayama, and M. Futamoto, IEEE Trans. Magn. **36**, 2402 (2000)
21. T.P. Nolan, Y. Hirayama, and M. Futamoto, J. Appl. Phys. **79**, 5359 (1996)
22. H. Gong, M. Rao, D. E. Laughlin, and D. N. Lambeth, J. Appl. Phys. **85**, 4699 (1999)
23. Y. Hirayama, Y. Honda, T. Takeuchi, and M. Futamoto, IEEE Trans. Magn. **35**, 2766 (1999)
24. T. Shimatsu, H. Komagome, I. Watanabe, H. Muraoka, Y. Sugita, and Y. Nakamura, J. Appl. Phys. **87**, 6367 (2000)
25. X. Song, J. Sivertsen, and J. Judy, J. Appl. Phys. **81**, 4387 (1997)
26. C. Yang, J.M. Sivertsen, and J. H. Judy, J. Appl. Phys. **81**, 4670 (1997)

27. K. Ishikawa, R. Taguchi, E. Miyashita, J. Numazawa, and H. Oshima, J. Appl. Phys. **81**, 4390 (1997)
28. K. Kuga, Y. Yoneda, J. Numazawa, and M. Naoe, IEEE Trans. Magn. **35**, 2745 (1999)
29. N. Honda, J. Ariake, K. Ouchi, and A. Iwasaki, IEEE Trans. Magn. **34**, 1651 (1998)
30. N. Honda, K. Ouchi, and S. Iwasaki, IEEE Trans. Magn. **33**, 3097 (1997)
31. J. Ariake, N. Honda, K. Ouchi, and S. Iwasaki, IEEE Trans. Magn. **36**, 2411 (2000)
32. Y. Ikeda, Y. Sonobe, and H. Uchida, IEEE Trans. Magn. **34**, 1636 (1998)
33. Y. Sonobe, Y. Ikeda, and Y. Tagashira, IEEE Trans. Magn. **35**, 2769 (1999)
34. Y. Sonobe, Y. Ikeda, H. Uchida, T. Toyooka, and Y. Tagashira, IEEE Trans. Magn. **33**, 3088 (1997)
35. T. D. Leonhardt, Y. Chen, M. Rao, D. E. Laughlin, D. N. Lambeth, and M. H. Kryder, J. Appl. Phys. **85**, 4307 (1999)
36. T. Suzuki, T. Kiya, N. Honda, and K. Ouchi, IEEE Trans. Magn. **36**, 2417 (2000)
37. B. N. Engel, C. D. England, R. A. Van Leeuwen, M. H. Wiedmann, and C. M. Falco, Phys. Rev. Lett. **67**, 1910 (1991)
38. R. F. C. Farrow, C. H. Lee, R. F. Marks, P. Harp, M. Toney, T. A. Rabedeau, D. Weller, and H. Brandle, in: *Magnetism and Structure in Systems of Reduced Dimensions*, R. F. C. Farrow, B. Dieny, M. H. Donath, A. Fert, and B. D. Hermsmeier (eds.), (Plenum, NY, 1993)
39. R. H. Victora and J. M. MacLaren, Phys. Rev. B **47**, 11583 (1993)
40. G. H. O. Daalderop, P. J. Kelly, and M. F. H. Schuurmans, in: *Science and Technology of Nanostructured Magnetic Materials*, G. C. Hadjipanayis and G. A. Prinz (eds.), (Plenum, NY, 1991)
41. R. H. Victora and J. M. MacLaren, J. Appl. Phys. **69**, 5652 (1991)
42. G. A. Bertero and R. Sinclair, Appl. Phys. Lett. **64**, 3997 (1994)
43. J. M. MacLaren and R.H. Victora, Appl. Phys. Lett. **66**, 3377 (1995)
44. J. Xue and R. H. Victora, J. Appl. Phys. **87**, 6361 (2000)
45. Y. S. Tyan, T. K. Hatwar, G. Farruggia, R. H. Victora, M. W. Culver, and C. F. Brucker, Jpn. J. Appl. Phys. **36**, 418 (1997)
46. K. Ho, B. M. Lairson, Y. K. Kim, G. I. Noyes, and S. Sun, IEEE Trans. Mag. **34**, 1854 (1998)
47. W. Peng, R. H. Victora, J. H. Judy, K. Gao, and J. M. Sivertsen, J. Appl. Phys. **87**, 6358 (2000)
48. W. Peng, R. H. Victora, and J. H. Judy, IEEE Trans. Magn. **37**, to appear
49. R. H. Victora, W. Peng, J. Xue, and J. H. Judy, J. Magn. Magn. Mat., to appear in 2001.
50. J. Xue and R. H. Victora, Appl. Phys. Lett. **77**, 3432 (2000)
51. L. Wu, S. Yanase, N. Honda, and K. Ouchi, IEEE Trans. Mag. **33**, 3094 (1997)
52. L. Wu, N. Honda, and K. Ouchi, IEEE Trans. Mag. **35**, 2775 (1999)
53. K. Takano, G. Zeltzer, D.K. Weller, and E. E. Fullerton, J. Appl. Phys. **87**, 6364 (2000)
54. G.-L. Chen, J. Appl. Phys. **87**, 6887 (2000)
55. A. Shukh, J. Magn. Magn. Mat., to appear 2001
56. T. Ando, M. Mizukami, and T. Nishihara, J. Magn. Soc. Jpn. **18**, S1, 87 (1994)
57. T. Ando, M. Mizukami, and T. Nishihara, IEICE Trans. Electron. **E78-C**, 1543 (1995)

58. T. Ando and T. Nishihara, IEEE Trans. Magn. **37**, to appear
59. S. Takahashi, Y. Yamakawa, and K. Ouchi, J. Mag. Soc. Jpn. **23**, S2, 63 (1999)
60. H. S. Jung and W. D. Doyle, IEEE Trans. Magn. **37**, to appear
61. H. Gong, B. Lu, D. E. Laughlin, and D. N. Lambeth, J. Appl. Phys. **87**, 6355 (1997)
62. Y. Kitamoto, M. H. Kim, S. Kantake, M. Abe, and M. Naoe, IEEE Trans. Mag. **33**, 3085 (1997)
63. T. Ichihara, S. Nakagawa, H. Matsumiya, and M. Naoe, IEEE Trans. Mag. **33**, 3004 (1997)
64. S. Iwatsubo, T. Takahashi, and M. Naoe, J. Magn. Soc. Jpn. **21**, S2, 193 (1997)
65. H. Hokkyo, S. Tsuboi, N. Ohshima, and K. Tagami, J. Magn. Soc. Jpn. **21**, S2, 517 (1997)

9 Self-Assembled Magnetic Nanoparticle Arrays

Shouheng Sun, Dieter Weller, and Christopher B. Murray

9.1 Introduction

Advances in magnetic recording technology have resulted primarily from proportional scaling of each individual storage bit [1]. To control the signal to noise ratio and other important recording parameters, it is essential to maintain a large number of independent ferromagnetic grains per bit. These trends have driven the development of new magnetic thin film media with smaller grains, high coercivity, low magnetization and minimal magnetic exchange coupling between neighboring grains [2–6]. Today, the grain size of Co-alloy based advanced recording media has been reduced to dimensions close to the onset of superparamagnetic behavior. At such dimensions the magnetization of individual grains tends to fluctuate randomly with ambient thermal energy, resulting in irreversible thermal decay of stored information [7] (Chaps. 5 and 6).

Two key technical barriers exist to the development of future high-density recording media. One is polydispersity of the magnetic grains in size, shape, crystallinity and magnetic easy axis orientation. The presence of these polydisperse magnetic grains has complicated the experimental observation of the magnetic properties of the granular thin films. Currently, the average size of the cobalt-based magnetic grains for advanced media has been reduced to about 8–10 nm. Further reduction of the grain size and the presence of broad grain size distribution enhance the influence of superparamagnetic effects, possibly leading to signal decay and loss of recorded information. The other technical barrier has been the difficulty in controlling the intergranular spacing. As a result, the grains produced in a typical thin film sputter process are often in physical contact, leading to intergranular exchange coupling and as a consequence media noise.

To overcome these shortcomings, lithographically patterned media (Chap. 7) or media based on monodisperse magnetic nanoparticle arrays, as discussed here, have been proposed. In lithographically patterned films, the magnetic bits are predefined lithographically. This could effectively eliminate the transition noise but fabrication costs and design of suitable read-write systems remain enormous challenges. In nanoparticle array films, grains are prepared with a uniform size distribution and a non-magnetic coating to reduce exchange coupling prior to their assembly into magnetic recording media. This new approach uses monodisperse magnetic nanoparticles as building blocks. By controlling deposition conditions, these nanoparticles can self-organize on a solid substrate into 2D or 3D assemblies [8–10]. This self-assembly approach is a low-cost alternative to fabricating nanostructured materials. Because nanoparticles can be synthesized with controlled dimensions as small as 2 nm, their self-assembly could also yield nanoparticle arrays with characteristic dimensions much smaller than conceivable with lithographic methods.

Solution phase synthesis for monodisperse magnetic nanoparticle materials has been developed. In this chapter, we focus on current chemical synthetic approaches to monodisperse magnetic nanoparticles, their assembly into magnetic superlattices and their potential applications in magnetic data storage media. Section 9.2 outlines chemical approaches to monodisperse nanoparticles, including solution phase nucleation and growth, particle dispersion stabilization and chemical synthesis. Section 9.3 discusses the self-assembly of magnetic nanoparticles and Sects. 9.4 and 9.5 illustrate two specific examples, Co and FePt, of magnetic nanoparticle assemblies. The potential application of magnetic FePt nanoparticle arrays as magnetic recording media is outlined in Sect. 9.6.

9.2 Chemical Synthesis of Monodisperse Nanoparticles

9.2.1 Solution Phase Nucleation and Growth

A general scheme for preparing monodisperse nanoparticle samples (with $<10\%$ standard deviation in diameter) requires a single, temporally short nucleation event followed by slower growth on the existing nuclei. This may be achieved by rapid addition of reagents into a reaction vessel to form a large number of critical nuclei that will grow simultaneously with material added to each nucleus at a similar rate. Under ideal conditions, the spread of particle sizes can become sharper with time. Methods based on this general principle have been developed for the preparation of a variety of monodisperse inorganic and polymer colloidal dispersions. The same principle has also been successfully used in obtaining monodisperse CdSe nanocrystals [11,12].

Another approach to produce monodisperse nanoparticle materials is using a matrix template to regulate particle growth. The most common matrix used for this purpose is reverse micelles (also called inverse micelles) (Fig. 9.1). These reverse micelles are formed by dispersing water droplets in the non-polar solvent in the presence of surfactant molecules, such as sodium bis(ethylhexyl)sulfosuccinate (AOT). By adjusting the water content, the size of the water droplets can be controlled. These reverse micelles act as restricted nanoreactors. Molecular precursors are concentrated in each of the nanoreactors, facilitating their growth into monodisperse nanoparticles.

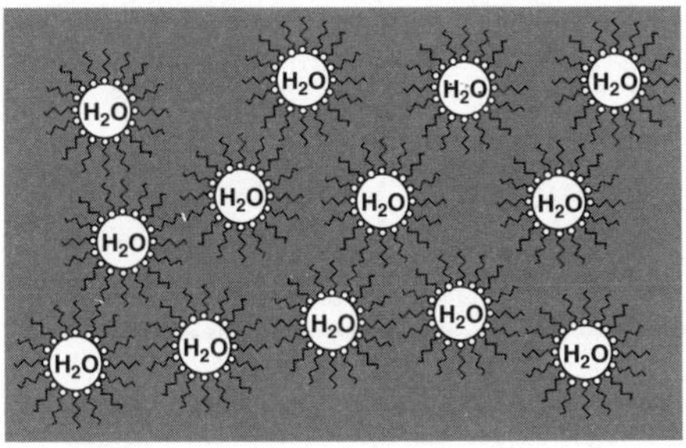

Fig. 9.1. Reverse micelles formed by dispersing water droplets in a non-polar solvent in the presence of hydrocarbon chained surfactants

Various nanoparticle dispersions have been prepared by the reverse micelle approach [13–18].

9.2.2 Stabilization of the Nanoparticle Dispersions

Nano-sized particles experience strong attractions between the particles that make them unstable and prone to aggregation. Therefore, one of the major problems in nanoparticle synthesis is particle aggregation resulting in undispersed fine powder, as illustrated in Fig. 9.2. The van der Waals forces of attraction exist in many nanoparticle systems while magnetic particles experience additional magnetic interactions, making their stabilization even more challenging. This has been a serious problem in magnetic nanoparticle synthesis, as most synthetic routes reported to date have yielded agglomerated powders. Strong magnetic interactions in Ni, Fe, and Co particle systems make it difficult to form stable colloids. Uncontrolled agglomeration of the magnetic particles often prohibits employing separation procedures that could narrow the size distribution. It also prevents the ready formation of the smooth, thin films required in magnetic recording applications. To stabilize nanoparticle dispersions, a repulsive or stabilizing force must be present to prevent ag-

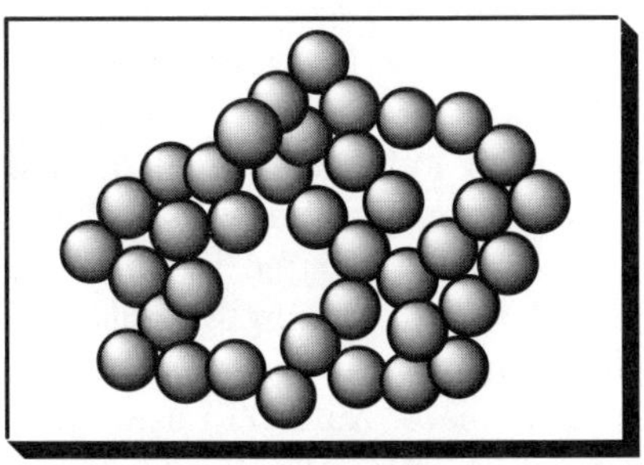

Fig. 9.2. An illustration of particle aggregation. Avoiding this kind of aggregation has been a major challenge in nanoparticle synthesis

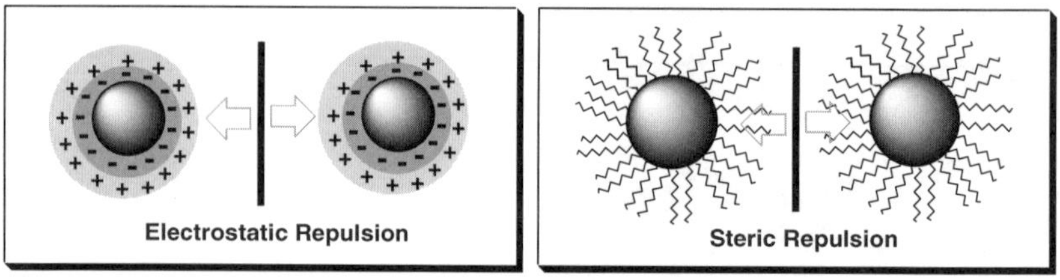

Fig. 9.3. Particle stabilization *via* coating of ionic compounds or long chain surfactants. The coating increases the electrostatic and steric repulsions among the particles, and can effectively stabilize the particle dispersion

gregation. This stabilization can be achieved by two methods: electrostatic repulsion and steric repulsion, as indicated in Fig. 9.3. Colloids remain stable with respect to aggregation only if there exists a repulsive force of sufficient strength and range to counteract the attractions. Coating the particles with ionic compounds will increase electrostatic repulsion when particles are approaching each other. The well-known gold sol is prepared by the reduction of $[AuCl_4]^-$ with sodium citrate. Each gold particle is coated with an electrical double layer of anions (such as citrate and Cl^-) and cations (such as Na^+). The electrostatic stabilization prevents gold colloids from aggregation, leading to stable gold sol. Another way of protecting particles from aggregation is to coat them with large molecules, such as polymers or surfactants. The long chain hydrocarbons in polymers or surfactants greatly increase the steric repulsion when particles get closer, efficiently preventing them from aggregation.

9.2.3 Synthetic Approaches to Magnetic Nanoparticles

9.2.3.1 Decomposition of Organometallic Precursors

Organometallic precursors, specifically metal carbonyls and metal arenes have proven useful precursors due to their inherent thermal instability. The most commonly used organometallic precursors for magnetic nanoparticle synthesis are dicobalt octacarbonyl, $Co_2(CO)_8$, and iron pentacarbonyl, $Fe(CO)_5$. The small molecule CO can be easily released under thermal conditions, leading to a variety of larger metal clusters that will further grow into desired magnetic nanoparticle materials in the presence of stabilizers (Fig. 9.4). This general synthetic method can be used to prepare Co, Fe and Ni based magnetic nanoparticles. Cobalt nanoparticles have been prepared by thermally decomposing $Co_2(CO)_8$ in refluxing toluene solution containing suitable polymeric materials, such as methylmethacrylate-ethylacrylate-vinyl pyrrolidone terpolymer or sodium bis(2-ethylhexyl)sulfosuccinate, Na(AOT) as stabilizers [19,20]. We recently found that decomposition of $Co_2(CO)_8$ at 200°C in diphenylether in the presence of trialkylphosphine/oleic acid gave monodisperse multiply-twinned fcc cobalt nanoparticles [21]. The size of the particles

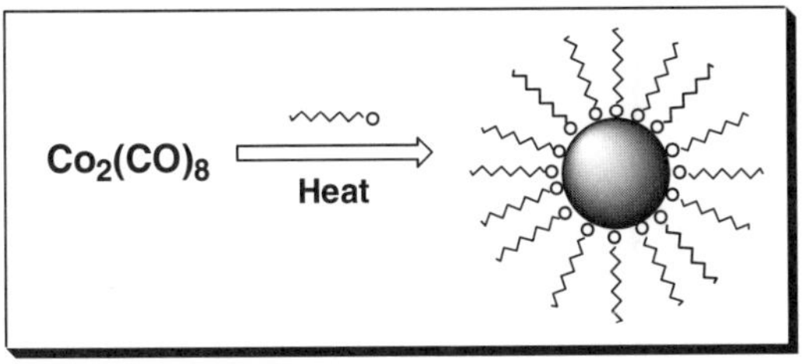

Fig. 9.4. Illustration of synthesis of cobalt nanoparticles *via* cobalt carbonyl decomposition in the presence of organic surfactant stabilizers

can be tuned from 3 nm to 12 nm by adjusting the ratio of stabilizers and cobalt carbonyl. Other cobalt organometallic precursors have also been used to prepare cobalt nanoparticles. For example, under 3 bar dihydrogen, decomposition of $Co(\eta^3–C_8H_{13})(\eta^4\text{-}C_8H_{12})$ in the presence of polyvinylpyrrolidone (PVP) at different temperatures 0, 20, 60°C yields differently sized (from less than 1 nm to about 1.5 nm) fcc cobalt nanoparticles [22]. Fe nanoparticles are prepared by decomposition of $Fe(CO)_5$ in the presence of a variety of polymer stabilizers [23], sarkosyl-O (n-oleyoyl sarcosine) [24–27] and TOPO/TOP [28]. $Fe(CO)_5$ is also subject to ultrasonic decomposition in solution phase to generate Fe nanoparticles [29–31]. Ultrasonic treatment of $Fe(CO)_5$ and $Co(CO)_3NO$ in the hydrocarbon solvent yields bimetallic Fe–Co nanoparticles [32]. In contrast to the synthesis of cobalt and iron particles, relatively few reports are dealing with the synthesis of nickel particles *via* organometallic precursor decomposition. This may be partly due to the high toxicity of $Ni(CO)_4$ precursors that prevents ready use as a starting material. Recently, decomposition of $Ni(COD)_2$ in methylene chloride in the presence of PVP has been used to prepare Ni nanoparticles [33,34].

9.2.3.2 Metal Salt Reduction

Metal salt reduction is the most versatile approach to produce magnetic nanoparticle materials. In solution, metal ions can be reduced by a variety of reducing agents. The reduction usually takes place instantly to give neutral metal species that further grow into magnetic nanoparticles in the presence of stabilizers. This process is illustrated in Fig. 9.5. The metal salts employed must be at least partially soluble in the solvent, and the reductions should be carried out under an inert atmosphere, as the neutral metal species are often very air sensitive. Reduction of the salt without any stabilizer present generally produces very reactive fine metal powders [35,36].

Over many years, BH_4^- has been used extensively to reduce metal salts and prepare metallic particles. However, it is also noticed that the metal particles produced from this reduction are often contaminated with metal boride. For example, in aqueous solution, the reduction of $CoCl_2$ by $NaBH_4$ yields a

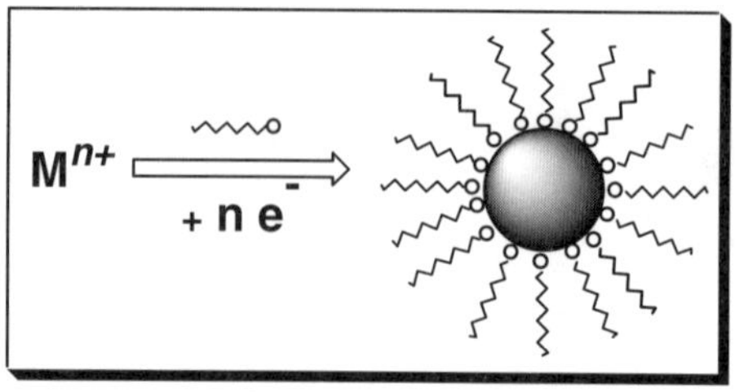

Fig. 9.5. Illustration of metal nanoparticle synthesis *via* metal salt reduction in the presence of organic surfactant stabilizers

product that is contaminated with Co_2B components [37–39]. Approaches to prepare pure magnetic nanoparticles include using different reaction media [40], different hydride reducing agents [41], and alternative families of reducing agents, such as Li, Na, K naphthalides [42–46]. The alkali metals are usually combined with naphthalene in an ether-type solvent to form metal naphthalide. This metal naphthalide can be easily dissolved in an ether-type solvent and the reduction is usually completed in a short period of time at relatively low temperature. However, these metal naphthalide-based reducing agents are extremely air sensitive and difficult to store. Organoborohydride type reducing agents are found to reduce metal salts in solution phase to give high purity magnetic nanoparticles in the presence of a variety of stabilizers, such as tetraalkylamonium salt, trialkylphosine/oleic acid, etc. Many metal elements can be prepared as pure particle materials by treating their metal salts with hydroorganoborides in either an ether or a hydrocarbon solvent in the temperature range from room temperature to 200°C [8,9,47,48].

Some organic compounds can be used as organic reducing agents to reduce metal salts. The sonication of aqueous Co^{2+} and hydrazine yields cobalt particle materials [49]. A better-known organic reduction process is called polyol process, in which organic diol is used as both a reducing agent and a solvent [50]. In this process, the metal salts are suspended in a liquid polyol (ethylene glycol, diethyleneglycol etc). The suspension is stirred and heated to a given temperature, which can reach the boiling point of the polyol for more electropositive metals, e.g., cobalt and nickel. The starting compound is at least moderately soluble in the diol. The general mechanism of metal particle formation through nucleation and growth from the solution is quite general for all metals obtained by this process. The authors have recently extended the use of short ethylene glycol to long chain diol, such as 1,2-hexanedecanediol and 1,2-dodecanediol, in almost stoichiometric manner in a high boiling ether solvent. Stabilized hcp cobalt nanoparticle dispersion has been separated [21]. This procedure has the advantage of allowing the particles to remain dispersed during growth, producing more uniform particles. In addition, careful adjustment of the ratio of the reducing agent and the metal precursor is possible to control the average particle size.

Reduction of metal salts *via* electrochemistry can produce magnetic nanoparticles. The synthesis employs a simple electrochemical cell containing only a metal anode and a metal or glassy carbon cathode. The electrolyte consists of organic solutions of tetra-alkylammonium halide that also serves as a stabilizer for the produced clusters. After application of a current, the anode undergoes oxidative dissolution forming metal complex. The subsequent reduction of these complexes leads to the formation of the desired particle dispersions [51,52]. This approach allows the synthesis of metal particles with narrow size distributions. The potential to control particle size by simply adjusting the electrochemical parameters is appealing.

The reverse micelle technique has been established as a simple method to obtain bulk quantities of nanoparticle materials, since the reduction of metal ions can be performed in reverse micelles, as shown in Fig. 9.1. The size of the particles synthesized inside the micelles can be varied by changing the [water]/[surfactant] ratio. An increase of the ratio increases the size of the water pool, and therefore allows bigger particles to form [53–59]. One disadvantage of the reverse micelle methods is that due to the low reaction temperatures, these methods generally yield particles that are poorly crystallized and irregular in their internal structure.

9.3 Magnetic Nanoparticle Assembly

As the control over particle size, shape and structure as well as the chemical nature and dimensions of the organic capping layer are established, engineering the properties of the individual nanoparticles becomes possible. The next major challenge is to rationally assemble these tailored building blocks into extended close packed solids to exploit this control and produce new materials and devices.

9.3.1 Self-Assembly

The tendency for monodisperse particle suspensions to form ordered arrays (self-assemble) is influenced by the nature of the interactions exhibited among the stabilized particles (Fig. 9.6). The synthetic strategy relies on a large num-

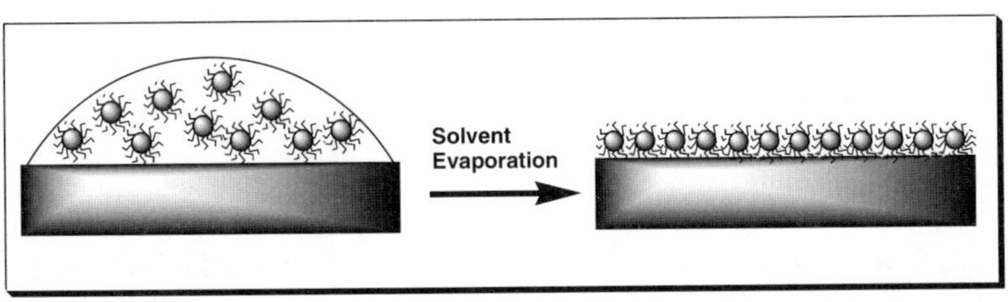

Fig. 9.6. An illustration of nanoparticle self-assembly *via* solvent evaporation

ber of weak and non-directional interactions, such as ionic bonds, hydrogen bonds and van der Waals interactions to organize the particles. Self-assembly into ordered arrays has been observed in a number of particle systems [60–68]. The key challenge to the success of this self-assembly method is the availability of well-controlled monodisperse nanoparticles. Extensive investigations have revealed that nanoparticle size, shape, as well as interparticle spacing, influence the assembled structures [69–73], and as a consequence the physical properties of the resulting assemblies. A particular challenge for this self-assembly approach is the preparation of robust thin nanoparticle assemblies for magnetic data storage applications. Self-assembled magnetic nanoparticle arrays are usually held together by very weak van der Waals and magnetic interactions. As a result, the assembly is readily destroyed. For example, the assembly from the particles protected by surfactants with long hydrocarbon tail groups can be re-dissolved back into alkane, aromatic, or chlorinated solvents. To make a mechanically robust particle assembly, thermal annealing or radiation exposure is usually required to induce chemical reactions among the stabilizers and interface reactions between the particles and the substrate.

9.3.2 Template-Assisted Assembly

Template-assisted assembly of nanoparticle materials refers to directionally assembling the particles on pre-treated substrates. These pre-treated substrates are functionalized either with molecules, to facilitate the chemical bonding between the particles and the substrate, or with porous membranes, to facilitate organization of the particles into regularly arrayed cores. Nanoparticles can be assembled in a controlled manner with robust mechanical properties.

9.3.2.1 Molecule-Assisted Assembly

Molecule-assisted approaches to robust nanoparticle assemblies on a solid substrate involve usage of ligand exchange reactions to exchange the stabilizers bound to the particles with the molecules that selectively bind to the substrate. The self-assembly of molecules on a solid substrate is a well-known technique to functionalize a substrate. Close-packed self-assembled monolayers are obtained by spontaneous adsorption of surfactant molecules from organic solution onto a solid substrate. In this molecular self-assembly approach, strong molecule-substrate interactions result in "chemisorption" of the monolayer on the substrate surface. An example of the self-assembly of organosilane compounds on a hydroxylated surface (e.g. glass, quartz or mica) is illustrated in Fig. 9.7. Hydrolysis of the Si–Cl bonds is caused by traces of water adsorbed on the hydrophilic substrate surface. The adsorption process proceeds with formation of polysiloxane that is connected to surface silanol groups *via* Si–O–Si bonds, leading to covalent attached siloxane with

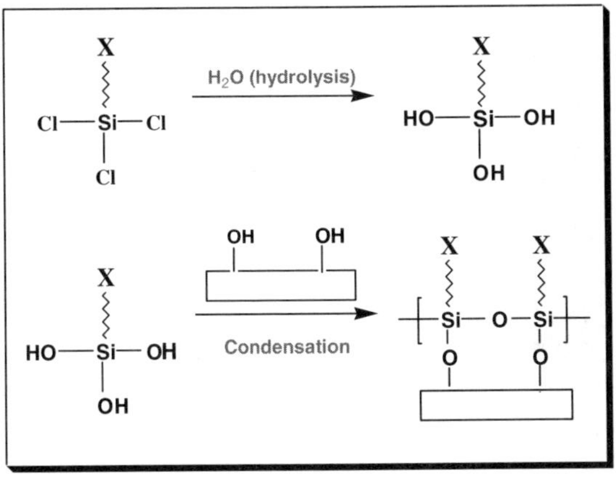

Fig. 9.7. An example of the self-assembly of organosilane compounds on a hydroxylated surface

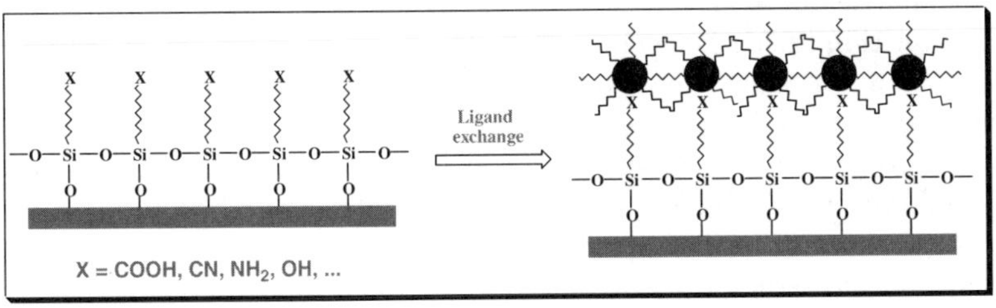

Fig. 9.8. Assembly of nanoparticles into a macroscopic two-dimensional array onto functionalized substrates *via* ligand exchange

the substrate [74–78]. For metal substrates, such as gold, silver and copper, sulfur containing surfactants (e.g., alkyl thiol, dialkyl sulfides and disulfides, alkyl and dialkyl dithiocarbamates) are readily adsorbed on the substrate to form monolayers [79–85].

The self-assembly of nanoparticles into a macroscopic two-dimensional array onto functionalized substrates is depicted in Fig. 9.8, in which the organosilane is used as an example. The assembly process includes: 1) surface-initiated polymerization of bifunctional organosilanes such as $(RO)_3Si(CH_2)_3X$ on a hydroxy-terminated substrate, e.g. a glass, SiO_2 or quartz surface; and 2) replacement of the particle stabilizers with a pendant functional group X. The pendant functional group X, chosen for its high affinity to metal surfaces, extends out into the solution. By dipping the polymer-derivatized substrate into the particle dispersion, ligand exchange occurs and a strong monolayer particle assembly is formed. This approach has been employed widely to assemble nanoparticles [86–90]. An example of such an assembly is shown in Fig. 9.9, where monodisperse cobalt nanoparticles are assembled onto a

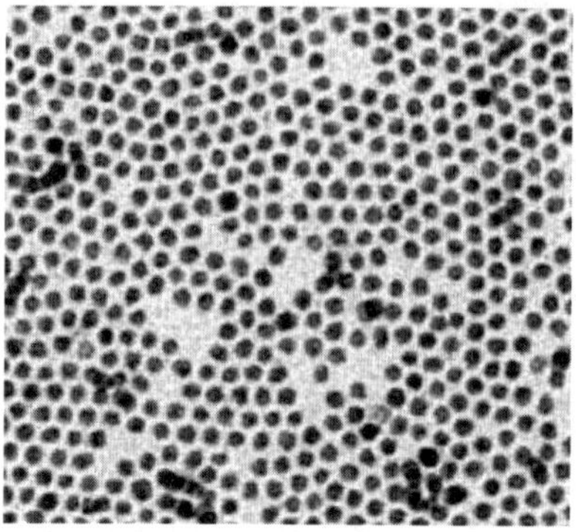

Fig. 9.9. TEM image of a monolayer of 8 nm cobalt nanoparticles on a $Cl_3Si(CH_2)_3COOH$-treated SiO_2 surface

$Cl_3Si(CH_2)_3COOH$-derivaitzed SiO_2 surface, yielding a uniformly covered cobalt nanoparticle film [91].

Multilayer thin films are usually fabricated using a layer-by-layer method. The fabrication of nanoparticles/polymer multilayer films can be carried out as follows: a substrate is immersed into a polymer solution, and then rinsed with fresh solvent. Subsequently, the slide is dipped into the nanoparticle solution followed by washing with more of the solvent used to prepare the particle dispersion. By repeating this simple two-step process in a cyclic fashion, layer-by-layer assembled polymer/nanoparticle multilayer films can be obtained [92–95].

Recently, a new approach of using biomolecules as a template for nanoparticle assembly has been tested. These biomolecules have shown a remarkable level of control over the assembly of functional molecules and other nanoscale building blocks into complex structures in biological systems. Their ability to direct the assembly of nanoscale components into controlled complex structures has spurred efforts to mimic and exploit their recognition capabilities found in biological systems. Both DNA [96,97] and protein [98] have been used to template assisted nanoparticle assembly.

9.3.2.2 Porous Membrane-Based Assembly

Although molecule functionalized template approaches allow precise control of the assembly thickness and smoothness, so far, it has not been successful to yield long range ordered nanoparticle assemblies. Porous membranes, in contrast, often contain well-organized pores. Each of these pores can be viewed as a nanoreactor in which a particle of the desired material can be synthesized. Loading the pores with magnetic materials can lead to the formation of well-organized magnetic dot arrays. Colloidal assemblies of latex, silica or block-copolymer particles have been employed to prepare porous membranes [99–105]. Larger mesoporous materials are prepared by using surfactants or emulsion droplets as templates. As a result, an ordered porous membrane could have a pore size from several nanometers up to a few microns. Porous alumina membranes are prepared from aluminum metal via anodization in an acidic electrolyte. The pores in these membranes are arranged in a regular hexagonal lattice. Pore densities as high as 10^{11} pores/cm^2 with a broad range of pore diameters can be achieved [106–111].

The advances in the chemistry of carbon nanotubes may allow these hollow tubes to act as templates for the formation of magnetic nanostructures. Nanotubes can have diameters of as small as $\sim$1 nm. Encapsulating the metal atoms within the tubes will have two benefits. The carbon coating will efficiently protect the reactive metal atoms from oxidation, and the size-limited tube will ensure the formation of well-controlled magnetic nanowires [112,113].

9.4 Monodisperse Cobalt Nanoparticles and Nanoparticle Superlattices

It was recently reported that decomposition of $Co_2(CO)_8$ and reduction of $CoCl_2$ by superhydride resulted in the production of a phase of cobalt not seen previously, ε-Co [8,9,114]. This synthesis coupled with size-selected precipitation yields monodisperse Co nanoparticles that can self-organize into Co nanoparticle superlattices. The ε-Co phase can be converted predominantly into the common hcp phase (with some residual fcc stacking faults) by annealing at 300–350°C. Particles with the common fcc crystal structure result from annealing the ε-Co at 450–500°C. These annealing processes have an added benefit in that the interparticle spacing can be controlled. Longer annealing times and higher annealing temperatures produce smaller average interparticle distances as the organic stabilizer is converted into an increasingly dense layer of hydrogenated amorphous carbon with separated particles in the final structure. Magneto-resistivity due to spin-dependent tunneling has been observed in these self-assembled nanoparticle superlattices [115].

9.4.1 Synthesis of Monodisperse ε-Co Nanoparticles

The ε-Co nanoparticles were synthesized by the injection of 2 M dioctylether superhydride solution into a hot (200°C) $CoCl_2$ dioctylether solution in the presence of oleic acid and trialkylphosphine (PR_3, $R = n - C_4H_9$, or $n - C_8H_{17}$) [8]. Reduction occurred instantly upon injection, leading to the simultaneous formation of many small metal clusters (nuclei). Continued heating at 200°C allowed steady growth of these clusters into nanometer sized, single crystals of cobalt. Extended heating did not readily increase the particle size, as side reactions, catalyzed by the cobalt particles, led to the progressive deterioration of the particles. However, the choice of alkylphosphine used in

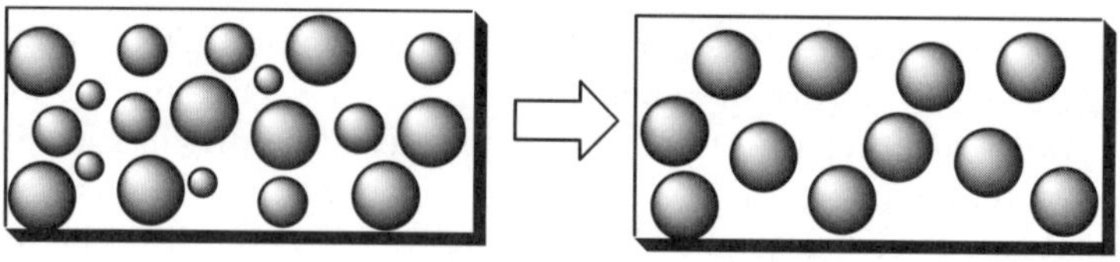

Fig. 9.10. Illustration of size selection from polydisperse nanoparticle dispersion. Monodisperse nanoparticles can be separated and collected

combination with oleic acid does allow the average particle size to be controlled in the synthesis. The steric bulk of alkylphosphine controls the rate of addition of materials to the particles. Short-chained alkylphosphines (e.g., tributylphosphine) allow faster growth, favoring production of large particles (7–11 nm) while bulkier species (e.g. trioctylphosphine) reduce the particle growth rate and favor the production of smaller particles (2–6 nm). Fractionating the initial dispersions and selecting the desired size fraction, as indicated in Fig. 9.10, provides the final level of size control. In this way a series of monodisperse particle samples can be produced. These organically stabilized cobalt particles are readily dispersed in aliphatic, aromatic and chlorinated solvents, facilitating their assembly on a solid substrate

9.4.2 Self-Assembled Co Nanoparticles

The Co particles are very uniform in size, allowing them to self-organize readily into 2-D and 3-D superlattices. Slow evaporation of a carrier solvent from cobalt dispersion spread on a flat substrate allows well-organized superlattice structures to be formed. The TEM images of these Co nanocrystal superlattices are shown in Fig. 9.11. The use of concentrated dispersion of Co nanoparticles in higher boiling solvents like dodecane allows slower evaporation at higher temperatures, producing well ordered multilayers. The added

Fig. 9.11. TEM images of (**a**) a 2D assembly, and (**b**) a 3D assembly of the 8 nm cobalt nanoparticles on an amorphous carbon surface

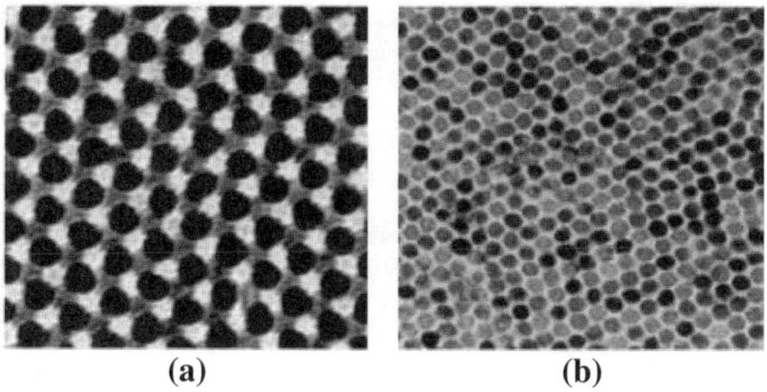

Fig. 9.12. TEM images of 10 nm cobalt nanoparticle assemblies of (a) before and (b) after thermal annealing at 350°C for 3 h

thermal energy permits the particles to diffuse to their lowest energy superlattice sites during evaporation, producing well-defined 3-D hcp superlattice structures, as shown in Fig. 9.11b. This hcp superlattice formation is driven by attractive van der Waals forces and magnetic interactions between the particles. However, this assembly process is reversible as the particles can be redispersed when exposed to additional solvent.

One way to form robust particle thin films is *via* thermal annealing. Thermal annealing induces the decomposition of the stabilizers around individual Co particles and moves particles closer to each other as denser amorphous carbon spacers are formed. For example, thermal treatment of a 3-D Co particle assembly at 350°C under vacuum for 3 h reduces the interparticle spacing from ~4 nm to ~2 nm (Fig. 9.12). This thermally controlled nanoparticle processing allows the fabrication of self-assembled cobalt nanoparticle-based devices [115]. A self-assembled superlattice of 10 nm cobalt nanoparticles has been deposited on a set of electron-beam patterned electrodes to allow the measurement of spin dependent transport in these assemblies. Annealing the superlattice under reducing conditions brings neighboring nanoparticles closer together to ~2 nm but retains enough hydrogen in the amorphous carbon to maintain insulating (tunnel) barriers between the nanoparticles. The H_2 annealing also reduces the surface oxide layer that forms on nanoparticles during handling. Current-voltage characteristics are well described by single-electron tunneling in a uniform array. At temperatures below 20K, device MR ratios are on the order of 10%, approaching the maximum predicted values for ensembles of cobalt islands with randomly oriented preferred magnetic axes [115]. The result demonstrates the potential utility of these ordered magnetic superlattice systems in magneto-transport systems. Granular materials consisting of nanometer size magnetic particles in a dielectric or non-magnetic metal matrix have been under intensive study recently because of their novel giant magneto-resistive properties [116–120].

9.5 Monodisperse FePt Nanoparticle and FePt Nanocrystal Superlattices

FePt-based nanostructured materials are promising candidates for future ultrahigh density recording media because of their good chemical stability and high magnetocrystalline anisotropy ($\sim 10^8$ erg/cm^3) observed in the ordered intermetallic phase [121,122]. This large magnetocrystalline anisotropy allows grains as small as $\sim$3 nm to be thermally stable over typical data storage periods of 10 years. A well-organized magnetic array of such particles should contribute to efforts to design magnetic media capable of recording densities well beyond 1 Tbit/in^2 [7].

Various vacuum deposition techniques have been developed to produce high coercivity FePt nanocrystalline films [121–125]. Post deposition annealing has proven essential to transform the as-deposited chemically disordered fcc structure into the chemically ordered fct phase that has high magnetocrystalline anisotropy. Random nucleation events in the initial stages of thin film growth, however, typically result in broad distributions of particle sizes, which may be further aggravated by agglomeration during annealing. To prevent this agglomeration, FePt particles have been embedded in a variety of insulating matrixes, such as SiO_2 [126] Al_2O_3 [127,128], B_2O_3 [129] and S_3N_4 [130].

Solution phase chemistry offers a promising alternative to fabricating FePt binary alloy nanoparticles. One synthetic procedure involves the reduction of metal salts with $NaBH_4$ in reverse micelles of cetyltrimethylammonium bromide [131]. Another procedure, which was developed by the present authors, involves decomposition of $Fe(CO)_5$ and polyol reduction of $Pt(acac)_2$ in solution with oleic acid and oleyl amine as stabilizers [10,132]. Controlled evaporation of the solvent from dispersed particles followed by thermal treatment leads to self-organized ferromagnetic nanoparticle arrays that can support high-density magnetic recording.

9.5.1 Synthesis of FePt Nanoparticles

The FePt nanoparticles were prepared by the combination of polyol reduction of $Pt(acac)_2$ and thermal decomposition of $Fe(CO)_5$ in the presence of oleic acid and oleyl amine. Under airless conditions, heating the mixture of $Fe(CO)_5$, $Pt(acac)_2$, oleic acid, oleyl amine, 1,2-hexadecanediol and dioctylether to 298°C produces monodisperse FePt nanoparticles that can be easily dispersed into alkane solvent [10,132]. The FePt particle size is tunable from 3 to 10 nm by first growing 3 nm monodisperse seed particles *in situ* and then adding more reagents to enlarge the existing seeds to the desired size. The composition of these FePt nanoparticles is controlled *via* the molar ratio of $Fe(CO)_5$ and $Pt(acac)_2$. Based on 0.5 mmol each of the $Pt(acac)_2$, oleic acid and oleyl amine, and 20 mL of dioctylether, the molar amount of $Fe(CO)_5$ and the resulting $Fe_xPt_{(100-x)}$ are shown in Fig. 9.13.

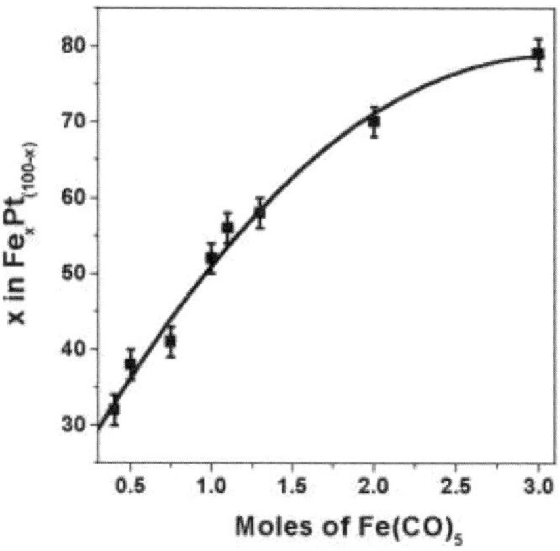

Fig. 9.13. Compositional relation between $Fe(CO)_5$ and x in $Fe_xPt_{(100-x)}$ based on 0.5 mmol of $Pt(acac)_2$

9.5.2 Assembly of FePt Nanoparticle Arrays

When the FePt nanoparticle dispersions are spread on a substrate, and the carrier solvent is allowed to slowly evaporate, FePt nanoparticles self-assemble into FePt nanoparticle superlattices. For TEM observation, a drop (~0.5 µL) of dilute FePt dispersion (~1 mg/mL) was deposited on a SiO_2-coated copper grid. Figure 9.14 is a TEM image of a cubic assembly of 6 nm $Fe_{50}Pt_{50}$ particles. Extensive studies on nanoparticle assemblies have shown that the superlattice symmetry and orientation with respect to a substrate depends on many factors, such as particle size, shape and relative dimensions of the particle core and the organic capping, and the chemical nature of the substrate [69–73].

Fig. 9.14. The TEM image of a 6 nm $Fe_{50}Pt_{50}$ nanoparticle assembly

To make chemically and mechanically robust thin films for magnetic recording applications, the nanoparticle superlattices are assembled on a substrate and then annealed. In a typical preparation, 0.05 mL of hexane dispersed 4 nm FePt nanoparticles (~10 mg/mL) are deposited on a solid substrate (~0.5 × 1 cm). The solvent is allowed to evaporate slowly (5-10 minutes) at room temperature. The as-deposited thin film is then transferred into a N_2 glove box (O_2 concentration <2 ppm) and annealed at 580°C for 30 minutes and removed from the furnace to cool in the N_2 environment. The resulting FePt nanoparticle assembly is a thin black film with a mirror like finish that can be handled and stored under ambient conditions. Rutherford backscattering measurements on these annealed 4 nm FePt nanoparticle assemblies indicate 40–50at% carbon content. This shows that annealing does not result in the loss of stabilizing ligands but rather their conversion to a hydrogenated amorphous carbon coating around each particle, effectively protecting the particles from agglomeration and rendering a strong surface for magnetic recording.

9.5.3 Magnetic Properties of FePt Nanoparticle Arrays

Controlled thermal annealing not only yields robust particle assemblies, but also facilitates internal particle structure transformations. The annealing induced internal particle structure change can be easily seen in FePt nanoparticle systems, as shown in wide-angle X-ray diffraction scans in Fig. 9.15 [10]. As-synthesized particles exhibit the chemically disordered fcc structure (Fig. 9.15a). Annealing induces the Fe and Pt atoms to rearrange into the long range chemically ordered fct structure, as indicated by the (111) peak shifts and evolution of the characteristic (001) and (110) chemical ordering peaks (Fig. 9.10b–e). At annealing temperatures below 500°C, only partial chemical ordering is observed (Fig. 9.15b,c). The chemical ordering can be increased by annealing at higher temperatures (Fig. 9.15d,e) or by increasing the annealing time (not shown). Diffraction patterns of the samples annealed above 540°C match those of fct FePt. It is worth mentioning, however, that the exact structure after thermal annealing depends strongly on the particle composition. XRD studies of annealed $Fe_{56}Pt_{44}$ assemblies indicate a high quality fct phase [132].

The role of the annealing temperature on magnetic properties has been probed by measuring temperature dependent magnetic properties. SQUID magnetometry measurements of as-synthesized FePt nanoparticles show that they are superparamagnetic at room temperature. The coercivity drops sharply as the temperature is raised from 5 K to 15 K, indicating that these small nanoparticles are thermally unstable. This is consistent with the low magnetocrystalline anisotropy of fcc FePt. Thermal annealing converts the internal particle structure from a chemically disordered fcc structure to the chemically ordered fct phase, and transforms the FePt nanoparticle superlattices into ferromagnetic assemblies. Hysteresis data of 580°C annealed FePt nanoparticle

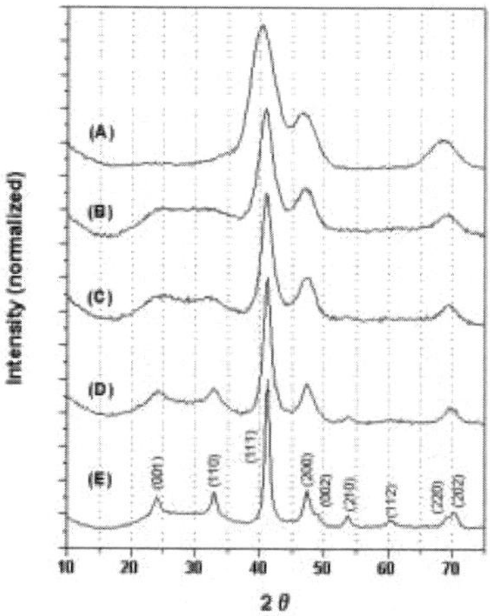

Fig. 9.15. X-ray diffraction patterns (**a**) of an as-synthesized 4 nm $Fe_{52}Pt_{48}$ nanoparticle assembly and a series of similar assemblies annealed under atmospheric N_2 gas for 30 minutes at temperatures of (**b**) 450°C, (**c**) 500°C, (**d**) 550°C and (**e**) 600°C. The indexing is based on tabulated fct FePt reflections. The diffraction patterns were collected with a Siemens D–500 diffractometer using Cu Kα radiation ($\lambda = 1.54056$ Å)

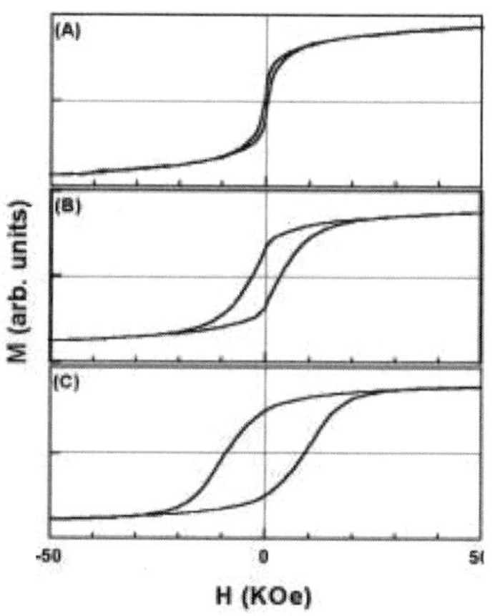

Fig. 9.16. Room temperature hysteresis loops of $Fe_{56}Pt_{44}$ nanoparticle assemblies annealed at (**a**) 500°C, (**b**) 550°C, and (**c**) 580°C

assemblies show that the coercivities of these annealed FePt nanoparticles depends strongly on FePt composition. The 'best' composition of FePt for highest coercivity is estimated to be $Fe_{55}Pt_{45}$ [132]. Shown in Fig. 9.16 are the room temperature hysteresis loops for 4 nm $Fe_{56}Pt_{44}$ nanoparticles that were annealed at 500, 550 and 580°C for 30 minutes. The coercivity, H_C, increases dramatically with annealing temperature showing the transition from

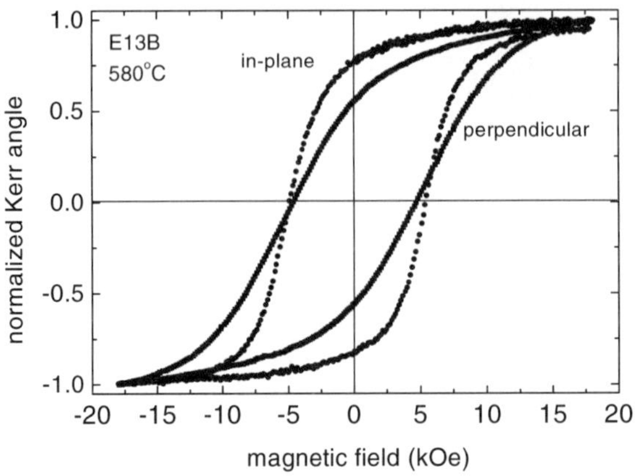

Fig. 9.17. Polar (perpendicular) and transverse (in-plane) Kerr hysteresis loops of a 140 nm thick, 4 nm-$Fe_{52}Pt_{48}$ nanoparticle superlattice annealed at 580°C [133]

superparamagnetic to ferromagnetic behavior. The sample annealed at 500°C appears nearly superparamagnetic at room temperature. An expanded view of the low-field part of the loop shows an H_C value of 330 Oe and the loop is hysteretic up to 6 kOe fields. This likely results from a minority fraction of the particles having sufficient anisotropy to be ferromagnetically ordered at room temperature. Increasing the annealing temperature to 550°C increases the ferromagnetic fraction of particles with the H_C-value increasing to 3200 Oe. However, there is still an inflection in the magnetization near $H = 0$, indicating some low H_C particles. Annealing at $T = 580°C$ results in a room temperature H_C value just over 9000 Oe with a loop shape characteristic of an isotropic distribution of high anisotropy particles.

Figure 9.17 shows polar (field perpendicular) and transverse (field in-plane) Kerr effect hysteresis measurements of a 140 nm thick assembly of 4 nm $Fe_{52}Pt_{48}$ nanoparticles after 30 min anneal at 580°C in dry N_2 (1 atm) atmosphere. The in-plane coercivity (5.1 kOe) is slightly larger than the out-of-plane coercivity (4.6 kOe). The remanence squareness is $S \cong 0.8$ and 0.56 for in-plane and out-of-plane orientation, respectively. The standard θ–2θ X-ray diffraction data, displayed in Fig. 9.10 show an intensity ratio $I_{(001)}/I_{(111)} \cong 0.13$, which is lower than the expected value of about 0.3 for a completely 3D random and fully chemically ordered powder, consistent with the notion of some possible in plane c-axis texturing. The magnetization per FePt nanoparticle volume is estimated to be $M_S \cong (1280 \pm 220)$ emu/cm^3 based on vibrating sample magnetometry [133].

9.6 Magnetic Recording Media Perspective

FePt nanoparticle arrays are prospective magnetic recording media candidates for longitudinal recording, perpendicular recording and patterned media recording. Whether or not such media will become practical depends on factors like large-scale disk coating ability, roughness control, corrosion robustness, control of the easy axis orientation and many others. The first magnetic recording media for hard disks were in fact 'particulate' media consisting of γ-Fe_2O_3 particles dispersed in polymeric binders. The coercivity was obtained from the shape anisotropy of the particles, which had large aspect ratios of 5:1 to 10:1. Disks were fabricated by spin-coating the particles onto Al–Mg substrates. The particle magnetization was oriented in the track direction by applying a circumferential magnetic field before the solvents had completely evaporated. The disks were then "cure-baked" to cross-link the epoxy-phenolic binder. Aluminum oxide was added to the coating for mechanical durability. The aluminum oxide particle diameters were larger than the coating thickness in order to provide separation between the head and the polymeric binder. The as-coated media were polished using fixed abrasives to obtain the final coating thickness and to improve the surface finish. This particulate technology was limited by the low volumetric magnetization. The particle volume concentration was kept below about 30% for adequate mechanical properties of the composite coating to be maintained. Therefore, relatively thick coatings were required to obtain sufficient read-back amplitude. Difficulties were encountered in thinning the coating uniformly and the non-magnetic aluminum oxide particles degraded the signal-to-noise performance of the system [139].

Thin-film technology offered the advantages of much thinner films and smoother surfaces allowing lower flying heights and tunable coercivities, depending on alloy composition. The first attempts at thin-film longitudinal media for disks used plating to deposit the magnetic alloy, but problems with corrosion and obtaining good overcoats for tribology soon made sputter-processing the obvious choice. Hexagonal Co-alloys were widely adopted because of their high uniaxial magnetocrystalline anisotropies.

Today, thermal stability imposed limitations make it unlikely that longitudinal Co-alloy based thin film media can be extended much beyond currently achieved areal densities (Chaps. 5 and 6). Areal densities of 20 Gbit/in^2 in products and 60 Gbit/in^2 in technology demonstrations have been reached [135] and 100–200 Gbit/in^2 may be feasible in longitudinal magnetic recording by continuing to scale head and media dimensions in proportion. Densities beyond 100–200 Gbit/in^2 will likely require a shift to new head and media technologies. Perpendicular magnetic recording appears least invasive promising Tbit/in^2 areal densities [140]. If one considers thermal stability alone it can be argued that 10–50 Tbit/in^2 is possible in single-particle-per-bit patterned media recording [133].

9.6.1 Medium Noise and Thermal Stability

The key impediment to achieving higher areal densities in magnetic recording is media noise, which limits the achievable signal-to-noise ratio SNR of the recording system. Media noise arises due to random placement and irregular shape and size of the magnetic grains. Trimming the grain size distribution and simultaneously reducing the average grain size will allow maintaining a larger number of grains per bit, which has traditionally contained media noise. Continued reduction of the grain size, however, will cause thermal instabilities (Chaps. 5 and 6) setting a lower bound for the grain size. The onset of such thermal instabilities depends besides temperature and time only on the magnetic hardness (anisotropy) of the storage material and on magnetic interactions between grains and across bit transitions. It is generally believed that today's Co-alloy-based media cannot support stable grain diameters smaller than 8–10 nm [137] (see Table 5.3 in Chap. 5). Smaller, stable grains can be obtained in magnetically harder materials, e.g., FePt-compounds, which support stable grain diameters down to the 2–3 nm range. Stability in these materials, however, comes at the expense of large switching coercivities, e.g., 50,000 Oe for FePt [141]. Such large coercivities by far exceed the write field capability of today's recording heads. This problem is somewhat eased in dual-layer perpendicular recording owing to a more efficient write geometry (Chap. 8). Generally, however, in order to record on high anisotropy alloys like FePt, more disruptive technologies, like thermally assisted recording or hybrid recording may be needed. Thermal stability requirements are discussed in detail in Chaps. 5 and 6. Relevant to the present discussion is the fact that a narrow grain size distribution, as observed in FePt nanoparticle arrays, relaxes the stability requirement and permits reduced average grain sizes without the penalty of increased media anisotropy constant and therefore write field of the head. The thermal stability requirement can be expressed as a product of a storage time dependent factor $\ln(\mathrm{f_0 t})$ ($\mathrm{f_0}$ = attempt frequency) and a linear density LD and distribution width σ_A dependent stability factor $\mathrm{r}_K(\mathrm{LD},\sigma_\mathrm{A})$

$$\frac{K_u V}{k_B T} > r_K \cdot \ln(f_0 \tau) \,. \tag{9.1}$$

The factor r_K increases with linear density in longitudinal recording whereas it decreases with linear density in perpendicular recording. A typical value is $r_K(800\ \mathrm{kfci}, \sigma_\mathrm{A} = 0.6) = 1.5$ and $\ln(\mathrm{f_0}\tau = 10\ \mathrm{years}) \cong 40$, hence $r_K \times \ln(\mathrm{f_0}\tau) = 60$, which has been used to estimate thermal stability 'cut-off' diameters (Chap. 5, [7,137]. Note that the logarithmic factor depends on the attempt frequency and ranges from 40–47 for $f_0 \cong 10^9 - 10^{12}\ \mathrm{s}^{-1}$ and $\tau = 10$ years storage time. An illustration of the effect of the area distribution width, σ_A, on thermal stability is given in Fig. 9.18 [142]. The dash dotted line separates the thermally stable region (to the right) from the thermally unstable region (to the left). A medium with broad grain area distribution,

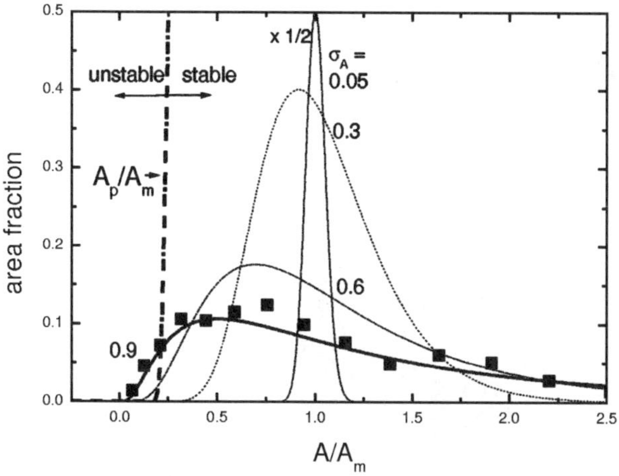

Fig. 9.18. Illustration of $\tau = 10$ year "superparamagnetic " cut-off as function of grain area (*dashed dotted line*). Lognormal grain area distributions for four different widths σ_A in comparison with measured TEM area fraction data [142]. The data are normalized to the median grain area A_m

e.g., $\sigma_A = 0.9$ has a considerable fraction of small grains falling below the thermal stability cut-off. Trimming the grain size distribution centers the grains on the median and moves all grains into the stable regime, as evident for $\sigma_A < 0.6$. A monodisperse nanoparticle array with $\sigma_A \cong 0.05$ allows shifting the median back toward the stability cut-off, i.e. reducing grain area without having to increase the anisotropy constant K_u.

9.7 Areal Density Perspective

The present FePt nanoparticle arrays have the potential of extending magnetic recording to ultimate recording densities in the 10–50 Tbit/in^2 range. They can serve as traditional "granular" media in either longitudinal or perpendicular recording or as 'naturally' patterned as opposed to lithographically patterned media, with extendibility to single particle per bit recording. Figure 9.19 illustrates this point, by showing a 160 nm × 160 nm array of 4-nm FePt-nanoparticles. The array comprises about 1600 particles, amounting to a particle density of about 40 Tparticles/in^2. A hypothetical bit cell at 100 Gbit/in^2 is 160 nm wide × 40 nm long (bit-aspect ratio BAR = 4) and comprises about 400 particles (left white dotted box). At 1 Tbit/in^2 one would still maintain 40 particles per bit cell (25.4 nm × 25.4 nm, BAR = 1, right white dotted box). Assuming that the signal-to-noise ratio (SNR) is determined by statistical averaging of N independent particles per bit, the SNR should scale as

$$\mathrm{SNR} \propto 10 \log(N) \,. \tag{9.2}$$

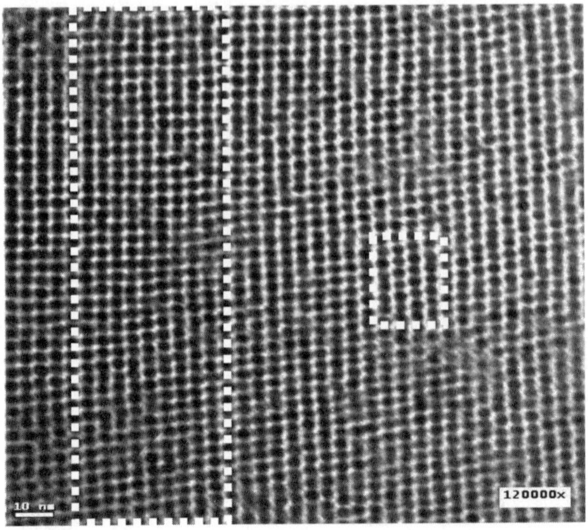

Fig. 9.19. TEM micrograph of a 3D assembly of chemically synthesized FePt nanoparticles with 4 nm diameter. Picture size is 160 × 160 nm comprising about 1600 particles at a particle density of about 40 Tparticles/in^2. Bit cell sizes for 100 Gbit/in^2 (160 nm × 40 nm, BAR = 4, 400 particles per bit) and 1 Tbit/in^2 (25 nm × 25 nm, BAR = 1, 40 particles per bit) are indicated

One obtains SNR = 26 dB for the above 100 Gbit/in^2 scenario and SNR = 16 dB for the 1 Tbit/in^2 scenario. By comparison, the perpendicular 1 Tbit/in^2 scheme described in [140] assumes about 10 grains per bit, hence SNR ≅ 10 dB.

9.7.1 Recording on Nanoparticle Array Media

Actual recording experiments on the present nanoparticle arrays have been performed at low density using an earlier product type merged write-read head with 2.45 μm write width. Figure 9.20 shows read-signal traces of square-wave patterns recorded at linear densities of 500, 1000, 2000, and 5000 fc/mm resulting in areal densities up to 1.3 Gbit/in^2. The experiments were performed using the contact write/read tester discussed in Chap. 5. The nanoparticle arrays used in this first recording demonstration were $t = 140$ nm thick and had an estimated demagnetization length of $M_r t/H_C \cong 26$ nm, using $M_r \cong 930$ emu/cm^3. Much smaller demagnetization lengths and therefore higher linear densities are expected in thinner assemblies. Considering for example a 5 nm FePt assembly of 10 nm overall thickness, i.e. consisting of two particle layers, one can estimate a demagnetization length $M_r t/H_C \cong 2$ nm. With an areal moment density of 0.5 memu/cm^2 and a coercivity of 5000 Oe we estimate a transition width parameter a ≅8.5 nm for reasonable assumptions on head-media spacing and head field gradient. The linear per-

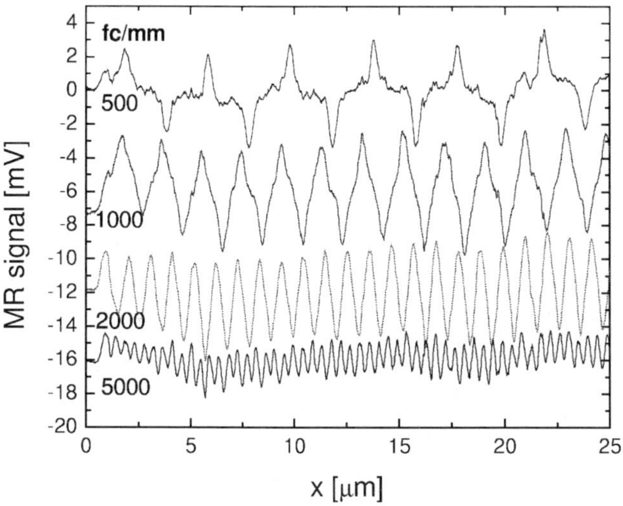

Fig. 9.20. Contact recording results of 140 nm thick 4 nm FePt nanoparticle assemblies (Fig. 4a of [10]

colation density would be 38,000 fc/mm, i.e., 7.6 times larger than shown in Fig. 9.20. Resulting densities would be 250 Gbit/in^2 with 100 nm wide tracks (160 grains/bit, BAR $\sim$ 5) and 1 Tbit/in^2 with 25 nm wide tracks (32 grains/bit, BAR $\sim$ 1).

9.8 Conclusions

Solution-phase chemistry and self-assembly procedures, in general, show great promise in producing high quality monodisperse magnetic nanoparticle materials and arrays. Self-assembled FePt nanoparticle arrays, in particular, as described in this chapter, may evolve into prospective alternative future ultrahigh density magnetic recording media with an areal density potential far beyond terabit per square inch. They may be viewed as either traditional "granular" media in either longitudinal or perpendicular recording or as 'naturally' patterned media, with extendibility to single particle per bit recording. Whether these magnetic nanoparticle arrays will become practical depends on numerous factors, such as large-scale coating ability, roughness control, assembly stability and control of the magnetic easy axis orientation.

Magnetic nanoparticle arrays may also have considerable promise in future nanoelectronic device applications. It is expected that well-controlled magnetic superlattice systems, e.g., monodisperse magnetic nanoparticles embedded in a nonmagnetic matrix will reveal novel magneto-transport effects.

References

1. D. A. Thompson, J. S. Best, IBM J. Res. Develop. **44** 311 (2000)
2. E. S. Murdock, R. F. Simmons, R. Davison, IEEE Trans. Mag. **28**, 3078 (1992)
3. T. Yogi, T. A. Nguyen, IEEE Trans. Mag. **29**, 307 (1993)
4. P.-L. Lu, S. H. Charap, IEEE Trans. Mag. **30**, 4230 (1994)
5. D. N. Lambeth, E. M. T. Velu, G. H. Bellesis, L. Lee, D. E. Laughlin, J. Appl. Phys. **79**, 4496 (1996)
6. J. Li, M. Mirzamaani, X. Bian, M. Doerner, S. Duan, K. Tang, M. Toney, T. Arnoldussen, M. Madison, J. Appl. Phys. **85**, 4286 (1999)
7. D. Weller, A. Moser, IEEE Trans. Magn. **35**, 4423 (1999)
8. S. Sun, C. B. Murray, J. Appl. Phys. **85**, 4325 (1999)
9. S. Sun, C. B. Murray, H. Doyle, Mat. Res. Soc. Symp. Proc. **577**, 385 (1999)
10. S. Sun, C. B. Murray, D. Weller, L. Folks, A. Moser, Science **287**, 1989 (2000)
11. C. B. Murray, D. J. Norris, M. G. Bawendi, J. Am. Chem. Soc. **115**, 8706 (1993)
12. X. Peng, L. Mana, W. Yang, J. Wickham, E. Scher, A. Kadavanich, A. P. Alivisatos, Nature **404**, 59 (2000)
13. I. Lisiecki, M. P. Pileni, J. Am. Chem. Soc. **115**, 3887 (1993)
14. T. Kawai, Y. Usui, K. Kon-No, Coll. Surf. A **149**, 39 (1999)
15. J. D. Hopwood, S. Mann, Chem. Mater. **9**, 1819 (1997)
16. T. Li, J. Moon, A. A. Morrone, J. J. Mecholsky, D. R. Talham, J. H. Adair, Langmuir **15**, 4328 (1999)
17. A. Taleb, C. Petit, M. P. Pileni, Chem. Mater. **9**, 950 (1997)
18. M. P. Pileni, B. W. Ninham, T. Gulik-Krzywicki, J. Tanori, I. Lisiecki, A. Filankembo, Adv. Mater. **11**, 1358 (1999)
19. J. R. Thomas, J. Appl. Phys. **37**, 2914 (1966)
20. J. S. Yin, Z. L. Wang, Phys. Rev. Lett. **79**, 2570 (1997)
21. C. B. Murray, S. Sun, W. Gaschler, H. Doyle, T. Betley, C. R. Kagan, IBM J. Res. Develop., in press (2001)
22. J. Osuna, D. de Caro, C. Amiens, B. Chaudret, E. Snoeck, M. Respaud, J.-M. Broto, A. Fert, J. Phys. Chem. **100**, 14571 (1996)
23. C. H. Griffiths, M. P. O'Horo, T. W. Smith, J. Appl. Phys. **50**, 7108 (1979)
24. J. van Wonterghem, S. Morup, S. W. Charles, S. Wells, J. Villadsen, Phys. Rev. Lett. **55**, 410 (1985)
25. J. van Wonterghem, S. Morup, S. W. Charles, S. Wells, J. Colloid Interface Science **121**, 558 (1988)
26. M. D. Bemtzon, J. van Wonterghem, S. Morup, A. Tholen, C. J. W. Koch, Phil. Mag. B **60**, 169 (1989)
27. C. Johansson, M. Hanson, P. V. Hendriksen, S. Morup, J. Magn Mag. Mater. **122**, 125 (1993)
28. S.-J. Park, S. Kim, S. Lee, Z. G. Khim, K. Char, T. Hyeon, J. Am. Chem. Soc. **122**, 8581 (2000)
29. K. S. Suslick, S.-B. Choe, A. A. Cichowlas, M. W. Grinstaff, Nature **353**, 414 (1991)
30. X. Cao, Y. Koltypin, G. Kataby, R. Prozorov, A. Gedanken, J. Mater. Res. **10**, 2952 (1995)
31. K. S. Suslick, M. Fang, T. Hyeon, J. Am. Chem. Soc. **118**, 11960 (1996)

32. K. S. Suslick, M. Fang, T. Hyeon, A. A. Cichowlas, Mat. Res. Soc. Symp. Proc. **351**, 443 (1994)
33. D. de Caro, J. S. Bradley, Langmuir **13**, 3067 (1997)
34. T. O. Ely, C. Amiens, B. Chaudret, E. Snoeck, M. Verelst, M. Respaud, J.-M. Broto, Chem. Mater. **11**, 526 (1999)
35. R. D. Rieke, Acc. Chem. Res. **10**, 301 (1977)
36. R. D. Rieke, Science **246**, 1260 (1989)
37. H. I. Schlesinger, H. C. Brown, A. E. Finholt, J. R. Gilbreath, H. R. Hockstrue, E. K. Hyde, J. Am. Chem. Soc. **84**, 1493 (1953)
38. L Yiping, G. C. Hadjipanayis, C. M. Sorensen, K. J. Klabunde, J. Magn. Mag. Mater. **79**, 321 (1989)
39. G. N. Glavee, K. J. Klabunde, C. M. Sorensen, G. C. Hadjipanayis, Langmuir **10**, 4726 (1994)
40. Y.-P. Sun, H. W. Rollins, R. Guduru, Chem. Mater. **11**, 7 (1999)
41. A. Duteil, G. Schmid, W. Meyer-Zaika, J. Chem. Soc., Chem. Commun., 31 (1995)
42. L. Yiping, G. C. Hadjipanayis, C. M. Sorensen, K. J. Klabunde, J. Appl. Phys. **67**, 4502 (1990)
43. K.L. Tsai, J. L. Dye, Chem. Mater. **5**, 540 (1993)
44. D. L. Leslie-Pelecky, X. Q. Zhang, R. D. Rieke, J. Appl. Phys. **79**, 5312 (1996)
45. D. L. Leslie-Pelecky, M. Bonder, T. Martin, E. M. Kirkpatrick, Y. Liu, X. Q. Zhang, S.-H. Kim, R. D. Rieke, Chem. Mater. **10**, 3732 (1998)
46. D. L. Leslie-Pelecky, X. Q. Zhang, S. H. Kim, M. Bonder, R. D. Rieke, Chem. Mater. **10**, 164 (1998)
47. H. Bonnemann, W. Brijoux, T. Joussen, Angew. Chem. Int. Ed. Engl. **29**, 273 (1990)
48. H. Bonnemann, R. Brinkmann, R. Koppler, P. Neiteler, J. Richter, Adv. Mater. **4**, 804 (1992)
49. C. P. Gibson, K. J. Putzer, Science **267**, 1338 (1995)
50. F. Fievet, J. P. Lagier, M. Figlarz, MRS Bulletin, 29 (1989)
51. M. T. Reetz, W. Helbig, J. Am. Chem. Soc. **116**, 7401 (1994)
52. J. A. Becker, R. Schafer, R. Festag, W. Ruland, J. H. Wendoff, J. Pebler, S. A. Quaiser, W. Helbig, M. T. Reetz, J. Chem. Phys. **103**, 2520 (1995)
53. J. P. Chen, C. M. Sorensen, K. J. Klabunde, G. C. Hadjipanayis, Phys. Rev. B **51**, 11527 (1995)
54. X. M. Lin, C. M. Sorensen, K. J. Klabunde, G. C. Hadjipanayis, Langmuir **14**, 7140 (1998)
55. C. Petit, A. Taleb, M. P. Pileni, Adv. Mater. **10**, 259 (1998)
56. X. M. Lin, C. M. Sorensen, K. J. Klabunde, G. C. Hadjipanayis, J. Mater. Res. **14**, 1542 (1999)
57. C. Petit, A. Taleb, M. P. Pileni, J. Phys. Chem. B **103**, 1805 (1999)
58. C. Petit, T. Cren, D. Roditchev, W. Sacks, J. Klein, M. P. Pileni, Adv. Mater. **11**, 1198 (1999)
59. D.-H. Chen, S.-H. Wu, Chem. Mater. **12**, 1354 (2000)
60. C. B. Murray, C. R. Kagan, M. G. Bawendi, Science **270**, 1335 (1995)
61. S. A. Harfenist, Z. L. Wang, M. M. Alvarez, I. Vezmar, R. L. Whetten, J. Phys. Chem. **100**, 13904 (1996)
62. S. A. Harfenist, Z. L. Wang, R. L. Whetten, I. Vezmar, M. M. Alvarez, Adv. Mater. **9**, 817 (1997)

63. M. P. Pileni, Langmuir **13**, 3266 (1997)
64. J. Fink, C. J. Kiely, D. Bethell, D. J. Schiffrin, Chem. Mater. **10**, 922 (1998)
65. B. A. Korgel, S. Fullam, S. Connolly, D. Fitzmaurice, J. Phys. Chem. B **102**, 8379 (1998)
66. Y. A. Vlasov, N. Yao, D. J. Norris, Adv. Mater. **11**, 165 (1999)
67. P. Jiang, J. F. Bertone, K. S. Hwang, V. L. Colvin, Chem. Mater. **11**, 2132 (1999)
68. Q.-H. Wei, D. M. Cupid, X. L. Wu, Appl. Phys. Lett. **77**, 1641 (2000)
69. Z. L. Wang, Adv. Mater. **10**, 13 (1998)
70. B. A. Korgel, S. Fullam, S. Connolly, D. Fitzmaurice, J. Phys. Chem. B **102**, 8379 (1998)
71. D. Zanchet, M. S. Moreno, D. Ugarte, Phys. Rev. Lett. **82**, 5277 (1999)
72. K. Pohl, M. C. Bartelt, J. de la Figuera, N. C. Bartelt, J. Hrbek, R. Q. Hwang, Nature **397**, 238 (1999)
73. Z. L. Wang, Z. R. Dai, S. Sun, Adv. Mater. **12**, 1944 (2000)
74. A. Ulman, Adv. Mater. **2**, 573 (1990)
75. C. Gros, F. Rabiet, F. Denat, S. Brandes, H. Chollet, R. Guilard, J. Chem. Soc., Dalton Trans., 1209 (1996)
76. Z. Yang, H. Yu, Adv. Mater. **9**, 426 (1997)
77. X. Yang, J. Shi, S. Johnson, B. Swanson, Langmiur **14**, 1505 (1998)
78. V. V. Tsukruk, I. Luzinov, D. Julthongpiput, Langmuir **15**, 3029 (1999)
79. R. G. Nuzzo, D. L. Allara, J. Am. Chem. Soc. **105**, 4481 (1983)
80. R. G. Nuzzo, F. A. Fusco, D. L. Allara, J. Am. Chem. Soc. **109**, 2358 (1987)
81. M. D. Porter, T. B. Bright, D. L. Allara, C. E. D. Chidsey, J. Am. Chem. Soc. **109**, 3559 (1987)
82. E. B. Troughton, C. D. Bain, G. M. Whitesides, R. G. Nuzzo, D. L. Allara, M. D. Porter, Langmuir **4**, 365 (1988)
83. C. D. Bain, E. B. Troughton, Y.-T. Tao, J. Evall, G. M. Whitesides, R. G. Nuzzo, J. Am. Chem. Soc. **111**, 321 (1989)
84. G. K. Jennings, P. E. Laibinis, J. Am. Chem. Soc. **119**, 5208 (1997)
85. C. K. Jennings, J. C. Munro, P. E. Laibinis, Adv. Mater. **11**, 1000 (1999)
86. R. G. Freeman, K. C. Grabar, K. J. Allison, R. M. Bright, J. A. Davis, A. P. Guthrie. M. B. Hommer, M. A. Jackson, P. C. Smith, D. G. Walter, M. J. Natan, Science **267**, 1629 (1995)
87. A. Doron, E. Katz, I. Willner, Langmuir **11**, 1313 (1995)
88. J. Zheng, Z. Zhu, H. Chen, Z. Liu, Langmuir **16**, 4409 (2000)
89. A. Gole, S. R. Sainkar, M. Sastry, Chem. Mater. **12**, 1234 (2000)
90. K. M. Chen, X. Jiang, L. C. Kimerling, P. T. Hammond, Langmuir **16**, 7825 (2000)
91. S. Sun, unpublished data.
92. Y. Liu, A. Wang, R. Claus, J. Phys. Chem. B **101**, 1385 (1997)
93. T. Cassagneau, T. E. Mallouk, J. H. Fendler, J. Am. Chem. Soc. **120**, 7848 (1998)
94. M. A. Correa-Duarte, M. Giersig, N. A. Kotov, L. M. Liz-Marzan, Langmuir **14**, 6430 (1998)
95. F. G. Aliev, M. A. Correa-Duarte, A. Mamedov, J. W. Ostrander, M. Giersig, L. M. Liz-Marzan, N. A. Kotov, Adv. Mater. **11**, 1006 (1999)
96. E. Braun, Y. Eichen, U. Sivan, G. Ben-Yoseph, Nature **391**, 775 (1998)

97. J. Richter, R. Seidel, R. Kirsch, M. Mertig, W. Pompe, J. Plaschke, H. K. Schackert, Adv. Mater. **12**, 507 (2000)
98. S. R. Whaley, D. S. English, E. L. Hu, P. F. Barbara, A. M. Belcher, Nature **405**, 665 (2000)
99. C. R. Martin, Chem. Mater. **8**, 1739 (1996)
100. D. Zhao, J. Feng, Q. Huo, N. Melosh, G. H. Fredrickson, B. F. Chmelka, G. D. Stucky, Science **279**, 548 (1998)
101. O. D. Velev, T. A. Jede, R. F. Lobo, A. M. Lenhoff, Chem. Mater. **10**, 3597 (1998)
102. S. H. Park, Y. Xia, Chem. Mater. **10**, 1745 (1998)
103. E. Kramer, S. Forster, C. Goltner, M. Antonietti, Langmuir **14**, 2027 (1998)
104. B. T. Holland, C. F. Blanford, T. Do, A. Stein, Chem. Mater. **11**, 795 (1999)
105. T. Tatsuma, A. Ikezawa, Y. Ohko, T. Miwa, T. Matsue, A. Fujishima, Adv. Mater. **12**, 643 (2000)
106. H. Masuda, K. Fukuda, Science **268**, 1466 (1995)
107. H. Masuda, H. Yamada, M. Satoh, H. Asoh, M. Nakao, T. Tamamura, Appl. Phys. Lett. **71**, 2770 (1997)
108. A. P. Li, F. Muller, A. Birner, K. Nielsch, U. Gosele, J. Appl. Phys. **84**, 6023 (1998)
109. D. Crouse, Y. H. Lo, A. E. Miller, M. Crouse, Appl. Phys. Lett. **76**, 49 (2000)
110. K. B. Shelimov, D. N. Davydov, M. Moskovits, Appl. Phys. Lett. **77**, 1722 (2000)
111. M. Zheng, L. Menon, H. Zeng, Y. Liu, S. Bandyopadhyay, R. D. Kirby, D. J. Sellmyer, Phys. Rev. B **62**, 12282 (2000)
112. N. Grobert, W. K. Hsu, Y. Q. Zhu, J. P. Hare, H. W. Kroto, M. Terrones, H. Terrones, Ph. Redlich, M. Ruhle, R. Escudero, F. Morales, Appl. Phys. Lett. **75**, 3363 (1999)
113. S. Liu, J. Zhu, Y. Mastai, I. Felner, A. Gedanken, Chem. Mater **12**, 2205 (2000)
114. D. Dinega, M. G. Bawendi, Angew. Chem Int. Ed. Eng. **38**, 1788 (1999)
115. C. T. Black, C. B. Murray, R. L. Sandstrom, S. Sun, Science **290**, 1131 (2000)
116. C. L. Chien, J. Appl. Phys. **69**, 5267 (1991)
117. A. E. Berkowitz, J. R. Mitchell, M. J. Carey, A. P. Young, S. Zhang, F. E. Spada, F. T. Parker, A. Hutten, G. Thomas, Phys. Rev. Lett. **68**, 3745 (1992)
118. J. Q. Xiao, J. S. Jiang, C. L. Chien, Phys. Rev. Lett. **68**, 3749 (1992)
119. Y. G. Pogorelov, M. M. P. de Azevedo, J. B. Sousa, Phys. Rev. B **58**, 425 (1998)
120. S. Takahashi, S. Maekawa, Phys. Rev. Lett. **80**, 1758 (1998)
121. A. Cebollada, D. Weller, J. Sticht, G. R. Harp, R. F. C. Farrow, R. F. Marks, R. Savoy, J. C. Scott., Phys. Rev. B **50** 3419 (1994)
122. R. F. C. Farrow, D. Weller, R. F. Marks, M. F. Toney, A. Cebollada, G. R. Harp, J. Appl. Phys. **79**, 5967 (1996)
123. Y. Ide, T. Goto, K. Kikuchi, K. Watanabe, J. Onagawa, H. Yoshida, J. M. Cadogan, J. Magn. Mag. Mater. **245**, 177 (1998)
124. N. Li and B. M. Lairson, IEEE Trans. Mag. **35**, 1077 (1999)
125. R. A. Ristau, K. Barmak, L. H. Lewis, K. R. Coffey, J. K. Howard, J. Appl. Phys. **86**, 4527 (1999)
126. C. P. Luo, S. H. Liou, D. J. Sellmyer, J. Appl. Phys. **87**, 6941 (2000)
127. B. Bian, K. Sato, Y. Hirotsu, A. Makino, Appl. Phys. Lett. **75**, 3686 (1999)

128. B. Bian, D. E. Laughlin, K. Sato, Y. Hirotsu, J. Appl. Phys. **87**, 6962 (2000)
129. C. P. Luo, S. H. Liou, L. Gao, Y. Liu, D. J. Sellmyer, Appl. Phys. Lett. **77**, 2225 (2000)
130. C.-M. Kuo, P. C. Kuo, J. Appl. Phys. **87**, 419 (2000)
131. E. E. Carpenter, J. A. Sims, J. A. Wienmann, W. L. Zhou, C. J. O'Connor, J. Appl. Phys. **87**, 5615 (2000)
132. S. Sun, E. E. Fullerton, D. Weller, C. B. Murray, MMM/Intermag, San Antonio, Jan 7–11, 2001, paper AA-06, IEEE Trans. Magn., in press (2001)
133. D. Weller, S. Sun, C. Murray, L. Folks, A. Moser, MMM/Intermag, San Antonio, Jan 7–11, 2001, paper DC-02, IEEE Trans. Magn., in press (2001)
134. S. H. Charap, P.-L. Lu, Y. He, IEEE Trans. Magn. **33**, 978 (1997)
135. Densities in the range 35–63 Gbit/in^2 have been demonstrated by Read-Rite, Fujitsu, Hitachi, Seagate and IBM in longitudinal recording. Hitachi has demonstrated 63 Gbit/in^2 in perpendicular recording
136. H. N. Bertram, M Williams, IEEE Trans. Magn. **36**, 4 (2000)
137. D. Weller, A. Moser, L. Folks, M.E. Best, W. Lee, M.F. Toney, J.-U. Thiele, and M. F. Doerner, IEEE Trans. Magn. **36**, 10 (2000)
138. Y. Ide, T. Goto, K. Kikuchi, K. Watanabe, J. Onagawa, H. Yoshida, and J. M. Cadogan, J. Magn. Magn. Mater. **245**, 177 (1998)
139. V. M. DePalma, R. J. Dilley, M. F. Doerner, D. A. Nepela, U.S. Patent No. 4,451,53 (1981)
140. R. Wood, IEEE Trans. Magn. **36**, 36 (2000)
141. Y. Ide, T. Goto, K. Kikuchi, K. Watanabe, J. Onagawa, H. Yoshida, and J. M. Cadogan, J. Magn. Magn. Mater. **245**, 177 (1998)
142. D. Weller and M.F. Doerner, Annual Review of Materials Science **30**, 611–644 (2000)

10 Theory of Magnetotransport for Magnetic Recording

William H. Butler, Olle Heinonen, and Xiaoguang Zhang

10.1 Playback of a Recorded Transition

The sensor in a disc drive read head must generate a signal large enough compared to noise levels that it can be processed with an acceptable error rate. In addition, the read head must satisfy geometrical constraints and also be manufacturable. Let us first estimate the orders of magnitudes involved in this process in order to develop a sense for the constraints imposed on the sensor. We shall use longitudinal recording as an example. Whether the estimates we shall obtain will change appreciably for perpendicular recording remains to be seen.

The read sensor is located between two shields (see the chapter Recording Head Design by Mallary in the present volume for basic head features). The geometrical constraints are set by the shield-to-shield spacing and the track pitch. The latter impacts the track width and so the read sensor width. The track pitch is largely determined by how narrow the write head can write the tracks and by how well the track-following servo system can keep the recording head on track. If we take as an example a 100 Gbit/in^2 system, a reasonable track pitch is 120 ktpi (kilo tracks per inch), corresponding to a track width of about 200 nm. The read sensor should be narrower than the track width in order to avoid the edges of recorded tracks as well as the influence of adjacent tracks, and also to allow for some play for the servo system which makes the read sensor follow a track, so a read sensor width of 100 nm is reasonable.

Since the areal density is given by the product of tracks per inch and bits per inch along the track, an areal density of 100 Gbit/in^2 would then require 833 kbpi (kilo bits per inch). The bit size in the track direction is then about 30 nm (this is on the average about three to four grains in a medium for 100 Gbit/in^2 applications!). The average bit size along the track determines the shield-to-shield spacing. The purpose of the shields is to suppress fields from transitions adjacent to the one the read sensor is trying to read; so we want to avoid having many transitions between the shields at the same time. (This is somewhat complicated by the fact that the actual bit size is larger at the outer diameter of the disc than at the inner diameter.) A rule of thumb for the shield-to-shield spacing is to set it at about twice the average bit size, so this would give us a shield-to-shield spacing of about 60 nm.

The signal is generated by the magnetostatic fringing fields emanating from written transitions in the medium. To get a simple estimate for the field strength from a written transition, we use the arctangent model [1] for a written transition. If we assume a medium remnant magnetization $M_r = 350$ emu/cm^3, corresponding to CoCrPtB medium (see Chap. 2 of this volume), a head-to-medium separation of 15 nm, medium thickness $\delta = 10$ nm, and a transition parameter $a = 7$ nm, the vertical component of the field at the air-bearing surface is about 400 Oe. Unfortunately, the field between the shields decays roughly linearly to zero at the upper edge of the read sensor, due to flux leakage to the shields, so the average maximum field at the read sensor is then about 200 Oe.

We would like our sensor to have an electronic signal-to-noise ratio of, say, 30 dB. For the sake of argument, let us lump all the noise sources into an equivalent resistor of resistance 200 Ω and a bandwidth of 500 MHz. At an operating temperature of 400 K, this gives a Johnson noise voltage of about 50 μV, so the output voltage of the read sensor should be about 2 mV in order to maintain a 30 dB signal-to-noise ratio. We can then summarize the above estimates as follows: the sensor for a 100 Gbit/in^2 application should be about 100 nm wide and no more than 60 nm thick, and at the same time be capable of producing a peak voltage of about 2 mV for an average field of less than 200 Oe. The sensor must be capable of producing this output at current densities less than something like 1×10^8 A/m^2 – this limit is set by the need to keep the Joule heating low enough that the operation of the sensor is not severely affected by elevated temperatures (this limit on current density may be eased somewhat for CPP GMR read sensor discussed later as these may provide better heat dissipation). These constraints on the design of a working read sensor severely restrict the degrees of freedom available to the design process so that producing a read sensor reliably and inexpensively is a formidable challenge.

Currently, and presumably also in the near future, magnetic sensors will rely on the physics of spin-dependent transport in order to generate a signal. Present devices based on this principle are spin valves. The relatively large signal amplitude from spin valve read sensors compared to earlier thin film heads, based on the anisotropic magnetoresistance, allowed for a rapid growth in areal density. As we look into the future, it appears that spin valves will continue to be the technology of choice for some time to come, but it also appears that heads based on tunneling magnetoresistance may emerge as strong contenders as the areal density continues to increase. In this chapter, we will look in some detail into the physics underlying spin valves and tunneling heads.

10.2 Two-Current Model for Transport in Transition Metals and Alloys

It has long been known that electrons in metals have two spin states. Often the coupling between these two spin states is ignored and they are treated as channels that conduct independently. In non-magnetic metals such as copper, the two spin channels are equivalent in the sense that they have the same Fermi energy[1] E_F, density of states and the same electron velocities, but in the ferromagnetic transition metals they are usually quite different. In order to understand spin-dependent transport we must first understand how applied electric fields lead to currents in metals. In the transition and noble metals that we will be concerned with here, the approximation of treating the two spin channels as independent is usually a good one. This approximation is called the "two current model" and is the subject of this section.

10.2.1 Electronic Structure of Transition Metals and Alloys

A basic understanding of the electronic structure of ferromagnetic transition metals and alloys is necessary for understanding electron transport in spin valves and spin tunneling devices. The electronic structures of Cu and fcc Co are presented in terms of their electronic densities of states (DOS) in Fig. 10.1. These DOS results were calculated using the local spin density approximation (LSDA) to density functional theory [2]. In these calculations we have neglected the relativistic coupling between the spin and orbital motions of the electrons and we have assumed that the magnetic moments of the Co atoms are all aligned.

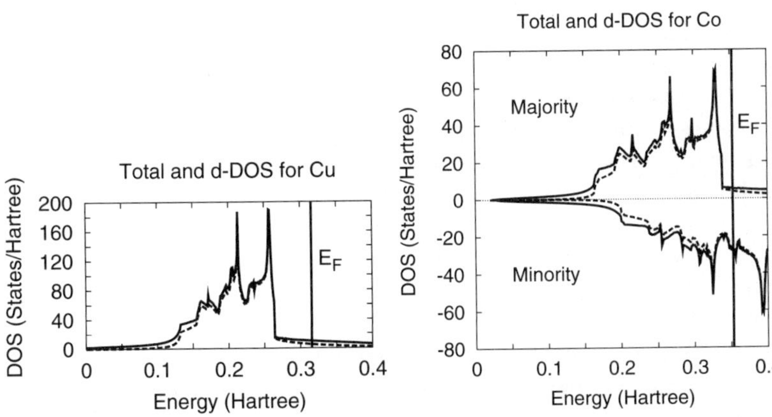

Fig. 10.1. Electronic density of states for copper (left panel) and for fcc cobalt (right panel). Dashed lines show the d-component of the density of states.

[1] The Fermi energy is the highest energy of all occupied electron states

For copper, the up and down spin DOS are identical so that the net spin moment is zero[2]. In this case a single curve is sufficient to display the DOS as shown in the left panel of Fig. 10.1. For Co, however, the majority and minority DOS curves are different as is shown in the right panel of Fig. 10.1. The area under the curve above the axis represents the majority or "up"-spin DOS and the area below represents the minority or "down" spin DOS. The majority DOS curve contains more states below the Fermi energy and can be viewed, at least approximately, as being shifted downward in energy relative to the minority DOS. The amount of this shift is sometimes called the spin-splitting energy. Depending on which part of the band structure is used as reference, one can obtain slightly different estimates for this quantity, but we estimate, based on Fig. 10.1 that the spin-splitting energy for Co is approximately 1.5 eV. The total area under the DOS curves up to the Fermi energy, E_F, gives the total number of valence electrons for each spin channel. The difference between the number of majority and minority valence electrons is the spin magnetic moment per atom in Bohr magnetons; in this case, 1.6 for Co (which may be compared to the experimental value of about 1.75 Bohr magnetons for the total moment, which includes both spin and orbital contributions).

As isolated atoms, all transition metals are magnetic in the sense that they have a net spin angular momentum and an associated magnetic moment. Near the center of the transition metal series these atomic spin moments can be quite high. These high moments come about because of the near degeneracy of the *d*-states in a transition metal atom. In fact, one of the results of atomic physics, known as Hund's rule [3], states that the lowest energy spin configuration will be one that maximizes the spin[3]. In contrast to the ubiquity of magnetic moments for isolated transition metal atoms, only Fe, Co, and Ni of the transition metals are ferromagnetic as bulk crystals[4]. The disappearance of magnetism when transition metal atoms are assembled to make a crystal occurs because the atomic *d*-level spreads into a band in the solid. (The part of the DOS with *d*-symmetry about the nucleus is shown by dashed lines in Fig. 10.1). For an atomic *d*-level, it costs no kinetic energy to put as many electrons as will fit (5 per atom) in the same spin state. In the solid, however, because the *d*-states are spread into a band, it costs additional kinetic energy if the up and down spin bands are not occupied equally. Usually, in a solid this additional kinetic energy cost overwhelms the Hund's rule energy that wants to align the spins. Among the transition metals, it is only at the end of the first transition metal series where the *d*-bands are relatively narrow that the Hund's rule energy still wins and the atoms have a net spin angular momentum and magnetic moment.

[2] Cu is weakly diamagnetic due to the orbital motion of its electrons.

[3] This rule arises because the high spin states keep the electrons farther away from each other and thus lower the repulsive Coulomb energy

[4] Mn and Cr are antiferromagnetic.

There are other magnetic elements, notably the rare earths and actinides in which the magnetism arises from very narrow f-bands. We will not discuss these systems here, however, because these bands are so narrow and so highly correlated that they do not contribute significantly to electron transport.

10.2.2 Two Current Model for Transport

The density of states curves shown in Fig. 10.1 arise from the underlying energy bands, $\varepsilon_{n,s}(\mathbf{k})$ through counting all of the states $|\psi_{n,s}(\mathbf{k})\rangle$ (i.e., all wave vectors $\mathbf{k}$, and all bands n) with energy E,

$$\mathcal{D}_s(E) = \sum_{n,k} \delta\left[E - \varepsilon_{n,s}(\mathbf{k})\right] . \tag{10.1}$$

Here n is the band index, s denotes the spin channel. The electron velocities at the Fermi energy will be very important for electron transport. The electron velocity is determined from the gradient of the energy bands with respect to the crystal momentum [4],

$$\hbar\mathbf{v}_{n,s}(\mathbf{k}) \equiv \frac{-i\hbar^2}{m} \langle\psi_{n,s}(\mathbf{k})|\nabla|\psi_{n,s}(\mathbf{k})\rangle = \nabla_{\mathbf{k}}\varepsilon_{n,s}(\mathbf{k}) \tag{10.2}$$

These velocities vary considerably from point to point on the Fermi surface, both in magnitude and in direction as can be seen in Figs. 10.2 and 10.3. We recall that the Fermi surface is an iso-energy surface in $\mathbf{k}$-space on which the electron states have band energies equal to the Fermi energy E_{F}.

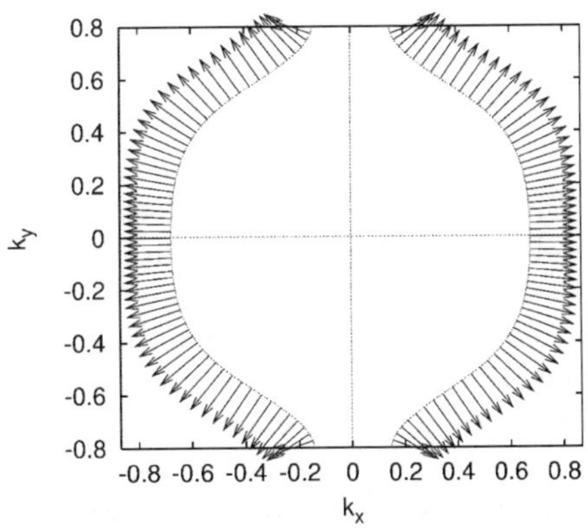

Fig. 10.2. Cut through the $k_z = 0$ plane of the Fermi surface of copper. The arrows indicate the magnitude and direction of the electron velocity. k_x and k_y are given in units of inverse Bohr radii (1.89 Å^{-1}). The k_y direction here is the (111) direction with respect to the conventional cubic axes

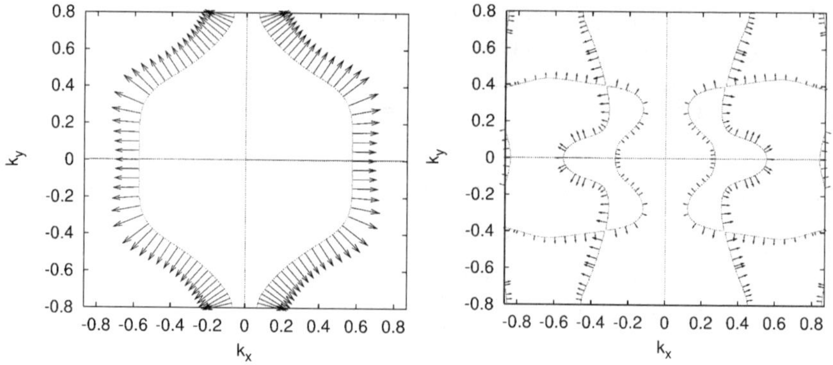

Fig. 10.3. Cuts through the $k_z = 0$ plane of the majority (left panel) and minority (right panel) Fermi surfaces of cobalt. The arrows representing the velocities are drawn to the same scale as for copper in Fig. 10.2. k_x and k_y are given in units of inverse Bohr radii

10.2.3 Corrections to the Two Current Model

It is important to remember that the two current model is an approximation. It is valid in a limit in which spin-orbit coupling is neglected and in which it is assumed that the direction of the magnetic moments of all of the atoms are aligned (parallel or antiparallel) along the same axis. Spin-orbit coupling is a relativistic effect that couples the spin and orbital motions of the electrons. It introduces a small additional term into the non-relativistic Schrödinger equation of the form,

$$H_{\mathrm{so}} = \frac{\hbar^2}{2m^2c^2r}\frac{\mathrm{d}V}{\mathrm{d}r}\mathbf{L}\cdot\mathbf{S}, \tag{10.3}$$

with $\mathbf{L}$ and $\mathbf{S}$ the orbital and spin angular momentum operators, respectively, and $V(r)$ the effective potential (assumed here to depend only on the distance to the nucleus). Because this term couples the spin and orbital motions of the electrons, their energy bands can no longer be described as purely up or down spin.

If all of the magnetic moments are not aligned parallel or antiparallel, the two current model also breaks down. In a ferromagnet (ignoring spin-orbit coupling) the majority and minority see different potentials, $V^{\uparrow}$ and $V^{\downarrow}$, respectively. If the moments in two nearby magnetic layers are aligned anti-parallel, the majority electrons in the first layer (where they experience potential $V^{\uparrow}$) will experience potential $V^{\downarrow}$ when they travel to the nearby layer with moments antiparallel to the first. Nevertheless, the two spin channels can be treated separately. If, on the other hand, the moments in the second ferromagnetic layer are aligned at some arbitrary angle relative to the first layer, the two spin channels defined in the first layer will be coupled in the second layer, and they can no longer be treated separately.

10.3 Giant Magnetoresistance and Spin Dependent Tunneling

The ability to sense small changes in magnetic field that occur over very small distances is critical to the design and operation of a read sensor for magnetic data storage. Currently, magnetotransport appears to be the technique that yields the best performance. Three magnetotransport effects are important in this regard, anisotropic magnetoresistance (AMR), giant magnetoresistance (GMR), and tunneling magnetoresistance (TMR) [5,6]. AMR is the dependence of the resistivity on the relative orientation of the magnetic field and the current. AMR arises from the relativistic coupling between the spin and orbital motions of the electrons discussed briefly in the last section [7]. Because AMR read sensors have largely been superseded by GMR sensors due to the larger signal provided by the latter, we will not discuss AMR sensors here. GMR sensors are currently used in most magnetic data recording devices and it appears likely that TMR sensors may be important for future devices.

GMR and TMR both require a system that is magnetically inhomogeneous on a fine scale. In both phenomena, the resistance typically decreases when an applied magnetic field aligns the magnetic moments in different regions of the material. Read sensor applications of GMR and TMR utilize a planar geometry. This is beneficial because magnetostatic interactions constrain the magnetization to lie in-plane so that a relatively small external field can rotate the magnetization to a different direction in the plane. For GMR, the layers are metallic. Ferromagnetic layers are separated by non-magnetic spacer layers. For TMR, the ferromagnetic layers are separated by a thin insulating barrier.

GMR can be observed in both a current-in-the-plane (CIP) geometry in which the current flows parallel to the layers and in a current-perpendicular-to-the-plane geometry (CPP) in which the current flows perpendicular to the layers. The GMR effect is generally larger for CPP in artificial periodic multilayer systems (multilayers for short), but applications so far have avoided the CPP geometry because of the extremely low resistances across the thickness of thin metallic films. Within the two-current model for magnetotransport, it is relatively easy to develop a zeroth-order model which contains the fundamental picture of these effects. We will do that in this section, and in subsequent chapters we will examine these effects from the point of view of semiclassical and quantum transport theory formulations.

The GMR effect was originally discovered in an artificial periodic Fe/Cr multilayer [9]. It was soon discovered that a large GMR effect could also be obtained in Co/Cu multilayers [10]. In multilayers, it can be arranged that consecutive magnetic layers have their magnetic moments aligned antiparallel in the absence of an applied magnetic field. This is accomplished by choosing the correct thickness of the non-magnetic (spacer) layer so that the exchange coupling between two magnetic layers across the spacer layer is negative. This exchange coupling is mediated by the conduction electrons in the spacer layer and oscillates with a period of $1/k^*$ where k^* is an extremal spanning vector of the Fermi surface of the spacer layer in the direction of stacking of the multilayer [11]. The origin of this oscillatory coupling can be traced to the same source as the familiar Ruderman–Kittel–Kasuya–Yosida (RKKY) coupling between magnetic impurities [12], namely the discontinuity in the occupation of the electronic states at the Fermi energy. The period and magnitude of the coupling depends on the growth direction of the spacer layer. For Co–Cu layers stacked along the (111) direction, for example, the coupling is strongly antiferromagnetic for a Cu thickness of about 8.5 Å.

By applying a sufficiently strong external field, the magnetization of all of the ferromagnetic layers can be aligned. With a current flowing either in CPP or CIP configuration, the resistance will change depending on the relative orientation of the magnetization in consecutive Co layers, with a concomitant change in current (for constant applied voltage).

10.3.1 CIP-GMR

For the CIP configuration, the electric field is uniform along the in-plane direction. One limit in which CIP GMR is relatively easy to understand is that in which the elastic mean free paths $\ell_\uparrow$ and $\ell_\downarrow$ for up- and down-spins, respectively, are both much larger then the thickness of the layers. The electrons will then, on the average, traverse many layers without being scattered as they drift along the in-plane direction. The resistance contributed by each spin channel is then proportional to the effective scattering rate for each channel. Let us denote the average scattering rate for each Co/Cu repeat unit by $1/\tau_\uparrow$ and $1/\tau_\downarrow$ for spins that are locally majority (up) and minority (down), respectively. In parallel (P) configuration, the resistance for the majority spin channel ($R_\uparrow$) and for the minority spin channel ($R_\downarrow$) is then proportional to $1/\tau_\uparrow$ and $1/\tau_\downarrow$ respectively. The current contributed by each spin channel adds in parallel so the total resistance of the system is then

$$R(P) = \frac{R_\uparrow R_\downarrow}{R_\uparrow + R_\downarrow} \, .$$

In the antiparallel (AP) configuration on the other hand, the average scattering rate for each spin channel will be $\frac{1}{2}(1/\tau_\uparrow + 1/\tau_\downarrow)$, since each spin channel is a majority spin in half the Co layers, and a minority spin in the other half. Again, due to the absence of spin-flip scattering, the two spin channels conduct in parallel, and the total resistance of the system is $R(AP) = \frac{1}{4}(R_\uparrow + R_\downarrow)$. The change in resistance in going from P to AP configuration is

$$\Delta R \equiv R(AP) - R(P) = \frac{(R_\downarrow - R_\uparrow)^2}{4(R_\downarrow + R_\uparrow)} \, ,$$

and the GMR ratio which is defined as

$$GMR = \frac{R(AP) - R(P)}{R(P)} ,$$

is given in the thin limit by

$$GMR_{\mathrm{thin-limit}} = \frac{1}{4}\frac{(R_{\downarrow} - R_{\uparrow})^2}{R_{\downarrow}R_{\uparrow}} .$$

Note that

$$\frac{R(AP) - R(P)}{R(P)} = \frac{G(P) - G(AP)}{G(AP)} = \frac{\Delta G}{G(AP)} ,$$

where $G(P)$ and $G(AP)$ are the conductances in P and AP configurations. This latter form tends to be more convenient to use in theoretical analyses, since by focusing on ΔG one can separate out shunting resistances which do not contribute to the GMR signal.

The opposite limit to the "thin" limit just described is the "thick" limit in which the mean free path is much shorter than the layer thicknesses. In the thick limit, the conductivity can be treated as local and there is no difference between the CIP conductance for parallel and antiparallel moment alignment. Thus $GMR_{\mathrm{thick-limit}} = 0$.

10.3.2 CPP-GMR

The theory for current flowing perpendicular to the planes (CPP) is more complicated than for CIP. For CPP, the current must be constant along the the direction perpendicular to the planes. Maintaining a constant current across material boundaries generally requires charge and spin accumulations at the interfaces because of the different properties of the layers. This does not happen in the CIP case because there is no *net* current along the direction perpendicular to the planes. As a consequence, there will be discontinuities in the electrostatic potential at the interfaces for CPP and a concomitant non-uniform electric field, as well as chemical potential barriers due to the charge and spin accumulations.

We shall deal with these complications in a later section. If one is willing to ignore them for the time being, it is possible to describe CPP GMR rather simply [13]. It is relatively easy to show, for example, that within the semi-classical two current model, if all layers have the same electronic structure, the resistance in either spin channel is given by

$$R_s = \frac{1}{A}\int dz \varrho_s(z) ,$$

where A is the area perpendicular to the current flow, and $\varrho_s(z)$ is the local resistivity which is proportional to the local scattering rate for electrons.

Thus consider a multilayer with N repeats of a ferromagnetic layer followed by a spacer layer. The resistance for parallel alignment of the magnetic layers will be given by

$$R(P) = \frac{R_\uparrow R_\downarrow}{R_\uparrow + R_\downarrow},$$

where R_s can be written in terms of the *local* resistivities of the ferromagnetic and nonmagnetic spacer layers, $\varrho_{\rm F}$, and $\varrho_{\rm N}$, and their thicknesses, $t_{\rm F}$ and $t_{\rm N}$,

$$AR_s = N[\varrho_{{\rm F}s} t_{\rm F} + \varrho_{{\rm N}s} t_{\rm N}] \,.$$

The resistance for antiparallel alignment will then be given by $R(AP) = \frac{1}{4}(R_\uparrow + R_\downarrow)$. This leads to a GMR of

$$GMR(CPP) = \frac{1}{4} \frac{(R_\uparrow - R_\downarrow)^2}{R_\uparrow R_\downarrow} \,.$$

We shall show later that the largest effect on GMR of including the effects of the different electronic structures is that a spin dependent interfacial resistance develops at the interfaces between the ferromagnet and spacer layers. This can be approximately accommodated by adding interfacial resistances, r_s to R_s. Thus

$$AR_s = N[\varrho_{Fs} t_{\rm F} + \varrho_{Ns} t_{\rm N} + 2r_s] \,.$$

Note that in contrast to CIP, (and for this simple model) the same result is obtained for layers that are thick compared to the electronic mean free path as for thin layers. This result breaks down, however, for layers that are sufficiently thick that the two current model breaks down because of spin-flip scattering or mixing of the spin channels. The simple two-current resistor picture gives some idea of the origin of the GMR. It must be the task of more fundamental theories to explain what the resistances and interface resistance are and how they are affected by material choices, layer thicknesses, etc. Furthermore, it is not quite correct that the GMR is independent of the number of layers for CIP and CPP multilayers, and this discrepancy also will have to be addressed by more fundamental theories.

Finally, it should be noted that multilayer systems may not be well suited for application as read sensors in disc drives. The anti-ferromagnetic RKKY coupling at 8.5 Å Cu is so strong that it takes a rather large external field (several thousands of Oe) to produce a response from the system – this is clearly not acceptable since the field a read sensor is exposed to is of the order of 100 Oe. There is a second, much weaker, anti-ferromagnetic RKKY peak at a Cu thickness of about 20 Å, but then the multilayer stacks run the risk of becoming so thick that they do not fit between the shields (in addition to the Co/Cu multilayers, there are seed and cap layers used to induce the correct growth and to protect the multilayer stack from the atmosphere). A multi-layer system also consists of a rather large volume of magnetic material that has to be rotated by the torque exerted by the field emanating from the medium. The net torque exerted by the magnetic field emanating from the transition is approximately inversely proportional to $M_s t$, where M_s and t are the saturation magnetization and total thickness of the magnetic layers in the reader element. Therefore, in order to gain sensitivity, it is desirable to have less magnetic material. These constraints led to the development of spin valve sensors, which we will consider in some detail in Sect. 10.5.

10.3.3 Tunneling Magnetoresistance

The prototypical TMR head consists of two ferromagnetic layers. One is a reference layer (R) with its magnetization kept in a fixed direction while the magnetization of the other layer (F) is free to rotate in response to an in-plane field. These two layers are separated by a thin insulating layer with a thickness of the order of 10 Å. A bias voltage is applied across the tunneling junction between the free layer F and the reference layer R. Electrons incident from F will have a finite probability amplitude for tunneling through the insulating layer into the R layer due to the exponential decay of the electron wavefunctions inside the insulating barrier. The conductance between two weakly coupled ferromagnets is, in the limit of vanishing bias voltage [14],

$$G = \frac{2\pi e^2}{\hbar} \sum_s \sum_{n_F,\mathbf{k}_F} \sum_{n_R,\mathbf{k}_R} |t(n_F,\mathbf{k}_F;n_R,\mathbf{k}_R;s)|^2 \times \\ A^F_{n_F,\mathbf{k}_F,s}(E^F_F) A^R_{n_R,\mathbf{k}_R,s}(E^R_F). \tag{10.4}$$

In this expression, $A_{n,\mathbf{k},s}(E_F)$ is the spectral weight function for band n at wavevector $\mathbf{k}$ in the F and R layers, respectively, at the Fermi energy E_F, and spin direction s, and $t(n_F,\mathbf{k}_F;n_R,\mathbf{k}_R;s)$ is the tunneling probability amplitude for state $(n_F,\mathbf{k}_F,s)$ to tunnel to state $(n_R,\mathbf{k}_R,s)$. In the absence of spin-flip scattering and at low temperatures, the spectral weight functions for metals are delta-functions, $A_{n,\mathbf{k},s}(E) = \delta(E_{n,\mathbf{k},s} - E)$.

It is often assumed that the tunneling amplitudes decouple into factors that depend separately on the band wavefunctions on opposite sides of the barrier, so that

$$|t(n_F,\mathbf{k}_F;n_R,\mathbf{k}_R;s)|^2 \approx |t(n_F,\mathbf{k}_F,s)||t(n_R,\mathbf{k}_R,s)|. \tag{10.5}$$

Under this assumption one can lump together the tunneling amplitudes and spectral weights into effective tunneling amplitudes and densities of states at the Fermi surface:

$$t_s \mathcal{D}_s \equiv \sum_{n;\mathbf{k}} |t_{n,\mathbf{k}}| \delta(E_F - E_{n,\mathbf{k},s}), \tag{10.6}$$

for the F and R sides, respectively. By defining effective polarizations P for the F and R sides, respectively, as

$$P = \frac{t_\uparrow \mathcal{D}_\uparrow - t_\downarrow \mathcal{D}_\downarrow}{t_\uparrow \mathcal{D}_\uparrow + t_\downarrow \mathcal{D}_\downarrow} \tag{10.7}$$

where the subscripts $\uparrow$ and $\downarrow$ denote majority and minority spin channels. We can then finally write the tunneling magnetoresistance as

$$\frac{G(P) - G(AP)}{G(AP)} = \frac{2P^{\mathrm{F}} P^{\mathrm{R}}}{1 - P^{\mathrm{F}} P^{\mathrm{R}}}. \tag{10.8}$$

Equation (10.8) is Jullière's formula [15] for TMR in terms of which is often used to analyze experimental results. Note that in this expression the effective polarizations are not simply densities-of-states, but more complicated objects involving tunneling amplitudes weighted over densities-of-states. Furthermore, the possibility of writing the TMR in the Jullière form of (10.8) depends crucially on the approximation (10.5) of factoring the tunneling amplitude in (10.4). This hypothesis would obviously have problems in epitaxial systems, where the transmission amplitude is determined in large part by the overlap of the lateral components to the incoming and outgoing Bloch wavefunctions, a factor not contained in (10.5). Indeed, a recent calculation of the transmission probabilities for an epitaxial system [16] showed that (10.5) is a poor approximation, at least for the particular system (Fe(100)|MgO(100)|Fe(100)) studied.

Again, the magnetoresistive response can be modeled in a rudimentary way using the Jullière formula, just as we can model CPP and CIP GMR using the two-current resistor models. However, just as with the case of GMR, the details of the TMR, the magnitude of the tunneling amplitudes, the densities of states etc. depend sensitively on band structures and material choices, and it must be the task of more fundamental approaches to try to understand the details of the tunneling magnetoresistance.

10.4 Semiclassical Theory of Spin-Dependent Transport

10.4.1 Boltzmann Equation

The semiclassical Boltzmann equation, with appropriate generalizations to include multilayers, provides a convenient approach for modeling CPP and CIP transport. In contrast to more fundamental, fully quantum-mechanical treatments of spin-dependent transport, it is very fast and lends itself more readily to interpretation of the results. Later in this section we shall describe in some detail an approach in which first-principle calculations are used to obtain some of the basic input into the Boltzmann equation, electronic band structure and transmission coefficients needed for boundary conditions at interfaces. This approach [17] uses realistic band structures and velocities for the electron states and yields results which are typically within 10% of experimental values (e.g., GMR and sheet resistance) without the need to 'tune' adjustable parameters. In addition, the first-principle calculations can be done once and for all for the materials and interfaces of interest, such as CoFe, Cu, and CoFe–Cu interfaces, providing a valuable interpretive and quantitatively predictive tool in device development.

The fundamental assumption underlying the Boltzmann equation is that the electrons can be characterized by a distribution function $f_{n,\mathbf{k}}(\mathbf{r},t)$. This function gives the probability density that a Bloch electron of wavevector $\mathbf{k}$ in band n is at position $\mathbf{r}$ at time t. Note that this concept, strictly speaking, violates the uncertainty principle which states that we cannot simultaneously specify both the momentum and the position of a particle to arbitrary precision. This objection can be overcome by treating the particles as wave packets centered on some momentum $\hbar\mathbf{k}_0$ and with a spread in momentum $\hbar\Delta\mathbf{k}$. This results in some spread $\Delta\mathbf{r}$ about the expected position, so that we give up some ability to specify position and momentum in order to comply with the uncertainty principle. However, the spreads $\hbar\Delta\mathbf{k}$ and $\Delta\mathbf{r}$ can be made small enough that the theory can be used to interpret the data from most commercial sputtered spin-valve systems.

An equation of motion for the distribution function is obtained by considering the continuity equation for $f_{n,\mathbf{k}}(\mathbf{r},t)$ in six-dimensional $(\mathbf{k},\mathbf{r})$-space [18]. In the steady state, the resulting so-called Boltzmann equation is

$$-\mathbf{v}\cdot\nabla_{\mathbf{r}}f_{n,\mathbf{k}}(\mathbf{r})-\frac{1}{\hbar}\mathbf{F}\cdot\nabla_{\mathbf{k}}f_{n,\mathbf{k}}(\mathbf{r})+\left(\frac{\partial f_{n,\mathbf{k}}(\mathbf{r})}{\partial t}\right)_{\text{collisions}}=0\,. \tag{10.9}$$

In this equation, $\mathbf{F}$ is the external force acting on the electrons. We will exclusively be considering the case where the external force $\mathbf{F}$ acting on a particle is due to an electric field $\mathbf{E}(\mathbf{r})$, so $\mathbf{F}=-e\mathbf{E}(\mathbf{r})$ for electrons of charge $-e$. The term $\left(\frac{\partial f_{n,\mathbf{k}}(\mathbf{r})}{\partial t}\right)_{\text{collisions}}$ is the time rate of change of the distribution function due to scattering processes. These scattering processes can involve any defect in the lattice, including phonons and even other electrons.

In most applications, and certainly the applications considered here, the applied external fields are small enough that the deviations of $f_{n,\mathbf{k}}(\mathbf{r})$ from its equilibrium value $f^0_{n,\mathbf{k}}$, which is independent of position for a homogeneous system, are small. We then write $f_{n,\mathbf{k}}(\mathbf{r})=f^0_{n,\mathbf{k}}+g_{n,\mathbf{k}}(\mathbf{r})$, where $g_{n,\mathbf{k}}(\mathbf{r})$ is of first order in the applied field. Inserting this into (10.9) and linearizing in $\mathbf{E}$, we obtain

$$-\mathbf{v}\cdot\nabla_{\mathbf{r}}g_{n,\mathbf{k}}(\mathbf{r})+\frac{e\mathbf{E}}{\hbar}\cdot\nabla_{\mathbf{k}}f^0_{n,\mathbf{k}}+\left(\frac{\partial g_{n,\mathbf{k}}(\mathbf{r})}{\partial t}\right)_{\text{collisions}}=0\,. \tag{10.10}$$

In this equation, $\nabla_{\mathbf{k}}f^0_{n,\mathbf{k}}$ depends on $\mathbf{k}$ only through the band energy $\varepsilon_n(\mathbf{k})$, $f^0_{n,\mathbf{k}}=f^0_{n,\mathbf{k}}[\varepsilon_n(\mathbf{k})-\mu]$, where μ is the chemical potential. We can then write

$$\nabla_{\mathbf{k}}f^0_{n,\mathbf{k}}=\frac{\partial f^0_{n,\mathbf{k}}[\varepsilon_n(\mathbf{k})-\mu]}{\partial\varepsilon_n(\mathbf{k})}\nabla_{\mathbf{k}}\varepsilon_n(\mathbf{k})=\frac{\partial f^0_{n,\mathbf{k}}[\varepsilon_n(\mathbf{k})-\mu]}{\partial\varepsilon_n(\mathbf{k})}\hbar\mathbf{v}_n(\mathbf{k})\,,$$

with $\mathbf{v}_n(\mathbf{k})$ the group velocity of state $(n, \mathbf{k})$. At absolute zero of temperature, the equilibrium distribution function $f^0_{n,\mathbf{k}}[\varepsilon_n(\mathbf{k})-\mu]$ is unity for band energies less than the Fermi energy ε_{F}, $\varepsilon_n(\mathbf{k}) < \mu(T=0) = \varepsilon_{\mathrm{F}}$. Finite temperatures introduce a thermal smear of the equilibrium distribution function near the Fermi energy when the thermal energy $k_{\mathrm{B}}T$ starts to approach the Fermi energy. However, for metals, $\varepsilon_{\mathrm{F}}/k_{\mathrm{B}} \sim 10^4$ K, so it is a very good approximation to set the equilibrium distribution function to its zero-temperature value, $f^0_{n,\mathbf{k}}[\varepsilon_n(\mathbf{k}) - \mu] \approx \theta[\varepsilon_n(\mathbf{k}) - \mu]$, from which we obtain

$$\frac{\partial f^0_{n,\mathbf{k}}[\varepsilon_n(\mathbf{k}) - \mu]}{\partial \varepsilon_n(\mathbf{k})} \approx -\delta[\varepsilon_n(\mathbf{k}) - \mu] \, .$$

This approximation, when inserted into the linearized Boltzmann equation, (10.10), shows that the first-order deviation $g_{n,\mathbf{k}}(\mathbf{r})$ is non-zero only on the Fermi surface. This is expected – transport properties of metals are Fermi-surface properties since the applied fields are weak enough that they can only cause excitations from immediately below to immediately above the Fermi surface.

We still need to produce a workable approximation for the scattering term $(\partial f_{n,\mathbf{k}}(\mathbf{r},t)/\partial t)_{\mathrm{collisions}}$. The simplest such approximation, and the one to which we will restrict ourselves here, is the relaxation time approximation [19,20]. The underlying assumption is that collisions cause a frictional force which tends to restore the equilibrium distribution. The only parameter necessary is a relaxation time τ which gives the average time between collisions for electrons at the Fermi surface. In the relaxation time approximation, we then write

$$(\partial f_{n,\mathbf{k}}(\mathbf{r},t)/\partial t)_{\mathrm{collisions}} = -\frac{f_{n,\mathbf{k}}(\mathbf{r}) - f^0_{n,\mathbf{k}}[\varepsilon_n(\mathbf{k}) - \mu]}{\tau} = -\frac{g_{n,\mathbf{k}}(\mathbf{r})}{\tau} \, . \tag{10.11}$$

It is clear that this is a rather crude approximation. Different states at the Fermi surface will, in general, have different lifetimes. This is particularly true if there are several bands crossing the Fermi surface. If one knew and were willing to keep track of the scattering processes in detail, it would be relatively straightforward to define a lifetime or scattering rate that was dependent on the state, $(n, \mathbf{k})$. Equally important, however, is the fact that electrons are being scattered into as well as out of the state $(n, \mathbf{k})$. This "scattering-in" complicates the solution of the Boltzmann equation because it couples different states.

Despite these limitations, the relaxation time approximation is a very useful and commonly used approximation when the details of the scattering mechanisms are either not known, or thought not to be of central importance to the system or issues at hand. Here we will typically treat the scattering rates as empirical adjustable parameters. Much of the error made in ignoring the state dependence of the scattering rate and the "scattering-in" term will be compensated by the choice of these parameters. We will here be interested in GMR effects in multilayers. As we shall see, scattering at the interfaces between different layers constitute a large and important contribution to the transport properties in such systems. Therefore, it is not so important to describe in detail the scattering due to impurities in the interior of the layers, and the relaxation time approximation suffices.

Within the relaxation time approximation, we then have the following form for the linearized Boltzmann equation:

$$-\mathbf{v}\cdot\nabla_{\mathbf{r}}g_{n,\mathbf{k}}(\mathbf{r})+\mathbf{v}_n(\mathbf{k})\cdot e\mathbf{E}\delta[\varepsilon_n(\mathbf{k})-\varepsilon_{\mathrm{F}}]=\frac{g_{n,\mathbf{k}}(\mathbf{r})}{\tau}\,. \tag{10.12}$$

Once a solution for $g_{n,\mathbf{k}}(\mathbf{r})$ has been found, the current density can then be calculated using,

$$\mathbf{j}(\mathbf{r})=-\frac{e}{(2\pi)^3}\sum_n\int\mathrm{d}^3k\mathbf{v}_n(\mathbf{k})g_{n,\mathbf{k}}(\mathbf{r})\,. \tag{10.13}$$

For a homogeneous system with a uniform applied field, the solution for $g_{n,\mathbf{k}}(\mathbf{r})$ is

$$g_{n,\mathbf{k}}(\mathbf{r})=e\tau\mathbf{v}_n(\mathbf{k})\cdot\mathbf{E}\delta[\varepsilon_n(\mathbf{k})-\varepsilon_{\mathrm{F}}]\,.$$

This shows that the first-order effect of the applied field is to cause a slight perturbation of the occupation of electron states at the Fermi surface. Moreover, the amplitude of the perturbation $g_{n,\mathbf{k}}(\mathbf{r})$ is larger for electrons with their Bloch velocities parallel or antiparallel to the applied field, and the longer the relaxation time, the larger the perturbation and the larger the resulting current response.

10.4.2 Inhomogeneous Systems

We must be more careful when we apply the linearized Boltzmann equation to inhomogeneous systems. The kind of inhomogeneous systems that we have in mind here are, of course, metallic multilayer systems. We first note that if the electric field is applied in the plane of the layers, nothing new has to be added. In this case, the electric field is uniform in the system. The current flows in-plane in the direction of the electric field and is constant in this direction. A different situation arises if the electric field is applied perpendicularly to the layers, i.e. the $\hat{z}$-direction. At the interface between two layers there will in general be an accumulation of charge and spin. This effect is to be distinguished from the band-bending and charge transfer near the interface that are necessary in order to equalize the Fermi energy on the two sides of the interface in the absence of an external field. Instead, band mismatches and disorder at the interface cause scattering which contributes additional charge and spin accumulation in the presence of an external field.

As a consequence, the total electric field, which is the sum of the applied external field plus induced electric fields, will not be constant along $\hat{z}$. We then have to modify the Boltzmann equation in the relaxation time approximation in order to ensure that the continuity equation is satisfied. The key point is that we assume that the distribution function relaxes not to the bulk equilibrium distribution function for each layer, but to a *local equilibrium distribution function* [20], $f_n^{\text{l.e.}}(\mathbf{k}, z)$. The local equilibrium distribution function describes the effects of charge accumulations at the interfaces. With the local equilibrium distribution function inserted into the linearized Boltzmann equation, we have

$$-v_z \frac{\partial g_n(\mathbf{k}, z)}{\partial z} + e v_z E(z) \delta[\varepsilon_n(\mathbf{k}) - \varepsilon_{\text{F}}] = -\frac{f_n(\mathbf{k}, z) - f_n^{\text{l.e.}}(\mathbf{k}, z)}{\tau} \,. \tag{10.14}$$

Here, $E(z)$ is the total electric field and satisfies Poisson's equation

$$\frac{\mathrm{d}E(z)}{\mathrm{d}z} = 4\pi e \left[n(z) - n_0(z) \right] \,,$$

with $n(z) = \sum_{n,\mathbf{k}} f_n(\mathbf{k}, z)$, $n_0(z) = \sum_{n,\mathbf{k}} f^0_{n,\mathbf{k}}$ the equilibrium density and with appropriate boundary conditions at the edges of the system. Because of the delta-function in the second term of (10.14), the difference between the local equilibrium distribution $f_n^{\text{l.e.}}(\mathbf{k}, z)$ and the equilibrium distribution function $f^0_{n,\mathbf{k}}$ must be constrained to the Fermi surface. We then write $f_n^{\text{l.e.}}(\mathbf{k}, z)$ as

$$f_n^{\text{l.e.}}(\mathbf{k}, z) = f^0_{n,\mathbf{k}} + \mu(z) \delta[\varepsilon_n(\mathbf{k}) - \varepsilon_{\text{F}}] \,.$$

Note that $f_n^{\text{l.e.}}(\mathbf{k}, z)$ then does not contribute to the net current, since it is just a position-dependent amplitude uniformly distributed over the Fermi surface. We determine $\mu(z)$ by requiring that the continuity equation be satisfied. For steady state, the continuity equation states that $\nabla \cdot \mathbf{j}(\mathbf{r}) = 0$, or

$$-e \sum_{n,\mathbf{k}} v_z(\mathbf{k}) \frac{\partial g_n(\mathbf{k}, z)}{\partial z} = 0 \,.$$

If we integrate (10.14) over $\mathbf{k}$ and sum over bands n, the first term on the left-hand side gives us $\nabla \cdot \mathbf{j}$, except for a factor of e. The second term on the left-hand side vanishes, since it is an integral of $\nabla_{\mathbf{k}} f^0_{n,\mathbf{k}}$ over $\mathbf{k}$. So the continuity equation is satisfied provided we demand

$$\begin{aligned} \sum_{n,\mathbf{k}} \left[f_n(\mathbf{k}, z) - f_n^{\text{l.e.}}(\mathbf{k}, z) \right] &= \sum_{n,\mathbf{k}} \left[g_n(\mathbf{k}, z) - \mu(z) \delta(\varepsilon_n(\mathbf{k}) - \varepsilon_{\text{F}}) \right] \\ &= n_1(z) - \mu(z) \mathcal{D}(\varepsilon_{\text{F}}) = 0, \end{aligned} \tag{10.15}$$

with $n_1(z) = \sum_{n,\mathbf{k}} g_n(\mathbf{k}, z)$ and $\mathcal{D}(\varepsilon_{\text{F}})$ the density of states at the Fermi surface. This shows that $\mu(z)$ is of first order in the applied field since $g_n(\mathbf{k}, z)$

is.[5] We can now combine all these equations into a set of equations which have to be solved self-consistently:

$$\begin{aligned} -v_z \frac{\partial g_n(\mathbf{k},z)}{\partial z} + e v_z E(z) \delta[\varepsilon_n(\mathbf{k}) - \varepsilon_\mathrm{F}] &= -\frac{g_n(\mathbf{k},z) - \mu(z)\delta[\varepsilon_n(\mathbf{k}) - \varepsilon_\mathrm{F}]}{\tau} \\ n_1(z) - \mu(z)\mathcal{D}(\varepsilon_\mathrm{F}) &= 0 \\ \frac{\mathrm{d}E(z)}{\mathrm{d}z} &= 4\pi e n_1(z). \end{aligned} \tag{10.16}$$

10.4.3 Boundary Conditions for Multilayered Systems

We will now discuss how to apply the linearized Boltzmann equation to multilayered systems. The fundamental assumption is that the electron states in each layer can be described by a bandstructure appropriate for a *bulk* system comprised of the material of that layer. This at first seems like a rather drastic approximation. The electrons in a metallic system will in general travel across several different layers, so each electrons state will not be a state characteristic of the bulk of only one layer.

We can, nevertheless, consider an electron wavepacket incident on an interface from the interior of one layer. The Schrödinger equation is a local equation, so here the wavepacket properties are essentially determined by the bulk properties of this layer. There may of course be local electric fields near interfaces in multilayer structures, but in metallic multilayers such fields are rapidly screened as one moves away from the interface so that the effects of these fields are quite naturally incorporated into the scattering properties of the interfaces. As the wavepacket approaches the interface, there will be some complicated interactions which distort the wavepacket during the time that it traverses the interface. After some time, well-defined asymptotic solutions appear in each layer, with reflected wavepackets in the initial layer and transmitted wavepackets propagating in the next layer. The reflected and transmitted wavepackets have properties intrinsic to the states in the bulk of each layer.

The properties of the interface are cast entirely in terms of transmission and reflection amplitudes and time delays. These can be obtained from scattering theory, which then provides the foundation for the approximation that the electrons can be described by a local bandstructure and distribution function in each layer, and it is actually not a bad approximation. We note, however, that it becomes questionable when the thickness of each layer becomes very small. In that case, a wavepacket incident on an interface may not develop its asymptotic solutions of reflection and transmission within the layers before it encounters another interface.

---

[5] Our local equilibrium distribution function is a different concept than that used in, e.g., [21], in which this term is used to reflect the large charge transfers that often occur in an inhomogeneous system even in the absence of an applied field

The approach then, consists of analytically solving the Boltzmann equation for each layer separately to obtain the distribution functions $g_{i,n}(\mathbf{k}, z)$ for each layer i. These solutions will have parameters that will be determined by the boundary conditions connecting the solutions in adjoining layers and by the boundary conditions at the bounding surfaces of the multilayer system. The view of wavepackets incident on interfaces and probabilities for reflection and transmission at the interfaces naturally gives us the boundary conditions we need here.

In order to describe the details of this approach we will need to introduce more notation. We denote by $g^+_{i,n}(\mathbf{k}, z)$ and $g^-_{i,n}(\mathbf{k}, z)$ the distribution functions in layer i with electrons moving in the $+z$ and $-z$ directions with z-velocities $v_z^{+,i}(n, \mathbf{k})$ and $v_z^{-,i}(n, \mathbf{k})$, respectively. Note that it is the z- component of the velocity rather than crystal momentum that determines the direction of travel for these semi-classical electrons. It can be shown that there are always an equal number of states with positive and negative v_z values. An electron in state $(n, \mathbf{k})$ in layer i traveling in the $+z$ direction with z-velocity $v_z^{+,i}(n, \mathbf{k})$ has a probability $T_i^{++}(n', \mathbf{k}'; n, \mathbf{k})$ of being transmitted to a state $(n', \mathbf{k}')$ (traveling in the $+z$ direction with z-velocity $v_z^{+,i+1}(n', \mathbf{k}')$) in layer $i + 1$. Similarly, $R_i^{+-}(n', \mathbf{k}'; n, \mathbf{k})$ denotes the probability that this state is reflected to a state $(n', \mathbf{k}')$ in layer i traveling in the $-z$ direction with z-velocity $v_z^{-,i}(n', \mathbf{k}')$. Similarly, $T_i^{--}(n', \mathbf{k}'; n, \mathbf{k})$ and $R_{i+1}^{-+}(n', \mathbf{k}'; n, \mathbf{k})$ denote the transmission and reflection probabilities, respectively, for electrons incident from layer $(i + 1)$. The boundary conditions are then obtained by demanding that the electron flux at each boundary is conserved. That is, at the boundary between layers i and $i + 1$, the outgoing flux in state $(n', \mathbf{k}')$ in layer $i + 1$ must equal the sum of the incoming flux that is reflected or transmitted into this state. Thus at the interface between layers i and $i + 1$, located at $z = z_i$, the outgoing flux on the $i + 1$ side is given by,

$$\begin{aligned} & g^+_{i+1,n'}(\mathbf{k}', z_i) v_z^{+,i+1}(n', \mathbf{k}') = \\ & \quad \sum_{n,\mathbf{k}} T_i^{++}(n', \mathbf{k}'; n, \mathbf{k}) g^+_{i,n}(\mathbf{k}, z_i) v_z^{+,i}(n, \mathbf{k}) \\ & \quad - \sum_{n,\mathbf{k}} R_{i+1}^{+-}(n', \mathbf{k}'; n, \mathbf{k}) g^-_{i+1,n}(\mathbf{k}, z_i) v_z^{-,i+1}(n, \mathbf{k}), \end{aligned} \tag{10.17}$$

and the outgoing flux on the i side is given by,

$$\begin{aligned} & -g^-_{i,n'}(\mathbf{k}', z_i) v_z^{-,i}(n', \mathbf{k}') = \\ & \quad - \sum_{n,\mathbf{k}} T_{i+1}^{--}(n', \mathbf{k}'; n, \mathbf{k}) g^-_{i+1,n}(\mathbf{k}, z + i) v_z^{-,i+1}(n, \mathbf{k}) \\ & \quad + \sum_{n,\mathbf{k}} R_i^{-+}(n', \mathbf{k}'; n, \mathbf{k}) g^+_{i,n}(\mathbf{k}, z_i) v_z^{+,i}(n, \mathbf{k}). \end{aligned} \tag{10.18}$$

We can further simplify this expression by recognizing explicitly that the deviation in the distribution function is confined to the Fermi surface,

$$g_{i,n}^{\pm}(\mathbf{k}, z) = h_{i,n}^{\pm}(\mathbf{k}, z)\delta[\varepsilon_n(\mathbf{k}) - \varepsilon_{\mathrm{F}}].$$

We then have

$$\begin{aligned}
&\sum_{\mathbf{k}} g_{i,n}^{\pm}(\mathbf{k}, z)v_z^{\pm}(n, \mathbf{k})G(n, \mathbf{k}) \\
&= \sum_{\mathbf{k}_{\|}} \frac{V}{2\pi A} \int dk_z h_{i,n}^{\pm}(\mathbf{k}, z)\delta[\varepsilon_n(\mathbf{k}) - \varepsilon_{\mathrm{F}}]v_z^{\pm}(n, \mathbf{k})G(n, \mathbf{k}) \\
&= \sum_{\mathbf{k}_{\|}} \frac{V}{2\pi A} \int dk_z h_{i,n}^{\pm}(\mathbf{k}, z)\delta\left[\frac{\partial\varepsilon_n(\mathbf{k})}{\partial k_z}(k_z - k_{z,n,F}(k_x, k_y))\right] v_z^{\pm}(n, \mathbf{k})G(n, \mathbf{k}) \\
&= \frac{V}{2\pi A}\sum_{\mathbf{k}_{\|}} h_{i,n}^{\pm}(k_x, k_y, z)\frac{v_z^{\pm}(n, k_x, k_y, k_{z,n,\mathrm{F}})G(n, k_x, k_y, k_{z,n,\mathrm{F}}(k_x, k_y))}{\left|\hbar v_z^{\pm}(n, k_x, k_y, k_{z,n,F})\right|},
\end{aligned} \tag{10.19}$$

where $k_{z,n,\mathrm{F}}$ are the values of k_z which satisfy $\varepsilon_n(k_x, k_y, k_z) = \varepsilon_{\mathrm{F}}$ for given (k_x, k_y), and $G(n, \mathbf{k})$ is an arbitrary function of the electronic structure. We also need to define the "on energy shell" transmission and reflection probabilities,

$$\begin{aligned}
T^{++}(n', \mathbf{k}'; n, \mathbf{k}) &= \frac{2\pi\hbar A}{V}v_z^{+}(n', \mathbf{k}_{\|}')t^{++}(n', \mathbf{k}_{\|}'; n, \mathbf{k}_{\|})\delta\left[\varepsilon_{n'}(\mathbf{k}') - \varepsilon_n(\mathbf{k}))\right] \\
R^{+-}(n', \mathbf{k}'; n, \mathbf{k}) &= \frac{2\pi\hbar A}{V}v_z^{+}(n', \mathbf{k}_{\|}')r^{+-}(n', \mathbf{k}_{\|}'; n, \mathbf{k}_{\|})\delta\left[\varepsilon_{n'}(\mathbf{k}') - \varepsilon_n(\mathbf{k}))\right] \\
T^{--}(n', \mathbf{k}'; n, \mathbf{k}) &= \frac{2\pi\hbar A}{V}v_z^{-}(n', \mathbf{k}_{\|}')t^{--}(n', \mathbf{k}_{\|}'; n, \mathbf{k}_{\|})\delta\left[\varepsilon_{n'}(\mathbf{k}') - \varepsilon_n(\mathbf{k}))\right] \\
R^{-+}(n', \mathbf{k}'; n, \mathbf{k}) &= \frac{2\pi\hbar A}{V}v_z^{-}(n', \mathbf{k}_{\|}')r^{-+}(n', \mathbf{k}_{\|}'; n, \mathbf{k}_{\|})\delta\left[\varepsilon_{n'}(\mathbf{k}') - \varepsilon_n(\mathbf{k})\right].
\end{aligned} \tag{10.20}$$

Thus, using (10.19) and (10.20) in (10.17) and (10.18) we obtain boundary matching conditions in terms of the transverse momentum that are restricted to the Fermi energy,

$$\begin{aligned}
h_{i+1,n}^{+}(\mathbf{k}_{\|}, z_i) &= \sum_{n', \mathbf{k}_{\|}'}^{i} t_i^{++}(n, \mathbf{k}_{\|}; n', \mathbf{k}_{\|}')h_{i,n'}^{+}(\mathbf{k}_{\|}', z_i) \\
&\quad + \sum_{n', \mathbf{k}_{\|}'}^{i+1} r_{i+1}^{+-}(n, \mathbf{k}_{\|}; n', \mathbf{k}_{\|}')h_{i+1}^{-}(\mathbf{k}_{\|}', z_i) \\
h_{i,n}^{-}(\mathbf{k}_{\|}, z_i) &= \sum_{n', \mathbf{k}_{\|}'}^{i} r_i^{-+}(n, \mathbf{k}_{\|}; n', \mathbf{k}_{\|}')h_{i,n'}^{+}(\mathbf{k}_{\|}', z_i) \\
&\quad + \sum_{n', \mathbf{k}_{\|}'}^{i+1} t_{i+1}^{--}(n, \mathbf{k}_{\|}; n', \mathbf{k}_{\|}')h_{i+1,n'}^{-}(\mathbf{k}_{\|}', z_i).
\end{aligned} \tag{10.21}$$

Here, the superscripts on the summations remind us that we are summing over states in different layers.

If we also assume that the interface has two-dimensional periodicity in the xy-plane, it follows that momentum $\hbar\mathbf{k}_{\|}$ parallel to the interface is conserved. Then the boundary conditions do not couple different values of $\mathbf{k}_{\|}$, so we obtain finally,

$$\begin{aligned} h^{+}_{i+1,n}(\mathbf{k}_{\|}, z_i) &= \sum_{n'} t^{++}_{i}(n, \mathbf{k}_{\|}; n', \mathbf{k}_{\|}) h^{+}_{i,n'}(\mathbf{k}_{\|}, z_i) \\ &\quad + \sum_{n'} r^{+-}_{i+1}(n, \mathbf{k}_{\|}; n', \mathbf{k}_{\|}) h^{-}_{i+1,n'}(\mathbf{k}_{\|}, z_i) \\ h^{-}_{i,n}(\mathbf{k}_{\|}, z_i) &= \sum_{n'} r^{-+}_{i}(n, \mathbf{k}_{\|}; n', \mathbf{k}_{\|}) h^{+}_{i,n'}(\mathbf{k}_{\|}, z_i) \\ &\quad + \sum_{n'} t^{--}_{i+1}(n, \mathbf{k}_{\|}; n', \mathbf{k}_{\|}) h^{-}_{i,n'}(\mathbf{k}_{\|}, z_i). \end{aligned} \tag{10.22}$$

We still have to deal with the outer boundaries of the multilayer system. One set of boundary conditions commonly used for thin films is due to Fuchs and Sondheimer [22]. The idea is that an electron incident on the boundary can be specularly reflected, in which case the momentum parallel to the bounding surface is conserved, with a probability p. On the other hand, the probability is $1 - p$ that the electron is diffusively reflected, in which case energy only is conserved after the reflection. We do not know what the distribution of diffusively reflected is, but a convenient approximation consists of assuming that the diffusively reflected electrons have a distribution equal to the equilibrium distribution of particles moving away from the surface. This way, the diffusively reflected electrons do not contribute to the current parallel to the surface and can be ignored in the calculation of the total current for CIP. This is not the case for CPP, which we will deal with later. Thus at the surfaces at $z = z_0 = 0$ and $z = z_N = L$ we have

$$\begin{aligned} h^{+}_{1,n}(\mathbf{k}_{\|}, z_0) &= \sum_{n'} p_0(n, \mathbf{k}_{\|}; n', \mathbf{k}_{\|}) h^{-}_{1,n'}(\mathbf{k}_{\|}, z_0) \\ h^{-}_{N,n}(\mathbf{k}_{\|}, z_\mathrm{N}) &= \sum_{n'} p_\mathrm{N}(n, \mathbf{k}_{\|}; n', \mathbf{k}_{\|}) h^{+}_{N,n'}(\mathbf{k}_{\|}, z_\mathrm{N}), \end{aligned} \tag{10.23}$$

where p_0 and p_N are unity for purely specular reflection, and zero for purely diffuse reflection.

The idea of specular and diffuse reflection at the outer boundaries is a useful one which we shall also adopt for the interfaces between different layers. For sputtered interfaces, there is always some degree of disorder due to the dynamics of the sputtering process[6]. There may also be additional disorder due to interdiffusion caused by subsequent anneals. Interfacial disorder

[6] disorder also occurs due to thermodynamic driving forces unrelated to the sputtering process.

decreases the probability of specular reflection. Following Hood and Falicov [23], we therefore introduce phenomenological specularity parameters S_i at interface i. The boundary conditions at an interface between two layers in the film then become for CIP

$$h^{+}_{i+1,n}(\mathbf{k}_{\|}, z_i) = S_i \left[\sum_{n'} t^{++}_{i}(n', \mathbf{k}_{\|}; n, \mathbf{k}_{\|}) h^{+}_{i,n}(\mathbf{k}_{\|}, z_i) \right.$$
$$\left. + \sum_{n'} r^{+-}_{i+1}(n', \mathbf{k}_{\|}; n, \mathbf{k}_{\|}) h^{-}_{i+1}(\mathbf{k}_{\|}, z_i) \right]$$
$$h^{-}_{i,n}(\mathbf{k}_{\|}, z_i) = S_i \left[\sum_{n'} r^{-+}_{i}(n', \mathbf{k}_{\|}; n, \mathbf{k}_{\|}) h^{+}_{i,n}(\mathbf{k}_{\|}, z_i) \right.$$
$$\left. + \sum_{n'} t^{--}_{i+1}(n', \mathbf{k}_{\|}; n, \mathbf{k}_{\|}) h^{-}_{i+1,n}(\mathbf{k}_{\|}, z_i) \right] , \qquad (10.24)$$

with S_i unity for purely specular reflection. Note that in this approach, the diffusely scattered electrons are simply neglected. For CIP, it can be shown that this is correct if the diffuse scattering is isotropic.

For CPP transport, the diffusely reflected electrons affect the local chemical potential and thus contribute to the voltage drop perpendicular to the interfaces even when the diffuse scattering is isotropic. The flux due to diffuse reflection must then be included in expressing flux conservation at the interfaces [24,25]. An incoming electron $(n, \mathbf{k})$ in layer i to the interface between layers i and $i+1$ can be specularly reflected and transmitted. The sum of all these probabilities adds up to the specularity coefficient S. The probability that the electron will be diffusely scattered is then $(1 - S)$. The diffuse scattering term should not contribute to the distribution function when v_z is zero. The lowest-order term in an expansion in v_z which satisfies this condition is a term in v_z^2. Using these assumptions the transmission and reflection matrices become

$$\frac{V}{(2\pi)^3 \hbar} T^{++}(n', \mathbf{k}'; n, \mathbf{k}) = \delta \left[\varepsilon_{n'}(\mathbf{k}') - \varepsilon_n(\mathbf{k})) \right]$$
$$\times \left[S\delta(\mathbf{k}_{\|} - \mathbf{k}'_{\|}) v_z^{+}(n', \mathbf{k}'_{\|}) t^{++}(n', \mathbf{k}_{\|}; n, \mathbf{k}_{\|}) + \frac{A}{(2\pi)^2}(1 - S) F \left[v_z^{+}(n', \mathbf{k}') \right]^2 \right]$$
$$\frac{V}{(2\pi)^3 \hbar} R^{+-}(n', \mathbf{k}'; n, \mathbf{k}) = \delta \left[\varepsilon_{n'}(\mathbf{k}') - \varepsilon_n(\mathbf{k})) \right]$$
$$\times \left[S\delta(\mathbf{k}_{\|} - \mathbf{k}'_{\|}) v_z^{+}(n', \mathbf{k}'_{\|}) r^{+-}(n', \mathbf{k}_{\|}; n, \mathbf{k}_{\|}) + \frac{A}{(2\pi)^2}(1 - S) F \left[v_z^{+}(n', \mathbf{k}') \right]^2 \right]$$

$$
\begin{aligned}
&\frac{V}{(2\pi)^3\hbar}T^{--}(n',\mathbf{k}';n,\mathbf{k}) = \delta\left[\varepsilon_{n'}(\mathbf{k}') - \varepsilon_n(\mathbf{k})\right] \\
&\times\left[S\delta(\mathbf{k}_\parallel - \mathbf{k}'_\parallel)v_z^-(n',\mathbf{k}'_\parallel)t^{--}(n',\mathbf{k}_\parallel;n,\mathbf{k}_\parallel) + \frac{A}{(2\pi)^2}(1-S)F\left[v_z^+(n',\mathbf{k}')\right]^2\right] \\
&\frac{V}{(2\pi)^3\hbar}R^{-+}(n',\mathbf{k}';n,\mathbf{k}) = \delta\left[\varepsilon_{n'}(\mathbf{k}') - \varepsilon_n(\mathbf{k})\right] \\
&\times\left[S\delta(\mathbf{k}_\parallel - \mathbf{k}'_\parallel)v_z^-(n',\mathbf{k}'_\parallel)r^{-+}(n',\mathbf{k}_\parallel;n,\mathbf{k}_\parallel) + \frac{A}{(2\pi)^2}(1-S)F\left[v_z^+(n',\mathbf{k}')\right]^2\right].
\end{aligned}
\tag{10.25}
$$

Using the same procedures that we used to derive (10.21) we obtain,

$$
\begin{aligned}
&h^+_{i+1,n}(\mathbf{k}_\parallel,z_i) = S_i\sum_{n'}^{i} t_i^{++}(n,\mathbf{k}_\parallel;n',\mathbf{k}_\parallel)h^+_{i,n'}(\mathbf{k}_\parallel,z_i) \\
&+S_i\sum_{n'}^{i+1} r^{+-}_{i+1}(n,\mathbf{k}_\parallel;n',\mathbf{k}_\parallel)h^-_{i+1,n'}(\mathbf{k}_\parallel,z_i) \\
&+(1-S_i)Fv^+_{z,i+1}(n,\mathbf{k}_\parallel)\left[\sum_{n',\mathbf{k}'_\parallel}^{i} h^+_{i,n'}(\mathbf{k}'_\parallel,z_i) + \sum_{n',\mathbf{k}'_\parallel}^{i+1} h^-_{i+1,n'}(\mathbf{k}'_\parallel,z_i)\right],
\end{aligned}
\tag{10.26}
$$

and

$$
\begin{aligned}
&h^-_{i,n}(\mathbf{k}_\parallel,z_i) = S_i\sum_{n'}^{i} r_i^{-+}(n,\mathbf{k}_\parallel;n',\mathbf{k}_\parallel)h^+_{i,n'}(\mathbf{k}_\parallel,z_i) \\
&+S_i\sum_{n'}^{i+1} t^{--}_{i+1}(n,\mathbf{k}_\parallel;n',\mathbf{k}_\parallel)h^-_{i+1,n'}(\mathbf{k}_\parallel,z_i) \\
&+(1-S_i)F|v^-_{z,i}(n,\mathbf{k}_\parallel)|\left[\sum_{n',\mathbf{k}'_\parallel}^{i} h^+_{i,n'}(\mathbf{k}'_\parallel,z_i) + \sum_{n',\mathbf{k}'_\parallel}^{i+1} h^-_{i+1,n'}(\mathbf{k}'_\parallel,z_i)\right].
\end{aligned}
\tag{10.27}
$$

The normalization constant, F, is obtained by requiring that the total flux leaving the interface is equal to the total flux entering it, $F = 1/\left(\mathcal{V}_i^- + \mathcal{V}_{i+1}^+\right)$. Here $\mathcal{V}_i^-$ and $\mathcal{V}_{i+1}^+$ are $\mathcal{V}_i^\pm = \sum_{n,\mathbf{k}_\parallel}|v^\pm_{z,i}(n,\mathbf{k}_\parallel)|$.

10.4.4 Application to Spin-Dependent Transport

We now have almost all of the formal tools we need to apply the Boltzmann formalism to spin-dependent transport in multilayers. The one remaining formality is to generalize the distribution functions to include spin indices. In the absence of spin-flip scattering, the spin channels decouple and all we need to do is to add another index for the spin to the distribution functions, to the Bloch velocities and band energies, to reflection and transmission coefficients, to specularity coefficients at interfaces and at outer boundaries, and to $\mu(z)$ in the local equilibrium distribution function. This makes the notation a little tedious but straightforward. Note that spin-flip scattering is more important for CPP than for CIP because as we pointed out earlier, for CPP it is the spin-diffusion length (determined by the spin-flip scattering rate) that determines how thick the layers can be before GMR vanishes while for CIP, the relevant length is the mean free path (normally determined by spin-conserving scattering). Spin-flip scattering can arise from thermally excited spin waves, from spin-orbit scattering or (especially in magnetic alloys) from non-collinear spins. We will here assume, however, that the spin diffusion length, which is a measure of how far an electron will travel with a well-defined spin direction, is long enough compared to the multilayer stack thickness that spin flip scattering can be neglected.

Next, we need to obtain Bloch velocities and band energies, as well as transmission and reflection probabilities. One approach consists of ignoring all band structure of the layers and assuming free electron states for all layers, but perhaps with different densities. In this approach, the reflection and transmission probabilities are treated as phenomenological numbers, constant over the Fermi surface (but perhaps different for each spin direction). A more rigorous approach consists of using Bloch velocities and band energies from first-principle calculations. The effect of disorder due to alloying can also be included in these calculations using approximations such as the coherent potential approximation [26]. These Bloch velocities and band energies can be calculated once and for all for the materials of interest and then stored. For epitaxial interfaces, the transmission and reflection coefficients can also be calculated from first principles using multiple-scattering techniques for each type of interface.

We will not discuss here the details of how such calculations are performed since this is a subject of its own. The main point is that the Bloch velocities, band energies, transmission and reflection coefficients can all be obtained from first principles and are then used as input to the Boltzmann approach without any adjustment or fitting. The only adjustable parameters left are then the specularity coefficients and bulk scattering times. For a non-magnetic material, such as Cu, there is only one scattering time, which can be fixed by demanding that the calculated bulk resistivity equals an experimental value. For magnetic materials, such as CoFe and NiFe, there are two scattering times, $\tau_\uparrow$ and $\tau_\downarrow$, one for each spin direction. The sum of these can again be fixed by the bulk resistivity, but the ratio of them is essentially an adjustable parameter. However, it cannot be completely arbitrary since the bulk scattering rate can in perturbation theory be expressed as a sum over squares of matrix elements of initial and final states, weighted by the density of final states. So some knowledge of the scattering potentials in the different spin channels, together with knowledge of the densities of states, can give

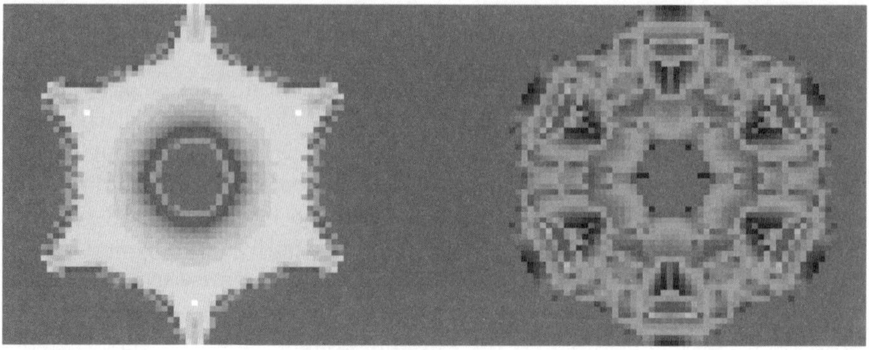

Fig. 10.4. Transmission probability on the Fermi surface in the k_x-k_y plane for majority (left) and minority (right) electrons incident from CoFe on Cu. The interface is in the (111) direction. The false color scheme goes from white (highest probability) to pink, green and blue in descending probability

an estimate for the ratio of the scattering times. Note that the scattering potential for a given nonmagnetic impurity need not be the same for spin-up and spin-down channels, so the ratio of the scattering times need not equal the inverse of the ratio of the densities of states. Indeed, in many magnetic materials, the scattering potential is often larger in the minority channel than in the majority channel for a given impurity. This tends to increase $\tau_\uparrow/\tau_\downarrow$ over the inverse of the ratio of the densities of states [27].

The remaining parameters are the specularity coefficients. Let us first consider the Co/Cu or CoFe/Cu interface. Detailed examination of the band structure of Co (or CoFe) and Cu shows that for the typical growth direction, (111) direction, there is an almost perfect matching of the bands for each $\mathbf{k}_\parallel$. This means that for an epitaxial interface, the majority electrons are transmitted almost perfectly through the interface. On the other hand, the minority spin Fermi surfaces of Co and CoFe are quite complicated and bear no resemblance to the rather simple Cu Fermi surface. As a consequence, there is a rather poor band match in the minority channel. Minority spins incident on the surface from CoFe thus experience a large effective potential and are rather likely to be reflected. Figure 10.4 depicts calculated transmission probabilities for minority and majority spins incident from CoFe on Cu, with the interface perpendicular to the (111) direction. The relatively good band match in the majority channel also has as a consequence that the majority electrons are rather insensitive to interdiffusion of Cu and Co across the interface – the scattering potential in the majority channel for a Cu impurity in Co or Co impurity in Cu is rather small provided, of course that the moments maintain their alignment. Therefore, interdiffusion of Co and Cu at the interface does not cause diffuse scattering at the interface. On the other hand, the scattering potential for Cu in Co or Co in Cu is rather large in the minority channel. As a consequence, interdiffusion of Cu and Co at the interface has a much larger effect on the minority spins and rapidly causes the scattering to be diffuse [28,29]. First-principle calculations of the specularity coefficients for majority and minority spins at disordered interfaces are in agreement with these arguments [30].

We are now in a position to qualitatively understand, from microscopic considerations, what drives the GMR in Co/Cu or CoFe/Cu multilayers. First, there is a contribution from the bulk scattering in the Co layers, since the scattering rates are different for majority and minority spins. This is what went into the simple two-current model earlier. In addition, there are important contributions from the spin-dependent scattering at the interfaces. Diffuse scattering suppresses contributions to the total current. In parallel configuration, this just gives additional resistance in the minority channel. Most of the current is shunted through the majority channel with its lower bulk resistivity and near-specular transmission at the Co/Cu interfaces. In the anti-parallel configuration, on the other hand, each electron is a minority spin at every other interface and in every other Co layer. Not only does the bulk scattering rate increase the resistivity of both channels, but both channels now suffer diffuse scattering at every other interface. This further increases the resistance in AP configuration and thus enhances the GMR over what would be obtained from bulk scattering alone.

10.5 CIP GMR in Spin Valve Sensors

There are two major technical challenges that must be overcome if antiferromagnetically coupled multilayer stacks are to be useful as read sensors. First, as we discussed in the introduction to this chapter, there is a constraint on the total thickness of the sensors set by the shield-to-shield spacing, which for a 100 Gbit/in^2 application is of the order of 60 nm. Second, the antiferromagnetic RKKY coupling used to keep consecutive ferromagnetic layers anti-parallel in the absence of an external magnetic field is usually so strong that a rather sizable external field must be applied in order to obtain a reasonably large response. This clearly makes it difficult to use in sensors in recording heads, which must be able to produce a large response for fields of the order of a few hundred Oe.

The spin valve is a design that uses CIP GMR to produce a usable response. The basic idea of a spin valve is to use only two ferromagnetic layers. The first of these, the reference layer (R), has the direction of its magnetization set in a fixed direction. Next is a metallic spacer layer (typically Cu), followed by the second ferromagnetic layer. This latter is free to move in response to an in-plane external magnetic field. This means that the Cu spacer thickness must not be such that there is a strong RKKY coupling between the free and reference layers. Normally, this is not an issue in industrial applications. In these, the layers are grown by a sputtering process that gives the layer interfaces enough roughness that the RKKY coupling is severely reduced. On the other hand, long wavelength roughness, or interface undulations, cause a so-called Néel orange peel coupling [31] between the magnetic layers and care has to be taken to ensure that this coupling is not unacceptably large.

A typical spin valve has the following layer structure. On top of the wafer substrate is first deposited a seed layer. As the layers are deposited, they usually develop a columnar grain structure along the growth direction. The purpose of the seed layer is to promote the growth of large grains, and sometimes also to provide the correct crystallographic texture (typically (111) texture along the growth direction). The seed layer also has to wet the substrate, i.e. grow uniformly across the substrate and not in disconnected globs or islands as it is deposited on the substrate. Ta is common seed layer, that gives a grain size of some tens of nm in the layers grown on top of it. For the basic structure known as bottom spin valves (BSV), the next layer is an antiferromagnet, such as NiO or PtMn. In a simple BSV, there are only two ferromagnetic layers, the first of which is the reference layer deposited on top of the antiferromagnet. Then come the Cu spacer, the free layer, and finally a cap, such as Ta, which protects the multilayer stack. Another basic structure is the Top Spin Valve (TSV), in which the free layer is first grown on the seed layer, followed by the Cu spacer, reference layer, antiferromagnet, and cap layer.

The purpose of the antiferromagnet is to pin the direction of the magnetization in the reference layer. In general, a ferromagnet deposited on top of an antiferromagnet and then annealed in an external field will experience a loop shift of the magnetization vs. external field. The magnitude of this shift is called the pinning field. The physics of this exchange bias effect could be subject of a chapter of its own, and we will not discuss the details [32] here.

In order to obtain a response that is as linear as possible, the free layer must be biased magnetically in a direction perpendicular to the reference layer in the absence of external fields. In addition, the device design has to be stable with a reversible response. Irreversible motion of the magnetization causes unacceptable noise and may also shift the electrical bias point. The magnetic bias is usually obtained by abutting permanent magnets to the free layer, with their magnetization in the desired free layer bias direction (cross-track). The strong magnetic field from the permanent magnets, and the demagnetizing field which tend to align the magnetization of the free layer along the edges, cause the response of the free layer in the device to be much stiffer than that of an unbiased sheet film structure. A figure of merit is the efficiency of the spin valve, which measures the average change in magnetization relative to saturation in the direction of the applied field (see Chapter 1 of this volume). In typical spin valves, the efficiency may be about 20%–25%.

From a transport point of view, the seed, cap and antiferromagnetic layers simply constitute parasitic conductances that tend to shunt some of the

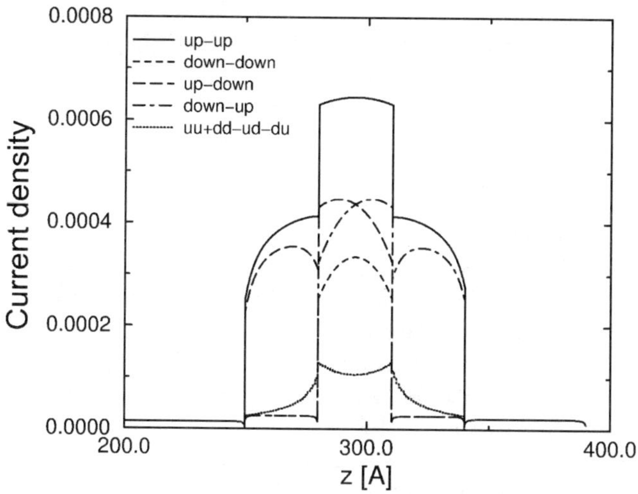

Fig. 10.5. Current densities for the different spin channels in P and AP configurations (in A/Å^2), and difference current density between P and AP configurations for a BSV spin valve Ta(5)/PtMn(20)/CoFe(3)/Cu(3)/CoFe(3)/Ta(5), with the layer thicknesses in nm

current away from the active Co/Cu layers. Therefore, it is desirable for these parasitic conductances to be as small as possible. Fortunately, typical antiferromagnets, such as PtMn, and seed and cap layers, such as Ta, have resistivities of a few hundred µΩ-cm and do not shunt the current very much. In order to apply the Boltzmann approach to a simple spin valve, we need parameters to describe these layers. First, the interfaces that they form with the active layers in the spin valve tend to be rather rough so it is a good approximation to set the specularity coefficients for both majority and minority electrons to zero at these interfaces. Thus the transport in these layers is then completely decoupled from the Co and Cu layers. In addition the electron mean free path tends to be quite short which implies that the transport is local in the parasitic layers. This allows us to use any model for the band structure and scattering times that gives the correct bulk resistivity.

Figure 10.5 depicts the current density in P and AP configurations in a CoFe/Cu spin valve. In these calculations, the bulk relaxation times of Ta and Cu were set to give bulk resistivities of about 240 µΩ-cm and 6.15 µΩ-cm, respectively. The majority and minority bulk spin relaxation times of CoFe were set to give a bulk resistivity of about 30 µΩ-cm, with a ratio of about 20. The interface specularity parameters were set to zero, except for majority spins at CoFe/Cu interfaces, for which the specularity parameter was set to unity. Since the actual parameters for sputtered spin valves may depend sensitively on seed and deposition conditions, the values used here may not necessarily reflect realistic parameters, but serve to give illustrative examples. The difference in current density between P and AP configuration shows directly from where the contributions to the GMR arise. In general, there are contributions both from the bulk of the CoFe and Cu layers. These contributions stem from the long scattering times for majority spins and the fact that majority spins can easily cross from CoFe into Cu without scattering. In addition, there are important contributions localized near the CoFe/Cu interfaces due to the spin-dependent interface scattering (the spin dependent specularity parameters).

In a patterned simple spin valve, there is a large magnetostatic coupling between the free layer and the reference layer, as well as edge effects due to magnetostatic interactions within the layers. In addition, the reference layer does respond to some degree to the external field. Both of these effects tend to reduce the GMR since they reduce the degree to which the free layer magnetization can rotate relative to that of the reference layer. The Synthetic Antiferromagnetic (SAF) spin valve is a design aimed at both reducing the response of the reference layer as well as the magnetostatic coupling between reference and free layers. The basic idea is to add another magnetic layer, the pinned layer (P), with a fixed magnetization direction and a very strong antiferromagnetic coupling to the reference layer. The magnetic field emanating from the pinned and reference layers then tend to form closed loops between the two layers, which substantially reduces the magnetostatic coupling to the free layer. In addition, the pinned and reference layer behave together as an antiferromagnet, with no net average response to an external field. This reduces the rotation of the reference layer in response to an external field.

The basic design of a SAF spin valve has the following structure. First there is the usual seed layer, followed by the antiferromagnet, which pins the pinned layer through the effect of exchange bias. After the pinned layer comes a very thin Ruthenium layer which separates the pinned and reference layers. Ruthenium has the property that it provides a very strong, of the order of 1 erg/cm^2, antiferromagnetic coupling [33] between the pinned and reference layers for a Ruthenium thickness of about 8 Å. Following the reference layer come the usual Cu spacer, free layer and cap.

While the SAF provides a magnetically improved design over the simple spin valve, it appears at a first glance that it should have a GMR signal lower than that of a simple spin valve. The reason is that both Ru and the pinned layer are extra conducting layers which shunt current. In addition, the pinned and reference layers are always in AP configuration, causing extra resistance for any electron no matter what the orientation of the reference and free layers. This tends to increase the resistance in P configuration more than in AP, so we would expect a net decrease in GMR, both due to the decrease in ΔG as well as the increase in $G(AP)$. However, in reality the GMR signal tends to be higher in SAF spin valves than in simple ones. Part of the reason for this appears to be the nature of the Ru-reference layer interface. There has been some speculation that the electrons incident on this interface from

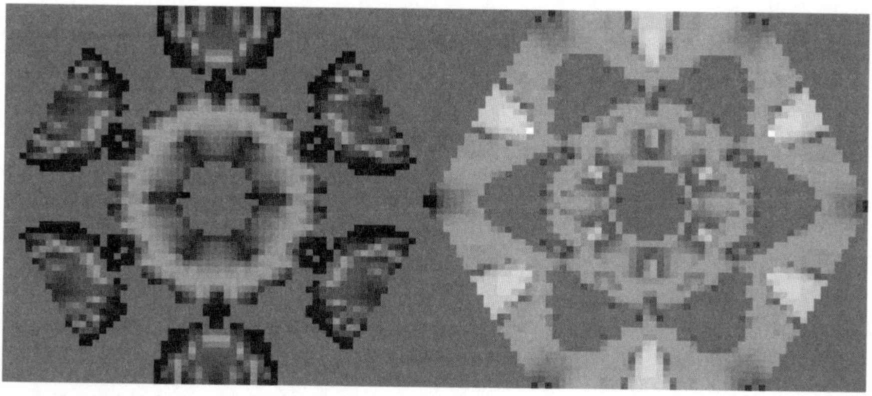

Fig. 10.6. This figure depicts the transmission probabilities on the Fermi surface in the k_x–k_y plane for majority (left) and minority (right) electrons incident from CoFe to Ru. The interface is perpendicular to the (111) direction. The same color scheme as in Fig. 10.4 is used

the reference layer are reflected specularly to some degree [34]. Such specular reflection would decouple the active reference, Cu, and free layers from the seed, AF, pinned and Ru layers and maximize ΔG since majority (minority) electrons from the R layer would not travel through the Ru to the P layer where they become minority (majority) electrons. There is indeed supporting evidence for this from first principle calculations. Figure 10.6 depicts the transmission probabilities for majority and minority carriers incident from CoFe on Ru for an epitaxial interface. The figure shows that the majority electrons tend to be reflected to a high degree across the Fermi surface, while the minority electrons are transmitted. This enhances ΔG of the spin valve since minority carriers entering the Ru and CoFe P layers will rapidly scatter.

For industrial applications it is obviously of great interest to maximize the signal from the device. Whether one wants specifically to maximize ΔR or the GMR itself depends on how the device is biased, constant current, constant voltage, or constant power. We will here not discuss optimization in these different bias schemes, but only briefly discuss in generic terms how to increase the GMR of a spin valve.

The left panel of Fig. 10.7 shows the GMR of a BSV as a function of Cu thickness, assuming a constant bulk resistivity of 4 µΩ-cm of the Cu film. The figure illustrates the general trend that the GMR increases monotonically with decreasing Cu thickness. The GMR also increases monotonically with decreasing Cu bulk resistivity, as is depicted in the right panel of Fig. 10.7. The reason is that the Cu itself does not directly contribute to the GMR. It is the traversal of electrons through the Cu from one CoFe layer to the other and the spin dependent scattering that occurs in the CoFe layers and at the CoFe/Cu interfaces that drives the GMR signal. Increased scattering in the Cu layer only increases spin *independent* scattering of all electrons going

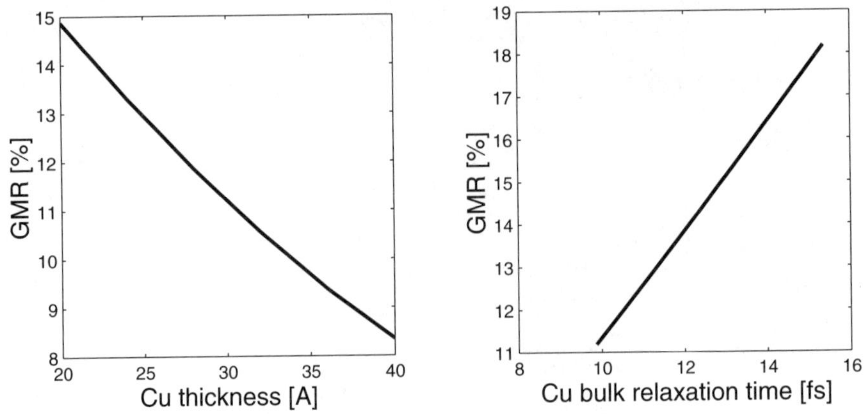

Fig. 10.7. The left panel depcits the GMR ratio as a function of Cu thickness for the BSV in Fig. 10.5. In the right panel is shown the GMR ratio as a function of the Cu bulk relaxation time for the same spin valve with a Cu thickness of 30 Å.

from one CoFe layer to the other. Therefore, the less Cu and the less spin independent scattering that occurs in the Cu, the higher the GMR signal.

In industrially manufactured sputtered thin films, the Cu thickness and electrical quality (bulk resistivity) are not independent. As we have discussed earlier, sputtering dynamics and diffusion during anneals cause some interdiffusion to occur at the Cu/CoFe interfaces. There is then a tail of Co and Fe impurities extending some distance into the Cu spacer layer from each interface. In addition, the interfaces themselves are not flat but have a certain roughness which depends on seed layer, multilayer stack thickness, and sputtering conditions. While some of the impurity scattering in the Cu may have a spin asymmetry with a higher scattering rate for minority carriers, the effect of the interdiffusion and roughness is inevitably to cause an increase in all scattering rates. As the Cu film is made thinner with unchanged interfaces, the net effect is to have higher impurity concentration and thus higher scattering rates in the Cu layer, which gives the appearance of a high bulk-like resistivity of the Cu layer. This tends to decrease the GMR. In addition, long wavelength undulations of the interface cause increased Néel coupling between the R and F layers, which makes it more difficult for the free layer to respond independently to an external field. Moreover, pinholes may appear in the Cu film due to incomplete coverage before the deposition of the next CoFe layer, and the pinholes may cause direct ferromagnetic coupling between the CoFe layers. Therefore, it is important to improve the quality of the interfaces as the Cu film is thinned down.

Another possible route to increase the GMR signal consists of improving the specularity at the outer boundary of the CoFe free layer [35–37]. Dirty metal caps, such as Ta, typically cause a very diffuse spin independent scattering at this interface, which in terms of the specularity coefficient means

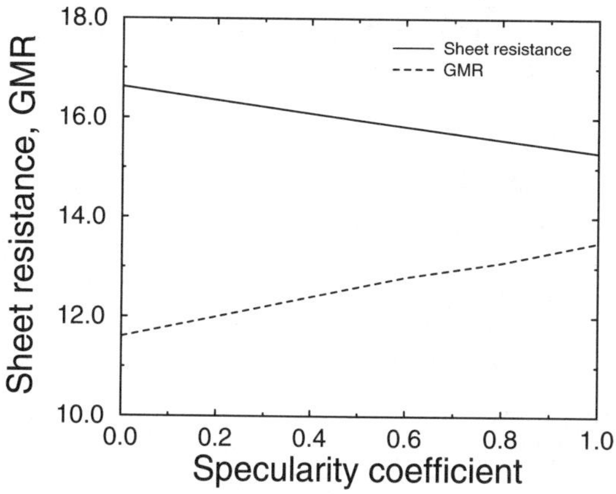

Fig. 10.8. Sheet resistance in Ω per square and GMR ratio in percent as functions of specularity coefficient of the free layer-cap interface for a BSV multilayer stack in Fig. 10.5, but with the Ta cap replaced by an insulator

$S \approx 0$. Current incident on this interface is then lost, causing an increased resistivity for free layers of the order of or thinner than the mean free paths of the spin species. This affects the majority spins the most, since they have a much longer mean free path than the minority spins (the mean free path of the minority spins is less than 10 Å). The result is to reduce the spin asymmetry of the transport in the free layer. The effect of the diffuse reflection is to set the effective mean free path of the majority spins to be approximately equal to the thickness of the free layer. If the reflection at this interface can be made more specular, the mean free path of the majority spins can be increased, which increases the spin asymmetry of the spin transport and so the GMR. Figure 10.8 depicts the sheet resistance and GMR for a spin valve as a function of the specularity coefficient at the free layer-cap interface. It is clear that a rather sizable increase in the GMR can be achieved with a high degree of specularity.

One possible way to achieve better specularity is to cap the free layer with an oxide. This can be done, for example, by in-situ or ex-situ oxidation of the free layer itself, or by deposition of a suitable oxide. Using the former method it may be difficult to accurately control the oxide thickness. Furthermore, ferromagnetic or antiferromagnetic oxides may result which adversely affect the magnetics of the free layer. Direct deposition of the oxide, or deposition of a nonmagnetic metal such as Al, which is subsequently oxidized, may provide better process control of the oxide thickness, but here the process and oxide have to be chosen such that the quality of the free layer/oxide interface is good enough that there is no appreciable diffuse scattering due to surface roughness. In addition, no matter which method of oxidation is used, the oxide has to be thermally stable so that oxygen does not diffuse into the multilayer structure during the lifetime of the device.

It may also be possible to introduce an oxide layer in the R layer to engineer specular reflection in the R layer before electrons reach the Ru and P layers. However, there are indications that the Ru/CoFe interface already provides a high degree of specular reflection for majority electrons incident on the Ru from the R layer, while minority electrons are transmitted into the Ru and subsequently rapidly scattered. So the Ru/CoFe interface already provides a high-degree of desirable spin-dependent scattering. In order for an oxide to improve on this, the oxide/CoFe interface has to provide better specular properties. This may not be easy to achieve. In addition, for BSVs the oxide then serves as a seed layer for the rest of the layers, which may degrade the growth. A similar problem occurs with free layer oxide caps in TSVs. The rest of the layers would have to be grown on top of the oxide, and this may result in degradation of grain size and layer quality.

Finally, we briefly discuss how far, that is, to what areal density, may CIP spin valve designs provide sensors in recording heads. A reasonable estimate of a potentially achievable GMR is about 20% [37,38] with a sheet resistance of about 20 Ω per square, assuming a suitable robust and stable structure can be prepared. This gives a ΔR of 4 Ω per square. Assuming a rectangular read sensor 100 nm × 50 nm for 100 Gbit/in^2, the device resistance is about 40 Ω. With a bias current of 1 mA and an efficiency of 25%, the peak signal output is about 2 mV. If we compare this to the rough estimate of 2 mV we obtained in Sect. 10.1, we conclude that current CIP GMR spin valves may perhaps be extendable to about 100 Gbit/in^2, but it is difficult to see how higher areal densities can be reached with these sensors.

10.6 CPP GMR Devices

The CPP GMR effect can be much larger than the CIP GMR, which is why devices based on this principle may provide attractive solutions for areal densities beyond 100 Gbit/in^2. As we have mentioned earlier, some technical issues related to bias and sensitivity must first be solved. Here, we will assume without further discussion that this can be done. An additional problem with CPP GMR sensors will remain, however. The actual resistance is rather low in these devices, because the current flows through a much larger cross-sectional area than in CIP devices. The product RA of resistance R and cross-sectional area A, which is independent of area, can be of the order of 10^{-3}–10^{-2} Ωμm^2 for a multilayer stack. Assuming a square sensor of side 100 nm, this gives a device resistance of 0.1–1 Ω. Even if the GMR is 30%, this gives a maximum ΔR of 0.03–0.3 Ω. It is not then feasible to achieve an acceptable output signal voltage for realistic bias currents in devices of this dimension. In order to increase the output signal voltage, the device resistance has to be increased (without a concomitant decrease in the CPP

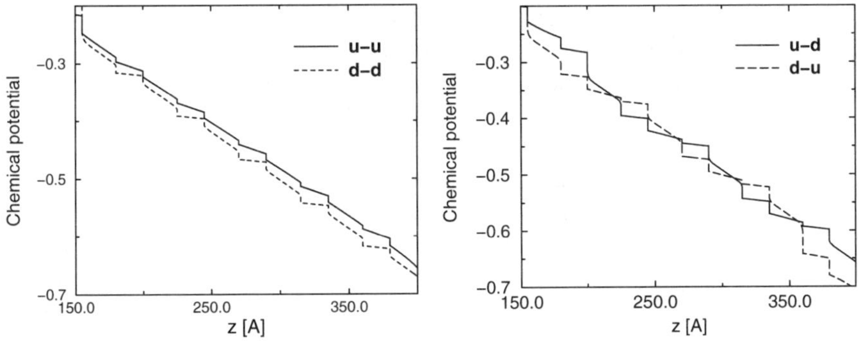

Fig. 10.9. Electrochemical potential along the current direction for the two channels in P, labeled u–u and d–d, (left panel) and AP, labeled u–d and d–u, (right panel) configurations. For clarity, only the section through the CoFe/Cu layers is shown

GMR). Conceptually, the easiest way to do this is by decreasing the size of the sensor. A smaller sensor implies a higher areal density, and we conclude that while CPP GMR sensors may not be useful for 100 Gbit/in^2, they may be important for higher areal densities, perhaps 150–200 Gbit/in^2 or beyond. Note that we said that decreasing the sensor size is the *conceptually* easiest way to produce working devices. There are of course enormous challenges related to mass-producing devices of dimensions 50 nm with a reasonable production yield of devices which will have to be overcome no matter what the specific read sensor will be.

Figure 10.9 depicts the chemical potential for a prototypical 6 bilayer CPP GMR multilayer structure, consisting of a 55 Å Ta seed, 100 Å Cu, 5 bilayers of CoFe(25 Å)/Cu(20 Å), capped with 100 Å Cu. The bulk scattering parameters were those used in the CIP calculations in the previous section, but for ease of computation, the interface scattering was assumed to be specular ($S = 1$) in all channels. The Cu pads at the ends of the structure are contacts. In practical realizations, the contacts may be made of Cu or Ta/Au, or the shields themselves may serve as contacts to the sensor. If we had ignored the charge build-up at the interfaces, represented by the quantity $\mu_\sigma(z)$ in the Boltzmann equation, the lines in the plot would simply be straight lines within each layer, the slope of which would give the resistance for each channel in that layer. Instead, there are large discontinuities at each Cu/CoFe interface. These discontinuities represent interface resistances and are due to charge and spin accumulation at the interfaces because of band mismatches and interface scattering. Note that the discontinuities are made necessary by the requirement that the continuity equation be satisfied. In addition to the interface resistances, there are exponential terms in the electrochemical potentials in the vicinity of an interface. These exponential terms decay at a length scale set by the component of the mean free path perpendicular to the interface, $\lambda_{\sigma,\perp} \approx v_z(n, \mathbf{k}_\parallel, \sigma)\tau_\sigma$. These terms can be viewed as a contribution to the interface resistance, but it should be pointed out that their contributions depend on the proximity to other interfaces and are not simple resistances of each interface that can independently be added in series. In addition, there are bulk-like contributions to the resistances which depend on the spin-dependent bulk scattering in the interior of the layers. Both the discontinuities and the exponential terms depend on the thicknesses and types of neighboring layers. For example, for parallel alignment, the discontinuous interface resistance may increase with the thickness of Cu spacer layer. For antiparallel alignment, on the other hand, the discontinuity may increase with Cu thickness on the minority side, but decrease on the majority side. Generically, the GMR increases when the resistance for the majority channel in parallel alignment decreases and the resistance for both channels in antiparallel alignment increase. From the arguments above, it is clear that optimizing the device output signal involves a careful balancing act.

10.7 Tunneling Read Sensors

A simple design for a tunneling read sensor consists of an AFM, reference layer, tunneling barrier, and free layer. The AFM pins the reference layer as in CIP spin valves, and the contacts can be made from separate leads, or the shields can be used as leads as can be the case for CPP read sensors. In addition, there must be some method of applying proper magnetic bias to the free layer. In contrast to CPP read sensors, one main technical challenge with tunneling read sensors is to get the resistance small enough. The resistance of a tunneling junction increases exponentially with the thickness of the oxide tunneling barrier. For practical applications, it is desirable to have a device resistance of less than 100 Ω. For a tunneling device of dimensions 100 nm × 100 nm, this implies an RA product of less than 1 $\Omega\mu m^2$. In order to achieve such low RA product, the tunneling barrier must be made as thin as possible. Current development work typically uses aluminum oxides as tunneling barriers, with thicknesses less than 10 Å. Since the lattice spacing of a solid is of the order of a few Å, this implies that the tunneling barrier is a few monolayers thick. As small variations in the barrier thickness lead to exponential variations in the device resistance, the barrier thickness has to be kept uniform to within a monolayer across an entire wafer in order to achieve acceptable yields. This obviously poses some technical challenges.

Tunneling read sensors have a noise characteristic which is different from that of CIP and CPP spin valves or multilayers. In the latter devices, the noise sources are broad band thermal Johnson noise, plus $1/f$-type noise due to magnetic fluctuations in the ferromagnetic layers. In tunneling read sensors, there is additional shot noise due to the the individual events of electrons tunneling through the barrier. For a simple tunneling device with a bandwidth Δf, the current-current correlation function is [39,40]

$$\langle I_n^2 \rangle = \frac{e^2}{h} \left[4k_B T \sum_{n,\mathbf{k}_\parallel} T_{n,\mathbf{k}_\parallel}^2 + \sum_{n,\mathbf{k}_\parallel} 2eV \coth\left(\frac{eV}{2k_B T}\right) T_{n,\mathbf{k}_\parallel} \left(1 - T_{n,\mathbf{k}_\parallel}\right) \right] \Delta f. \tag{10.28}$$

Here, V is the bias voltage across the junction, T the operating temperature, and $T_{n,\mathbf{k}_\parallel}$ the tunneling probability for a state at the Fermi surface in band n and wavevector $\mathbf{k}_\parallel$ parallel to the junction interface. The noise power P_n can then be obtained by multiplying (10.28) with the device resistance. Equation (10.28) describes the cross-over from Johnson noise to shot noise when $k_B T \sim eV$. For operating temperatures of 300–400 K, and bias voltages of a few tenths of a volt, $eV \gg k_B T$ and the Johnson noise can typically br neglected compared to the shot noise. In this limit, the expression becomes the classical shot noise $P_n = 2e|I|R\Delta f$ for very small tunneling probabilities $T_{n,\mathbf{k}_\parallel} \ll 1$. In order to obtain an estimate of the noise voltage, consider a tunneling device with a resistance $R = 100\ \Omega$, $\Delta R/R = 10\%$, $V = 0.1$ V, and a device bandwidth of 500 MHz. We then obtain a signal-to-noise power ratio of about 57 dB. This result depends sensitively on achieving a low value for the product of area and resistance with a reasonably high $\Delta R/R$. If we instead assume $RA = 50\ \Omega\mu\mathrm{m}^2$ we obtain a signal-to-noise power ratio of about 38 dB for a $100 \times 100\ \mathrm{nm}^2$ device with $\Delta R/R = 5\%$. While this is a crude estimate, it illustrates the point that shot noise can be a potential source of problems for achieving a reasonable signal to noise ratio unless the RA product can be made small enough for working devices. We note that suppression of shot noise below its classical limit can occur if the tunneling probabilities are appreciable. Whether this can occur for tunneling devices with thin oxide tunneling barriers is an open question.

In the Jullière model, (10.8), the tunneling magnetoresistance is expressed only in terms of the spin polarization ratios of the tunneling density of states at both sides of the junction. There is no reference to band structure of the ferromagnetic materials nor to the electronic structure of the tunneling barrier. This clearly cannot be the whole story. In a simple WKB approximation of tunneling through a sharp (epitaxial interfaces) barrier of thickness t, the tunneling probability amplitude can be expressed as [41]

$$T(n, \mathbf{k}_\parallel, s) = \sum_n T^0(n', \mathbf{k}_\parallel; n, \mathbf{k}_\parallel, s) \exp\left[-\int_0^t k_\mathrm{b}(z, n', \mathbf{k}_\parallel, s) dz \right], \tag{10.29}$$

where the prefactor T^0 depends on properties of the ferromagnetic layers, and $-\mathrm{i}k_\mathrm{b}(z, n', \mathbf{k}_\parallel, s)$ is the complex wavevector in the barrier for an electron at the Fermi surface in band n' with wavevector $\mathbf{k}'_\parallel$ and spin s. In contrast to the Jullière formula, (10.29) clearly depends on electronic properties of both ferromagnets and barrier. This is also demonstrated in first-principle calculations of the tunneling probabilities for epitaxial Fe/vacuum/Fe and Fe/MgO/Fe junctions [42]. In addition to fine details, there are some generic characteristics that emerge from such calculations. The majority electrons incident on the barrier tend to be s-like and can penetrate a barrier more readily than p- or d-like electrons, which have a higher curvature of the wave function in the plane parallel to the interface. Minority spin electrons, on the other hand, tend to be more d-like. Higher curvature in the plane parallel to the interface leads to a shorter decay length in the tunneling barrier. If the curvature is low, states may have a long decay length, with an ensuing tunneling probability which can be relatively large.

References

1. See, for example, *Introduction to Magnetic Recording*, Robert M. White (ed.), (IEEE Press, New York, NY, 1985), or H. Neil Bertram, *Theory of Magnetic Recording* (Cambridge University Press, Cambridge, UK, 1994)
2. See, for example, R. M. Dreizler and E. K, U. Gross, *Density Functional Theory*, (Springer-Verlag, Berlin, 1990)
3. N. W. Ashcroft, N. D. Mermin, *Solid State Physics* (Holt, Rinehart and Winston, New York, NY, 1976)
4. See, for example, Appendix E in [3]
5. J. S. Moodera, L. R. Kinder, T. M. Wong, and R. Meservey, Phys. Rev. Lett. **74**, 3273 (1995)
6. T. Miyazaki, N. Tezuka, J. Mag. Magn. Mat. **151**, 403 (1995)
7. For a review, see, for example, T.R. McGuire and R.I. Potter, IEEE Trans. Mag. **MAG-11**, 1018 (1975)
8. See, for example, A. Messiah, *Quantum Mechanics* (North Holland, New York, 1961)
9. M. N. Baibich, J. M. Broto, A. Fert, F. Nguyen Van Dau, F. Petroff, P. Etienne, G. Creuzet, A. Friederich, and J. Chaezelas, Phys. Rev. Lett. **61**, 2472, (1988)
10. S. S. P. Parkin, Z. G. Li, and D. J. Smith, Appl. Phys. Lett. **23**, 2710 (1991)
11. P. Bruno, J. Phys. Condensed Matter **11**, 9403 (1999)
12. M. A. Ruderman and C. Kittel, Phys. Rev. **96**, 99 (1954); T. Kasuya, Prog. Theor. Phys. **16**, 45 (1956); K. Yosida, Phys. Rev. **106**, 893 (1957)
13. T. Valet and A. Fert, Phys. Rev. B **48**, 7099 (1993)
14. A. H. MacDonald, T. Junwirth, and M. Kasner, Phys. Rev. Lett. **81**, 705 (1998)
15. M. Julliere, Phys. Lett. A **54**, 225 (1975)
16. X.-G. Zhang and W. H. Butler, unpublished
17. For a review, see, for example, W. H. Butler, X.-G. Zhang, and J. M. MacLaren, J. Supercond. **13**, 221 (2000)
18. See, for example, K. Huang, *Statistical Mechanics* (Wiley, New York, NY 1967)
19. An excellent treatise on the Boltzmann equation and its application to electron transport can be found in J. Ziman, *Electrons and Phonons* (Oxford University Press, New York, NY, 1960)
20. For a more modern brief discussion and applications, see for example, H. Haug and A.-P. Jauho, *Quantum Kinetics and Transport in Optics and Semiconductors* (Springer Verlag, Berlin, 1996)

21. H. U. Baranger and J. W. Wilkins, Phys. Rev. B **30**, 7349 (1984)
22. K. Fuchs, Proc. Philos. Camb. Soc. **34**, 100 (1938); E. H. Sondheimer, Adv. Phys. **1**, 1 (1952).
23. R. Q. Hood and L. M. Falicov, Phys. Rev. B **46**, 8287 (1992); L. M. Falicov and R. Q. Hood, J. Appl. Phys. **76**, 6595 (1994)
24. S. Zhang and P. M. Levy, Phys. Rev B **57**, 5336 (1998)
25. A. Shpiro and P.M. Levy, Phys. Rev. B **63**, 14 419 (2000)
26. See, for example, A. Gonis, *Greens Functions for Ordered and Disordered Systems* (Elsevier Science, Amsterdam, 1992)
27. For a review, see R. K. Nesbet, IBM J. Res. Dev. **42**, 1, (1998)
28. W. H. Butler, X.-G. Zhang, D. M. C. Nicholson, and J. M. MacLaren, Mater. Res. Soc. Symp. Proc. **384**, 291 (1995)
29. W. H. Butler, X.-G. Zhang, D. M. C. Nicholson, and J. M. MacLaren, Phys. Rev. B **52**, 399 (1995)
30. X.-G. Zhang and W. B. Butler, unpublished
31. L. Néel, Comptes. Rendus Ser. R **255**, 1545 (1962); *ibid* 1676 (1962); J. C. S. Kools, W. Kula, D. Mauri, and T. Lin, J. Appl. Phys. **85**, 4466 (1999)
32. For a recent review, see J. Nogués and I. K. Schuller, J. Mag. Mag. Mat. **192**, 203 (1999)
33. S. S. P. Parkin, N. More, and K. P. Roche, Phys. Rev. Lett. **64**, 2304 (1990)
34. B. Dieny, M. Li, S. H. Liao, C. Horng, and K. Ju, J. Appl. Phys. **87**, 3415 (2000)
35. H. J. M. Swagten, G. J. Strijkers, R. H. J. N. Bitter, and W. J. M. de Jonge, IEEE Trans. Mag. **34**, 948 (1998)
36. W.F. Egelhoff et al., J. Vac. Sci. Technol. B **17**, 1702 (1999).
37. S. Araki et al., J. Appl. Phys. **87**, 5377 (2000).
38. W.F. Egelhoff et al., J. Appl. Phys. **82**, 6142 (1997).
39. M. Büttiker, Phys. Rev. B **46**, 12 485 (1992).
40. Th. Martin and R. Landauer, Phys. Rev. B **45**, 1742 (1992).
41. Khairurrijal, W. Mizubayashi, S. Miyazaki, and M. Hirose, Appl. Phys. Lett. **77**, 3580 (2000).
42. W.H. Butler, X.-G. Zhang, T.C. Schulthess, and J.M. MacLaren, Phys. Rev. B**63**, 054416 (2001); W. H. Butler, X.-G. Zhang, and T.C. Schulthess, Phys. Rev. B**63** March 1, (2001).

11 Recording Head Design

Michael Mallary

11.1 Introduction

Magnetic recording is literally as old as the hills. Recently a satellite, flying high over the polar regions of Mars, detected stripes of magnetisation spaced at more than 50 miles. This is proof that Mars had a significant magnetic field and a liquid core in the distant past. In order to detect finer structures, the height of the orbit would have to be reduced. At fly heights, h, greater than one third of the plus to minus spacing of the stripes, S_b, the field intensity from a periodic source falls as:

$$B \propto e^{-\pi h/Sb} \tag{11.1}$$

This is known as the Wallace spacing loss formula. At this time the fly height in disk drives is about 25 nm. At this height the spacing loss factor is about 1/5 for a 50 nm bit space. This is approaching the minimum that a modern partial response maximum likelihood (PRML) channel detector can cope with, for a reasonable minimum signal to noise (SNR) requirement. The fly height in a disk drive is the key variable from which all the others scale. Recently density has doubled each year. This has been achieved by scaling all of the salient system dimensions, including tolerances, down by 30% per year.

11.2 Superparamagnetic Limit

At this time the recorded bits in all commercial media consist of a collection of several hundred randomly located switching units. Maintenance of adequate signal to noise requires that the size of these switching units (and therefore their total number per bit) be scaled with the size of the bits. Therefore the volume of the switching units, V, has been going down at 65% per annum. For decoupled grains, the energy barrier to thermally induced demagnetisation, E, is proportional to the volume (e.g., $E = K_u V$, K_u is the magnetic anisotropy energy per unit volume). For constant K_u, E is also going down at 65% per year.

In the near future, we will not be able to continue scaling down the size of these switching units. If their switching energy barrier falls below 40 kT they will not retain the written magnetic state for the requisite ten years (see

Chap. 5 for more detail on this). Approximately every nanosecond, thermal agitation attempts to switch grains' magnetic orientation. The probability that such an event will occur per attempt is:

$$P_{\mathrm{switch}} = e^{-E/\mathrm{kT}} . \tag{11.2}$$

Here T is the temperature in degrees Kelvin, and k is Boltzman's constant. In ten years there will be e40 attempts (1 billion per second), hence the need for $E > 40$ kT. This is known as the superparamagnetic limit for magnetic recording.

The superparamagnetic effect sets a lower limit to the switching volume below which we must not go. With scaling as usual, this would bring the growth of areal density to a screeching halt. However there are several ways to work around this problem which will be exploited in the near future. They require adjustments in the projected evolution of head and media design. The most obvious approach is to increase the magnetic anisotropy energy density so that a smaller grain volume can be used. However this has stringent limits because:

$$K_u = H_K M_S / 2 . \tag{11.3}$$

H_K is the anisotropy field (Oe) and M_S is the saturation magnetisation (emu/cc).

Adequate overwrite of old data in longitudinal recording limits the usable H_K to a value that is approximately 60% of the saturation magnetisation of the pole tips expressed in Gauss (see below). The energy barrier reducing effects of demagnetisation fields, H_d, limit the usable M_S thickness product. This limit is proportional to H_k (see below). Therefore the usable K_u is proportional to the square of the saturation magnetisation of the pole tips. At this time $Ni_{45}Fe_{55}$ pole tips with a saturation magnetisation of about 16 kG are commonly used. The use of $Co_{35}Fe_{65}$ layers in the tips, with a saturation magnetisation of 25 kG, could allow for a factor of 2.4 (25/16-squared) increase in usable K_u. This in turn will allow for a proportionate reduction in grain volume. However the area of a grain is proportional to the 2/3 power of the volume so the increase in areal density from high saturation moment pole tips will be a maximum factor of 1.8 (e.g., $1.8 = 2.4^{2/3}$). At the present rate of areal density growth this is about one years increase. It should also be kept in mind that this cobalt alloy is susceptible to corrosion and it has a high coercivity. Third element additions to mitigate these issues will lower its saturation moment so this improvement should be viewed as an upper limit on the benefits of increasing the saturation field of the pole tips. For example, 22 kG pole tips (CoFeX) will allow for a 50% increase in areal density over 16 kG poles.

11.3 Reduced Bit Aspect Ratio

A second way to buy time against the superparamagnetic limit is to reduce the ratio of bit density to track density. This ratio is known as the bit aspect ratio or BAR. It will be shown below that low values of BAR can yield systems with equivalent media noise at higher areal density. For the last forty-five years of disk recording the BAR has been between 16 and 20. If we reduce the BAR we are no longer scaling by the same rules that have been used in the past. This impacts head design by requiring less reduction in fly height (per generation) but tighter limits on fringing and the manufacturing errors that effect the read and write widths.

The areal density benefit of reduced BAR can be seen from the following thought experiment which neglects write fringing, width errors, and tracking errors. For a given signal processing channel design there is a minimum signal to noise ratio, SNR, requirement to achieve a required raw error rate target. In this thought experiment we wish to design a new recording system with a small BAR (relative to the old system). The new system must have the same SNR and the same thermal stability (its grain volume and demag fields) as the old one.

If the media is partially oriented and fully saturated by the write process, the DC erase noise can (in principle) be reduced to a negligible level. If the transition noise is the dominant source, the amplitude SNR is proportional to the bit space, S_b, divided by the transition jitter, σ_t. The transition jitter is in turn proportional to the grain diameter, D, divided by the square root of the number of grains, N, that can be lined up across the read width of the head, W. Noting that the number of grains is the diameter divided by the read width then gives:

$$\mathrm{SNR} \propto S_b/\sigma_a \propto S_b/(D/N^{1/2}) \propto S_b W^{1/2}/D^{3/2}\,. \tag{11.4}$$

For longitudinal recording the demagnetising field, H_d, in the centre of a dibit is proportional to the media thickness, δ, divided by the bit space (for $W \gg S_b$). Therefore the constancy of the demagnetising field requires that their ratio be constant ($\delta/S_b = C =$ constant). This principle also holds for perpendicular recording in that the minimum writeable bit space is proportional to the media thickness. This is because the write field gradient goes down as the thickness increases. With this principle (noting that BAR $\propto W/S_b$ and that $V \propto \delta D^2$) we can rewrite the proportionality in Eq. (11.4) as:

copy of Eq. (11.4)	$\mathrm{SNR} \propto S_b W^{1/2}/D^{3/2}$	
sub. $D \propto (V/\delta)^{1/2}$	$\mathrm{SNR} \propto S_b W^{1/2}/(V/\delta)^{3/4}$	
sub. $\mathrm{W} \propto S_b\mathrm{BAR}$	$\mathrm{SNR} \propto S_b^{3/2}\mathrm{BAR}^{1/2}/(V/\delta)^{3/4}$	
sub. $\delta = CS_b$	$\mathrm{SNR} \propto S_b^{9/4}\mathrm{BAR}^{1/2}/(V/C)^{3/4}$	(11.5)
V and C constant	$\mathrm{SNR} \propto S_b^{9/4}\mathrm{BAR}^{1/2}$	
$x(S_bW)9/8/(S_bW)^{9/8}$	$\mathrm{SNR} \propto (S_bW)^{9/8}Sb^{9/8}\mathrm{BAR}1/2/W^{9/8}$	
sub. $\mathrm{BAR} \propto W/Sb$	$\mathrm{SNR} \propto (S_bW)^{9/8}/\mathrm{BAR}^{5/8}$	

Noting that areal density $\propto 1/S_b W$, constant SNR gives:

$$\text{Areal Density} \propto \text{BAR}^{-5/9} . \tag{11.6}$$

Therefore, from Eq. (11.6) it can be seen that reducing the BAR from 20 down to 4 would allow for an areal density increase of a factor of 2.45.

An additional benefit is gained from the reduction of self-demagnetisation when the BAR is small. At a distance of $y = S_b/2$, the field from a line charge of length W_{wr} (the write width), is proportional to the integral of $(y/(x^2 + y^2))dx$ from zero to $W_{wr}/2$. This is then proportional to the Arctan(W_{wr}/S_b). The write width is typically 80% of the track pitch, P. So for a BAR of 4 (e.g., $P/S_b = 4$ so $W_{wr}/S_b = 3.2$) the demagnetisation field, H_d, is reduced by 16%. This would allow for $x1.18$ thicker media relative to the bit space. Therefore the constant grain volume requirement allows the bit space to be reduced by 5.4%. This is the 1/3 power of this demagnetisation factor (e.g., $V \propto \delta S_b W$; $\delta \propto S_b H_d$; $S_b \propto W$; so $V \propto H_d S_b^3$). Therefore areal density is inversely proportional to the 2/3 power of the demagnetisation factor.

$$\text{Areal Density} \propto [1/\arctan(W_{wr}/)]^{2/3} \propto [1/\arctan(0.8\ \text{BAR})]^{2/3} \tag{11.7}$$

In the example of reducing the BAR from 20 down to 4, this allows a 12% boost in areal density to a net gain of 2.75. If the BAR were reduced from 20 down to 1.0 this demagnetising effect becomes stronger (e.g., density gain of **1.71**) and it boosts a density gain of 5.3 up to 9.0. But if the adjacent tracks have coherent data (the same bit pattern) this benefit is washed out unless the adjacent data is timed to be half a bit cell out of phase. The difficulty of achieving this phase shift timing is comparable to the timing problem of writing on patterned media (see Sect. 11.12 below).

In order to achieve a density gain from low BAR, the pitch overhead associated with width errors, tracking errors, and fringing must be kept at a constant percentage of track pitch. This represents a very significant demand on head design, head manufacture and track following. Undoubtedly micropositioners will be part of the head gimbal assembly (HGA) by the time the industry is working at low BAR. The extremely small width dimensions and the requisite tolerances will go beyond the capabilities of standard photolithography as well. Therefore on wafer or air bearing surface (ABS) focussed ion beam milling or electron beam lithography will probably be required. It should be kept in mind that the very small exposure area that is required at the tips of a recording head will make these approaches economic in the recording head industry long before they are in the semiconductor industry.

On the bright side of a low BAR is that low head media separations will not be encountered in high density product quite as soon. The fly height, h, is proportional to the bit space for a constant percentage of magnetic separation overhead (e.g., head and media overcoats and pole tip recession). Then for a given areal density, fly height is inversely proportional to the square root of the BAR. In the 20 vs 4 example, the low BAR system can fly 120% higher (if the areal density and media grain geometry is kept constant). For a read width that is 60% of the track pitch, the media SNR would improve by 3.5 dB and the demagnetisation field would go down. However if the grain volume and demagnetisation is kept constant (same thermal stability) then Eq. (11.5) applies (not including the Hd factor benefit) and the net media SNR gain is 8.7 dB.

11.4 Longitudinal Recording

Since the introduction of the RAMAC disk drive in the late 1950's the recording industry has been dominated by recording on media in which the magnetisation was oriented in the plane of the disk. This is almost a historical accident in that perpendicular recording was a contender for this first product [1]. Serious reconsideration of perpendicular recording did not occur for another two decades with the movement started by S.-I. Iwasaki. After some initial excitement it became clear that whatever advantage perpendicular might have had the entrenchment of the longitudinal approach out weighed it. Even though perpendicular recording will allow for smaller thermally stable grains (e.g., higher areal density) longitudinal recording will remain dominant for the near future. The industry has a remarkable track record based on the evolution of old technologies so it will stick with longitudinal recording until every bit is squeezed out of it.

A typical longitudinal recording head media system is shown in Fig. 11.1. It consists of a read/write head structure mounted on a slider that flies over a thin film disk at a height of about a microinch. The magnetic anisotropy axis of the disk grains is largely random in the plane. Even in circumferentially "oriented media" there is a large dispersion of the easy axis of the grains so that most of them will switch at a write field of about $H_K/2$ as they would without orientation. The write transducer has the ring structure of the old inductive read/write head. The read head consists of a giant magneto-resistive (GMR) sensor that is located in a narrow gap between two ferromagnetic

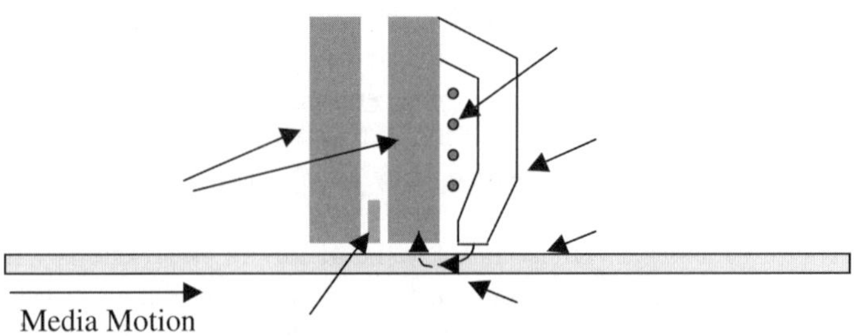

Fig. 11.1. Typical longitudinal recording head media system

shields. Flux from the transitions in the media induces a change in the resistance of the sensor that is quite linear within its saturation limits.

The shields absorb flux with low spatial frequency and thus slim the read back pulse. Equivalent slimming in an equaliser would boost high frequency noise excessively. The early rejection of low frequency flux also reduces the dynamic range requirements on the GMR sensor. This then allows it to be designed to maximise sensitivity on the low flux levels of the highest density data. Without the shields, the excess low frequency flux would drive a typical sensor into saturation and thus defeat the fundamental linearity assumptions of partial response multilevel (PRML) recording channel designs. PRML has provided very significant advantages (e.g. >3 dB equivalent SNR boost) over the previous peak detect schemes. Therefore sensors will be required to be approximately linear for the foreseeable future. We will see (below) that this can be a problem for perpendicular recording sensors.

The top shield is usually part of the write structure as indicated in Fig. 11.1. It is therefore called the shared shield. However it is becoming popular to provide a separate bottom pole for the write structure. This avoids driving the shared shield material near the GMR sensor and the sensor itself with extreme magnetisation excursions (during the write process) which can cause "domain instabilities" that distort the read back wave form.

Prior to the introduction of magneto-resistive read back sensors in 1990 all heads integrated the read and write functions into a single inductive read-write transducer. This made the write design more challenginging since it had to incorporate very soft magnetic materials with an absolute minimum of domain walls that were parallel to the flux path (which will induce domain wall instability). Now that the functions are separated they can be optimized separately. The write width is now wider than the read width to forgive servo error and the poor writability at the edge of the down stream write pole. Typically the writer width is 80% of the track pitch while the reader is about 60%. The tolerable sigma on these dimensions has ususally been about 5% of the track pitch. Therefore the write width is larger than the read width by a compound error that is just under three sigma (0.5% of the heads have $W_r > W_{wr}$). In fact all shipped heads are tested for these magnetic widths so the wings of the distribution have been truncated.

11.5 Longitudinal Recording Writer Tips

For both the shared shield and the separate writer designs, the high track densities of todays' products have required that the bottom pole of the writer be track trimmed to the width of the top pole. An ABS view of three possible pole tip designs is shown in Fig. 11.2. The original untrimmed shared shield approach is shown in Fig. 11.2a. This has process simplicity but the write fringing (beyond the geometric edge of the top pole) is about twice that of the track trimmed design shown in Fig. 11.2b. In this design the top pole

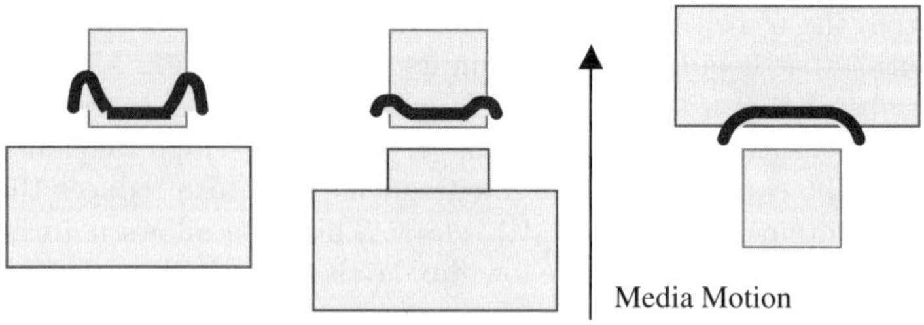

Fig. 11.2. Air Bearing Surface view of pole tip designs showing the coercivity field contours (note that media moves from bottom to top): **a** untrimmed shield; **b** trimmed shield; **c** shields over the write pole

is plated extra thick and is used as a selfaligned mask to ion mill trim the shared shield [2]. Good performance requires that: the cut into the shared shield be more than about 1.5 gap lengths; the shield be trimmed away at a lateral distance > several gaps; and that the material removal depth behind the ABS be a couple of gaps.

With these criteria, at a distance of half a gap from the pole edge, the write field will fall to about 1/3 of its maximum on track value. The maximum on track field should be about twice the media switching coercivity [3] to minimise non-linear transition shift (NLTS). Therefore the field at the half gap distance is 2/3 of the switching field, H_{co}. This is the threshold where repeated (e.g., hundreds) hits from adjacent track writes will produce a measurable increase in NLTS by a kind of room temperature thermally assisted erase process. After many writes on track N, tracks $N \pm 1$ will have their edges nibbled away if their servo error biased them towards track N. Therefore the high quality region of a written track (e.g., between the edge hooks) must be margined away from the half gap point by the 4 sigma servo error (compounded error of track N and $N + 1$ is then 2.8 sigma) in order for the fringe field of an adjacent track write to be $< 0.66 H_{co}$ 99.5% of the time in the sweet zone between the hooks.

We can shed crocodile tears over the loss of the track edges provided that it does not go too far. The write process there is comparatively poor. This is due to: a) the lower maximum field strength (80% of on track maximum when at the gap centre and at the geometric edge); b) the lower field gradient; c) the write contour is phase shifted relative to on track; and d) the large transverse field component there produces transverse recording. The contours where the write field falls below the media coercivity are drawn schematically onto the various down stream poles in Fig. 11.2. The "hooks" at the track edge contain essentially unusable flux because of the above list of pathologies. The track trim design in Fig. 11.2b is half as bad in this respect. The up side down design in Fig. 11.2c is a little better yet but its process is more difficult to execute. Yamaha shipped inductive heads with this design to the Quantum Fireball drive program in 1993 and Das Technologies built prototype MR heads in this

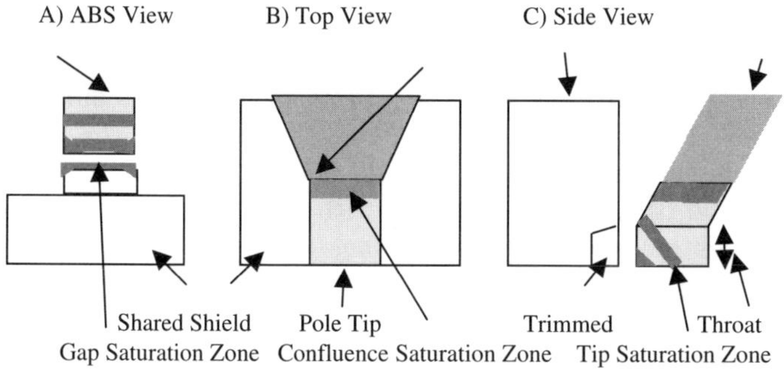

Fig. 11.3. Track trimmed pole tip region showing saturation zones

way [4]. Later we will see that this approach is theoretically advantageous for perpendicular recording heads of the "shielded pole design" [5]. At this time the tips shown in Fig. 11.2b are common so we will concentrate on this design.

Three views of track trimmed tips are shown in Fig. 11.3. This figure defines the various geometric terms that will be used here and it also shows three types of saturation zones that can occur at or near the pole tips. We will see that the successful performance of any advanced writer design depends strongly on the control of saturation in the gap, tip, and confluence zones. Loose talk leads to the impression that all saturation is bad. This is not the case. Controlled saturation at the confluence point is essential for good performance (see below). Maximisation of the write field (for thermally stable high coercivity media) requires that some significant degree of saturation in the gap zone occur within acceptable limits. In fact, in the region adjacent the write gap there is always saturation zones for any non-zero value of the gap field. This is due to the singular nature of the field intensity in the vicinity of the corner where the gap surface of the pole meets its ABS surface. Here the field grows as the inverse 1/3 power of the distance from the corner. It would go to infinity if the saturation magnetisation was infinite but of course it is not. Therefore at some distance it maximises at the saturation field of the tips. Inverting this spatial relationship predicts a saturation zone size that grows as the cube of the deep gap field.

When the gap saturation zone has grown to approximately the magnetic separation (e.g. $h + \delta/2 =$ the fly height plus overcoats plus pole tip recession plus lubrication plus half the media thickness) it starts to impact the write process in a number of ways: a) the peak value of the write field starts to max out, thus saturating the improvement in overwrite (OVW); b) the write field gradient starts to degrade thus increasing the transition width and the pulse width (PW50); c) the image of the transition magnetic charges in the pole tip weaken thus increasing self demagnetisation and therefore increasing the transition width; d) the write fringing field grows rapidly; and e) the pole tip image of the previously written transition in a dibit weakens so that its field strength increases and thus non-linear transition shift (NLTS) increases.

This laundry basket of gap zone saturation pathologies tumbles out suddenly when the deep gap field, B_g is raised above about 75% of $M_{S,\mathrm{poletip}}$. There are two things the head designer can do about this: a) use a high $M_{S,\mathrm{poletip}}$ material; and b) rigorously limit B_g with write current control or control of saturation in the confluence zone (see below). In the high $M_{S,\mathrm{poletip}}$ approach there is a contradiction between the materials that are good for the yoke (e.g., high permeability, low coercivity, transverse domain structure, high resistivity, electroplateable, etc) and the maximisation of $M_{S,\mathrm{poletip}}$. Therefore advanced heads greatly benefit from separate tip and yoke materials optimisation. As stated in Sect. 11.2 above, CoFe has the highest available $M_{S,\mathrm{poletip}}$ but it corrodes easily and it is magnetically hard. If it is used as a thin (e.g., less than a gap thick) "dusting layer", adjacent the gap, its hard magnetic properties are of little consequence (e.g. permeability >50 is good enough). As long as the overlying pole tip material has an $M_{S,\mathrm{poletip}}$ that is at least 75% of that of the dusting layer, saturation will not occur in it either if the geometry is well designed. However if the dusting layer has electrical contact (Gigaohms is way too low a resistance so contact must be assumed) with an electrolytically dissimilar overlying pole tip material, the battery effect is fatal if there is any water present. Therefore any head that uses this material at the tips must rigorously exclude water from the lapping process (e.g. non-aqueous lapping fluids) and from the head disk assembly (HDA) itself. Third elements can be added that reduce corrosion but all of these additions will lower $M_{S,\mathrm{poletip}}$. Inevitably an engineering compromises result in $4\pi M_{S,\mathrm{poletip}} < 25$ kG.

Having selected a material with the highest possible $M_{S,\mathrm{poletip}}$, the designer now must make sure that $B_g < 0.75\%$ of 4π Mspoletip, at least until the written transition has moved out of the high field region. Controlling the write current, I_w, for this is of limited practicality. Fast rise times inevitably cause overshoot of the current due to the inductance of the head and leads and the capacitance of the leads and write driver. In fact some current overshoot is of great assistance in getting the yoke structure to respond rapidly (see below) in order to minimise NLTS. Therefore the burden of limiting the write field falls upon saturation at the confluence point. If media switching speed is not a problem [6], the best possible B_g vs I_w function rises linearly up to the optimum B_g and abruptly saturates at that point [3]. With saturation dependence like this, the write driver designer has great freedom to overdrive the current in order to get a high dB/dt and ringing has little consequence. The yoke designer's job is also made easier in that an over driven poor yoke can in fact switch faster than a marginally driven good one. The read designers life is also improved by alleviation of the read instability problem that excess flux in the shared shield often causes. Finally, the servo engineer can now reduce the track pitch because the write fringing of the

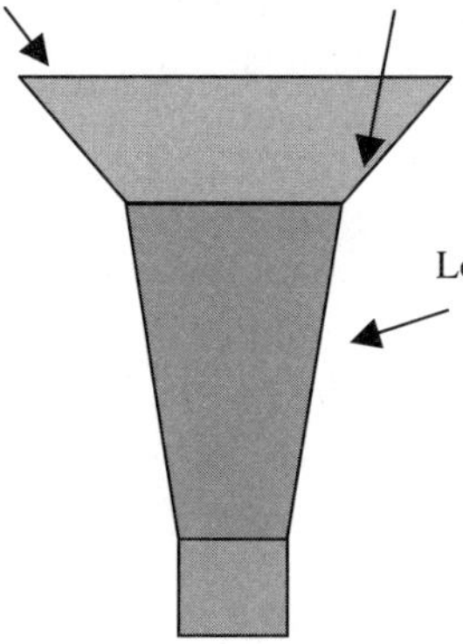

Fig. 11.4. Double confluence design for sharp saturation

head is not excessive. Experiments on thermally stable media show that the threshold field for corrupting adjacent track data with repeated exposures to a fringe field is 65% of the switching coercivity. Media with marginal thermal stability requires even lower fringe fields.

With all these advantages to a sharp saturation curve why doesn't every head behave this way. Part of the answer is that if the saturation dependence is too sharp, and at the wrong B_g, then adjusting the write current will not compensate. Fortunately there are no proposed designs that are this sharp. The sharpest saturation curves are obtained from the design shown in Fig. 11.4. This is a double confluence point design [7]. With this type of approach, tuning the initial confluence angle in the range of 6 to 11 degrees (depending on gap, throat, ramp angle, etc.) can yield a slope of the B_g vs I_w curve after saturation that is less than 1/6 of the initial slope. This is more than twice as good as that obtained from a single confluence design. The slope can be made even flatter by lengthening the saturation region and slightly grading down the opening angle far from the ABS (e.g., an 8 degree initial break angle graded smoothly down to 7 degrees moving away from the ABS). However most of the benefit is obtained with the double break design and an excessively long saturation zone can require excessive write current and throat height control.

One problem with selecting sharp B_g vs I curve is that one must then decide on the optimum B_g for saturation to produce. In general the optimum is that which produces the greatest write field gradient at the written transition at the moment it is being written [3]. This minimises transition width and NLTS. For a head in which the gap is small (this is not realistic but it is easy to discuss), this is obtained when the maximum field in the media is twice the coerciivty and the writing point is one fly height down stream of the gap [3].

However the need to write the highest possible coercivity (for thermal stability) will push the engineering optimum write point closer to the gap centre and to fields that are closer to the maximum field that the head can produce. At the thermal stability limit the trade off is between SNR (fine high coercivity grains) on the one hand and writeability (OVW, NLTS, PW50 etc) on the other.

As the write head gap is increased, the ratio between the deep gap field and the field in the media improves while the maximum of the normalised gradient goes down a little (e.g., 15%). Therefore the gap to fly height ratio (G/h) is typically large (e.g., $G/h > 4$). Though large ratios improve manufacturing tolerance on throat height they can reduce the deep gap field at which the confluence point saturates. Larger gaps need more electromotive force (amp-turns) for a given field. Therefore more flux leaks out the sides of the poles relative to the gap flux. In addition, the range of the fringe field of the head increases with gap so the track pitch must increase. For low BAR this can be quite important.

In the limit of large gap, narrow tracks, steep ramps and high fields, a tip saturation zone (see Fig. 11.3c) can develop at a significant distance down stream of the tip to gap corner. As with gap zone saturation, the write field gradient and transition imaging in the pole are degraded. At higher currents these zones melt together and the write process is very poor. Because of the inherent geometric complexity in these saturation issues it is quite difficult to make general statements about optimum design. The goal here is to point out design options and the pit falls that the designer must look for in the finite element simulations that are necessary for this kind of work.

11.6 Longitudinal Write Yoke Design

The longitudinal write yoke should provide a rapid ramp of flux to the tips. Ideally the effective electromotive force at the confluence point should flatten out suddenly after some controlled overshoot. This can be achieved with the write current waveform provided saturation does not occur elsewhere (other than at the confluence point) first. The interaction of the driver current ramp with the LRC of the driver-leads-head system can be designed for this over shoot or it can be built into the driver current ramp design. This overshoot helps drive the rapid switching of the magnetisation in the tip regions. In general the high speed switching of the tip region involves the complex interaction of micromagnet, gyromagnet and microeddy current effects [8] all of which inhibit switching. Therefore overshoot followed by a flat plateau (for the reasons listed above) is ideal. However good saturation properties at the confluence point can forgive some degree of long term ramp up of the EMF (typical yoke behavior).

As with the tips, the switching behaviour of the yoke involves micromagnet, gyromagnet and microeddy current effects. However they are somewhat

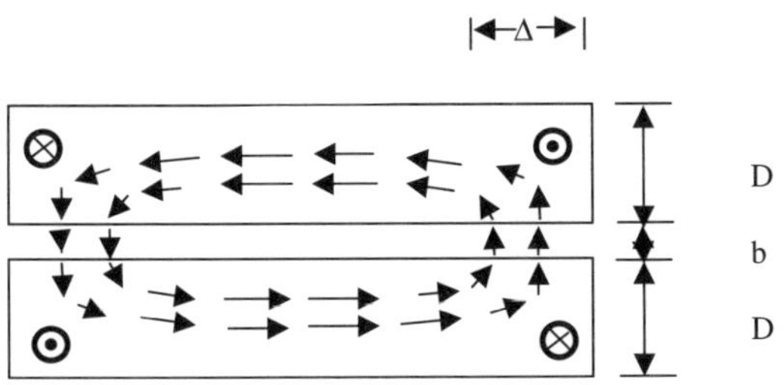

Fig. 11.5. Static domain flux closure in a pair of closely spaced magnetic films

more tractable in that the magnetisation changes are typically small by comparison to the saturation moment $M_{S,\text{yoke}}$. As mentioned above, the value of $M_{S,\text{yoke}}$ is not very important in that geometry can compensate for it. Though the domain structure, resistivity, and soft magnetic properties of the tips are not of overriding importance (because the path length is short), this is not the case in the yoke. Transverse domains, high permeability, low coercivity, and high resistivity are all-important features of fast yokes [9–12].

A domain structure in which the majority of the flux path allows for conduction by the rotation of transversely oriented domains reduces the negative impact of low domain wall mobility. In the case of inductive read/write yokes, conduction by wall motion will also lead to Barkhausen noise during read back when walls snap free of pinning sites. This will also occur in the shields of GMR heads but with less impact on the read back wave form. Fine lamination can eliminate closure domains and their associated walls thus allowing rotational conduction in narrow regions (e.g., tips, confluence region, or narrow yokes) [13]. The effect of fine lamination (e.g., magnetic layer thickness, D, less than 100 nm and interlayer thickness, b, less than 5 nm) on domain structure is shown in Fig. 11.5. There is still an edge closure region but it circulates domain flux between layers and it is very narrow by comparison to edge domains in single layer films. The width of the edge closure region [14] for a single pair of magnetic layers is given by (neglecting exchange coupling inside of each layer):

$$\Delta = (2\pi b D M_{S,\text{yoke}}/H_{K,\text{yoke}})^{1/2} . \tag{11.8}$$

So for $D = 5$ nm, $b = 1$ nm, $H_{K,\text{yoke}} = 100$ Oe, and $M_{S,\text{yoke}} = 1500$ emu/cc, (e.g., CoFeNi alloys) the edge closure size will be 22 nm. Therefore this analysis indicates that domain wall free pole tips, with a write track width of 50nm (e.g. 400 KTPI with a write width of 80% of the pitch) can be fabricated in this way. However exchange coupling within a layer would be a dominant source of potential energy for these dimensions even if the interlayer were non-conductive so that exchange coupling between the layers is prevented.

In the above example the exchange energy in each layer (with perfect isolation) would be orders of magnitude larger than the anisotropy energy for longitudinal domain lock up. Therefore this design needs fixing. An alternate

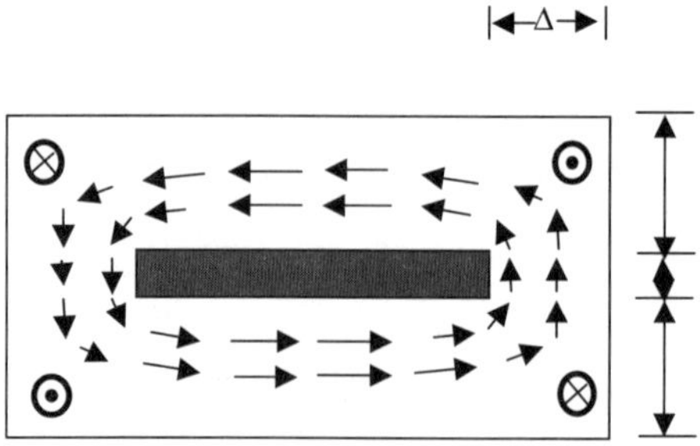

Fig. 11.6. Static domain flux closure in a pair of magnetic films with a partial ruthenium antiferromagnetic exchange layer

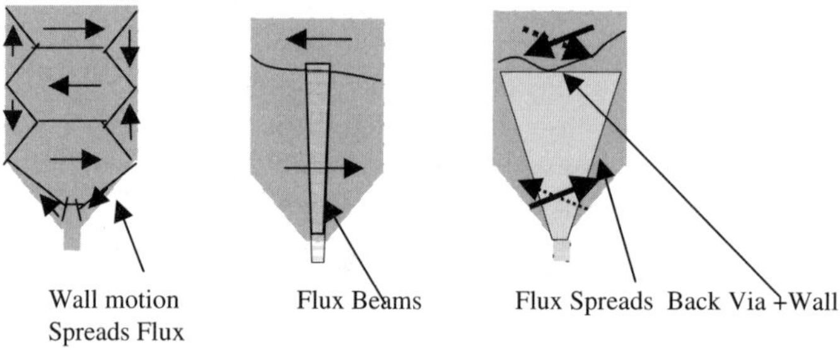

Fig. 11.7. Yoke domain patterns and resulting flux beams

design is shown in Fig. 11.6. It uses a thick laminate (e.g., 25 nm) and a wider pole (e.g., 70 nm) to reduce the exchange energy associated with curling of the magnetisation at the edges. Brute force in the form of very high H_K (e.g., 500 Oe), or antiferromagnetic coupling across a 6 Angstrom ruthenium layer can mitigate the exchange energy problem. However this causes low permeability. If the write throat height (distance from the ABS to the confluence point) is small enough (it must be for high write field), this will not be a fatal problem. Also short throats will give exchange coupling to the wider yoke structure, which can be more easily maintained in a transverse orientation.

Typically, transverse 180-degree walls are greatly reduced by using these kinds of lamination (see Fig. 11.7b). This reduces problems with domain wall resonance [15] that occur when the yoke thickness is less than about two microns. If the interlayer material is an insulator and care is taken to avoid shorting at the edge, eddy current suppression will improve the frequency cut off in direct proportion to the number of layers (until gyromagnetic effects cut in). However, total elimination of domain walls can be a problem in finely laminated heads with uniaxial anisotropy. This is because conduction of flux in the lateral direction, by domain wall motion, is usually needed to spread out the flux in the width direction. This reduces magnetic impedance [9–12],

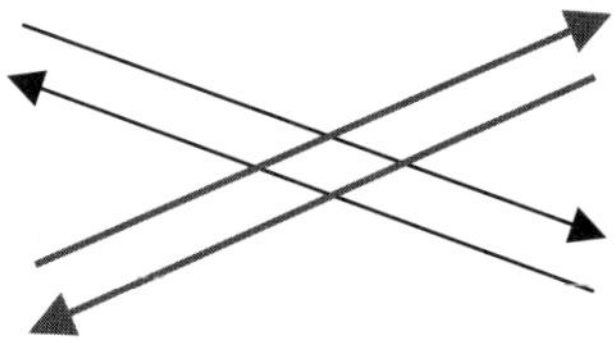

Fig. 11.8. Magnetization in a multilaminant with pairwise easy axis cant

which is inverse with the flux path width (see Eq. (11.11) below). Without the flux spreading, provided by domain wall motion, uniaxial laminants beam the flux in a narrow high impedance path whose width increases only gradually with its length. This problem is evident in a head that was made with fine lamination [16]. It is illustrated in Fig. 11.7b.

This problem can be fixed by canting the easy axis up and down by about 10 degrees in every other pair of layers in the laminate [17]. Thus layers 1,2,3,4,5,6 ... are oriented at +, +, −, −, +, + ... 10 degrees to the transverse direction, respectively. This is illustrated in Fig. 11.8. This provides a rotational transverse flux conduction mechanism without resorting to domain wall motion. Each pair of layers conducts flux perpendicular to its own easy axis so the effective transverse permeability is the sine of the cant angle times the longitudinal value. A transverse permeability of 1/6 the longitudinal is more than adequate for flux spreading [12]. The spreading of flux has a wide short path (vs the long narrow direct path) so a low permeability (e.g., 200 but not 1.0) is adequate. This approach is also applicable to the soft underlayer of perpendicular recording media. Good flux spreading there can allow for high H_K, which reduces domain wall noise.

In choosing an optimum yoke material there is a conflict between high permeability ($\mu = 4\pi M_{S,\mathrm{yoke}}/H_K$) and rapid gyromagnetic switching. From [18], the resonant angular frequency (note that $\gamma = 17.6 \cdot 10^6$ radians/s/Oe) for a plate is:

$$\omega_r = \gamma H_K (1+\mu)^{1/2} = \gamma 4\pi M_{S,\mathrm{pole}} (1/\mu + 1/\mu^2)^{1/2} \; . \tag{11.9}$$

Therefore high permeability implies a low resonant frequency. However high permeability helps high speed switching by giving high head efficiency (the ratio of the gap impedance to the total impedance). In a simple impedance analysis the head efficiency is:

$$\varepsilon = 1/(1 + F/\mu) \; , \tag{11.10}$$

where the geometric factor, F, is:

$$F = (L_{\mathrm{yokepath}}/\text{Yoke Crossection Area})/(\text{Gap}/\text{GapArea}) \; . \tag{11.11}$$

The impulse response can be analysed by determining the frequency dependence of μ and the yoke crossection (both complex) from the skin effect and/or the gyromagnetic behaviour of the yoke material depending on the relative importance of these two phenomena.

In the simplest analysis the important question is: what is the frequency, $f_{c,eff}$, of the -3dB point for the efficiency? If an exponential form is fit to the step response, the fitted time constant will be approximately given by $\tau = 1/(2\pi f_{c,eff})$. The 10–90% rise time (or −80% to +80%) is then about 2.2τ. Using numerical analysis or some algebra one can verify this. For the critically damped, gyromagnetically limited situation, the real and imaginary permeability are given by [19]:

$$\begin{aligned} \mu' &= \mu_i/(1+(f/f_r)^2) \\ \mu'' &= \mu_i(f/f_r)/(1+(f/f_r)^2) \,, \end{aligned} \tag{11.12}$$

where f_r is the resonant frequency (e.g., $\omega_r/2\pi$) and μ_i is the initial permeability. Then the −3dB point in the efficiency vs frequency curve is related to the −3dB point in the absolute permeability vs frequency curve, f_r, by:

$$f_{c,eff} \sim f_r/(1-\varepsilon) \,, \tag{11.13}$$

Therefore, for nearly critical damping (typical in geometrically inhomogeneous systems like pole tips), the rise time of the head is related to the relaxation time of the permeable material by:

$$\tau_{\text{head}} \sim \tau_u(1-\varepsilon) \,, \tag{11.14}$$

and for high efficiency, Eq. (11.10) gives:

$$\tau_{\text{head}} \sim \tau_u F/\mu \,. \tag{11.15}$$

Finally, applying Eq. (11.9) gives:

$$\tau_{\text{head}} \sim F/(\gamma 4\pi M_{S,\text{pole}}\mu^{1/2}) \,. \tag{11.16}$$

Therefore, even though low permeability implies a high resonant frequency, this is not necessarily the way to go with head design. The rise time in fact gets worse because of the benefit of high efficiency as expressed in Eqs. (11.14)–(11.16) above. This is also why NiFe is still used in heads even though “better” materials have been developed. It is also why ferrite heads have hung in there for so long. The main caveat on this is that “critical “ damping is assumed here and this may not be the case in the uniform regions of a yoke (though it is in the tips [8]). At very high data rates (e.g., $\gg$1 Gb/s), the lamination style shown in Figs. 11.7c and 11.8, which is quite homogeneous, will become advantageous. In this context, the low resonant frequency and damping of NiFe will become a problem and other alloys (e.g. CoFeHfO has $f_r = 4$ GHz for $\mu = 150$) will become important.

From Eqs. (11.16) and (11.11) it can also be seen that the rise time is directly dependent on the geometric factor, F, and therefore on the yoke length. Therefore the trend in modern heads towards very short yokes is well founded. This has also become more feasible in that fewer amp turns are required for the smaller scaled down gaps even though the coercivity has been going up.

It should also be pointed out that the above type of analysis applies to a skin depth limited situation. The conclusion is about the same. In this context the skin depth goes down with the inverse square root of the frequency times the permeability and F increases in proportion. Therefore the rise time is still inverse with the square root of the permeability. However in this case the designer has more knobs to turn. The value of $f_{c,eff}$ increases directly with the resistivity or with the square of the number of layers in a laminant.

11.7 Longitudinal Read Design

Presently all longitudinal readers are shielded flux sensors as illustrated above in Fig. 11.1 (see Chap. 10 for more detail). As mention before, the shields slim the read back pulse by absorbing low frequency components. This also improves the dynamic range of the sensor. The pulse slimming can be accomplished with a tapped delay line filter (FIR) but at the expense of amplified high frequency noise. The shields also provide improvement in read fringing in that low frequency adjacent track flux generates flux-cancelling image charges in the shields. Therefore the cross talk falls inversely with the distance instead of with the log of the distance. In addition, the shields perform the important but much ignored function of conducting away the heat generated by the sense current. This creates a conflict between high frequency magnetic performance ($f_{c,u} \propto \rho$, the resistivity) and thermal conductance ($C \propto 1/\rho$ for metals). In general the high thermal conductivity is more important in that the magnetically soft materials with the best thermal conductance (e.g., $Ni_{80}Fe_{20}$) contribute about 1/3 of the total thermal impedance. Frequency response is less important because an effective permeability of 100 is still adequate for shields.

The ABS and plan views of a typical shielded flux sensor are shown in Fig. 11.9. In the design that is shown here the current carrying leads define the read width of the device. Resistance changes due to read flux outside of this region have little effect because they are not part of the current path. This approach was used on the first MR heads because of its simplicity but was later replaced by a design in which the permanent magnet to MR junction carried the current.

When the read width is defined by the permanent magnet junction, PMJ, a sharper off track sensitivity roll-off results. The effective read width is then the geometric distance between junctions minus the dead regions in which the field from the permanent magnets stiffens the sensor magnetisation so

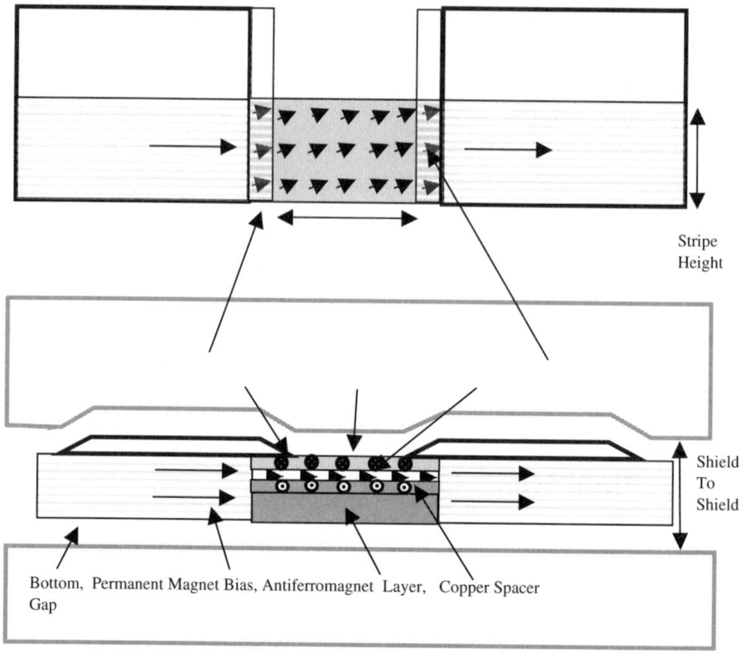

Fig. 11.9. Plan view and air bearing surface views of a typical current in plain (CIP) shielded GMR sensor

that it will not respond well to read flux. For a very sharp junction, the permanent magnet field falls exponentially with distance with a spatial constant of Gss/π (shield to shield gap/π). This field starts from the coercivity of the permanent magnet (e.g., about 1500 Oe) and must fall to about 30 Oe before the sensitivity to read flux at the edge approaches the central value. This factor of 50 change in bias field is four powers of e so the dead zone is about 4 Gss/π wide on each side.

Though, in principle, PMJ width definition can yield better roll-off behaviour, it is becoming increasingly difficult to achieve for very narrow track widths. To see this consider a conceptual design for a 60 Gb/in^2 longitudinal head based on a permutation of a recently reported 25 Gb/in^2-demonstration [20]. The authors demonstrated 10^{-8} bit error rate with: a magnetic spacing of 21 nm (h); a bit aspect ratio (BAR) of 15 (e.g., 610 KBPI × 41 KTPI); a switching field (H_{co}) of 5500 Oe (VSM H_C of 3900); average thermal stability factor (K_uV/kT) of 122 at 77 C; media areal moment density of 0.3 memu/cm^2; $Ni_{45}Fe_{55}$ pole tips; and a shield to shield spacing (Gss) of 120 nm.

In order to preserve thermal stability and media SNR, a BAR of about 6 is appropriate for our 60 Gb/in^2 design (e.g. 600 kBPI × 100 kTPI). Using Eq. (11.6) we can get from 25 to 43 Gb/in^2 by going from the demonstrated

BAR of 15 down to 6 (e.g., 507 kBPI $\times$ 85 kTPI). The moment of the media goes up in proportion to the bit space to 0.36 memu/cm^2. Now to get up to the density goal, we need to use rigorous scaling with a scaling factor of 0.85.

This brings the BPI and the media moment back to the demonstrated values so all of the reported head dimensions and the spacing can be used as is. However the stability factor at 77C will go down to 75 (e.g., 122x.853 = 75). This might be adequate if the media is exchange coupled because then the width of the energy barrier distribution is reduced below that of the grain area distribution [21]. However if it is not exchange coupled so that the lognormal energy barrier width is that of the grains (e.g. 70%) then 10% of the media moment will come from grains with a stability factor of less than 38. This is further reduced to 27 by the 1100 Oe demagnetising field of the dibit (see Eq. (11.26) below) at this density (e.g. demag factor = $(1-H_d/H_{co})^{3/2}$ – see Eq. (11.27) below). Therefore it is necessary to increase the stability factor by 40/27 = 1.5.

Increasing the coercivity increases the stability factor of a field free grain in direct proportion, if the media thickness is not increased as well. However we have that option, because the demagnetisation factor and the transition width will remain constant, if we increase thickness in proportion to coercivity. With this type of scaling the stability factor improves as the 4/3 power of the coercivity. This gets us a factor of 1.36. Therefore, use of a $2T$ high $M_{S,\text{poletip}}$ layer adjacent the gap would allow for comfortable writing on 5200 Oe media. This would have adequate stability for the more stable 90% of the magnetic moment (e.g., 10% dibit amplitude loss after 10 years).

So, after all of this, we now have a 60 Gb/in2 head media system that looks very much like the 25 Gb/in^2 demonstration in side view except that the media moment is now 0.4 memu/cm^2, the VSM coercivity is 5200 Oe, and a high $M_{S,\text{poletip}}$ layer has been added to the write pole tip. The shield to shield space is the demonstrated 120 nm. But in the plan view, the width is only 40% of the demo width. The PMJ to PMJ width is then 140 nm (demo used 350 nm). However the dead zones of 4 Gss/π at each junction is 150 nm. The permeability of the free layer, even in the centre, is suppressed by the excessive PM field there. This is the problem with PMJ track width definition at low BAR.

However, before reverting to the old current lead or antiferromagnetic pinning definition of track width and the associated fringing problems, we should take a closer look at the PMJ approach. The dead zone effect is really due to over kill on the PM thickness. Typically its $M_r\delta$ is many times that of the free layer in order to guarantee magnetic stability (single domain) even when the manufacturing process is imperfect. The PMJ is never as shown in Fig. 11.9. Actual PMJs are usually intermediate between the ideal geometry and the poor one shown in Fig. 11.10. Here the PMJ is far from vertical. This smears out its associated magnetic charges. As a result the local field can actually be opposite to the PM field. If a perfect PMJ could be produced

Fig. 11.10. a Ideal and **b** actual permanent magnet junctions

then the PM $M_r\delta$ would need to be only slightly greater than that of the free layer so the dead zone would be greatly reduced. In all likelihood, the design of the future will be a hybrid between a good PMJ and leads that lap over the PMJ slightly to exclude whatever dead zone is left. Another consideration that biases toward this type of design is that it avoids using the PMJ to conduct high current densities. In addition to the difficulty of achieving a steep PMJ (using ion beam PM deposition and/or baffles to get normal incidence), it is also difficult to avoid a large contact resistance there. The ion milling process which defines the edge of the free layer (prior to PM deposition) also deposits insulating materials on that edge that interferes with good electrical contact there.

For the purpose of minimising read fringing, other designs can be considered. Focused Ion Beam (FIB) milling can be employed at the ABS to trim back the sensor edges to a depth that is much less than the shield to shield spacing. This will significantly suppress the flux gathering capability there. As long as the cut depth is much less than the exchange length, the untrimmed portion of the free layer, that defines the read width, should remain in a single domain state (almost). A step further in this direction is to provide a very short (to preserve efficiency) laminated flux pipe to a recessed sensor [22]. This is essential in the case of spin tunnelling sensors that cannot be located at the ABS without excessive risk of smear induced shorts.

Another promising approach to the narrow track reader problem is to use a current perpendicular to the plain (CPP) GMR sensor [23] shown schematically in Fig. 11.11. This sensor is in effect a finely laminated flux pipe. It consists of alternating layers of copper (e.g., 2 nm) and CoFe (e.g., 2 nm). The current flows perpendicular to the layers through electrodes attached to the shields. With this design, flux closure occurs between every other layer of CoFe. The CPP configuration approximately doubles the GMR effect. The termination of the sandwich at the read width cleanly defines the read track width. The perpendicular flow of current through shield contacts obviates the shield to sensor shorting issues. The lack of an AF bias layer allows a small shield-to-shield gap.

With all these advantages, why isn't CPP GMR in use? The answer is that the resistance of the device at this time (e.g., about one ohm for the 60 Gb/in$^{2\prime\prime}$ design here) is far too low to be matched to that of the leads (e.g., 2 ohms) and the equivalent noise resistance (e.g., 30 Ohms) of the preamplifier. However, for a fixed BAR, the resistance is proportional to the density so the lead resistance issue (device noise resistance must include leads) will

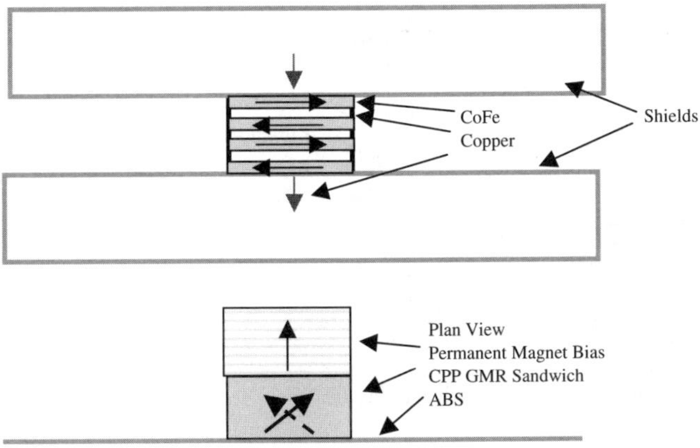

Fig. 11.11. Current perpendicular to plain GMR: ABS and plan views

fade. However, the preamplifier noise is more of a problem. Both, the lead resistance and the preamplifier noise problems can be fixed with an on slider transformer that can make the low resistance CPP GMR look like a high impedance device. Similarly, reduction of the lead resistance would allow for the use of a macroscopic transformer at the preamplifier front end. Alternately, at the cost of increased power dissipation, a very low noise preamplifier with a large base area and I_C would have a reduced equivalent noise resistance.

As areal density increases it is important that the read sensor sensitivity also increases. To some extent this happens automatically because scaling down the device dimensions allows for an increase in current density for the same temperature rise. Another phenomenon that reduces the demand for high sensitivity is that the thermal decay limit on longitudinal recording will force a situation where system noise is dominated by media noise from grains that cannot be shrunk any further. Nonetheless improvements in sensitivity will help. This implies that the GMR effect should be enhanced because the electronic SNR is proportional to $\Delta R/R$.

Many small increases in $\Delta R/R$ have been obtained through refinements of the basic GMR sensor. Since its introduction, shipped GMR heads have gone from 4% to 13% $\Delta R/R$. Buried layers of oxides in both the free and the pinned layers have been used to get multiple passes of the electrons across the spacer layer. This is also enhanced with the use of specular reflection layers above the free layer. An additional copper layer above the free layer provides more conductivity (spin valve design). An additional spacer layer and pinned layer above the free layer also helps (>22% $\Delta R/R$ has been reported for cobalt free and spacer layers [24]). The CPP approach achieves this in spades and a doubling of $\Delta R/R$ has been achieved with many layer cobalt copper sandwiches [25]. Therefore it is expected that GMR based devices will support areal densities >100 Gb/in^2.

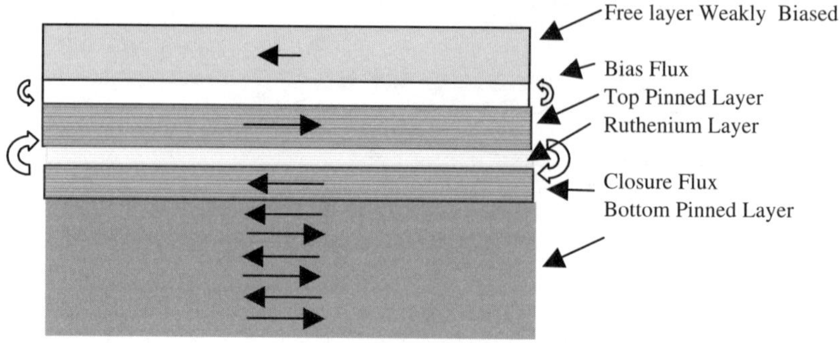

Fig. 11.12. Synthetic antiferromagnetic pinned layer

An additional refinement to the GMR sensors is the use of "synthetic antiferromagnetic" (SAF) pinned layers [26]. The magnetisation of the free layer in Fig. 11.12 is not shown to be in the ideal purely transverse direction because in practice it never is. Therefore asymmetry between the positive and negative going pulses results. The rest state of the free layer magnetisation is determined by many competing sources of torque. These are: a) its uniaxial anisotropy (induced and magnetostrictive); b) the field from the sense current; c) the PM field; d) exchange coupling to the pinned layer through the spacer (controlled by spacer thickness); e) roughness induced magnetostatic coupling to the pinned layer (orange peal effect) through the spacer; and f) the pinned layer edge flux which is almost entirely absorbed by the very proximate free layer. This last source is particularly damaging in that it maximises at the ABS where it adds directly to the signal flux to drive the free layer towards saturation every half cycle. Thinning the pinned layer can cut this down, but $\Delta R/R$ suffers.

The SAF approach (see Fig. 11.12) greatly reduces this problem by providing two opposed pinned layers separated by a 0.7 nm thick ruthenium layer. Very strong (e.g., many kOe) antiferromagnetic coupling across this ruthenium layer locks the pair of ferromagnetic layers into the opposed orientation. The magnetostatic flux from these two layers nearly cancels in the ideal case. Therefore its flux loading effect on the free layer is minimised. This near cancellation also means that very large external fields are needed to torque the pair even if the coupling to the underlying AF layer is not very strong. This AF coupling is not effected by the SAF approach while the coupling of external fields goes to zero in the case of perfect balance. However some imbalance is necessary in order to set the pinning direction with anneals in an external field.

Besides the GMR effect a number of other magnetoresistive phenomena have been proposed for read head sensors. Though very high $\Delta R/R$ values have been reported, these alternatives have signal to noise problems that will be hard to overcome. Magnetic semiconductor materials (based on lanthanum manganese oxides) display the "Colossal Magneto Resistance" (CMR) effect in which $\Delta R/R$ can be much greater than 100%. However these materials have resistivities that are many orders of magnitude too high (note that noise power is proportional to resistance) and they require huge fields (e.g. $> 1T$) to achieve the effect [27]. A donut geometry for an InSb device also achieves very high ΔR/R but I expect the noise to be enhanced as well by the geometry. In a basic Hall effect type of geometry, InSb breaks even with the old AMR technology for an AMR current density of 10^7 Amp/cm^2 when both sensors are remotely located [22]. Therefore it is very unlikely that a geometric permutation can compete with ABS GMR at current densities $>10^8$ A/cm^2.

Finally the spin tunnelling magnetoresistive (TMR) effect has been proposed for sensors. This phenomenon has achieved $\Delta R/R$ values comparable to equivalent (with Co free layers) CPP GMR but with the opposite impedance problem. Initial devices had grossly excessive resistances but recent advances have almost fixed this. As with the CPP GMR effect, the current flows perpendicular to the plain. So as density increases the impedance per unit area must come down. At this time numbers like 50 ohm-micron^2 can be achieved with very thin (e.g. 0.7 nm) tunnel barriers [28]. However further thinning of the barrier to reduce impedance so far has cut deeply into $\Delta R/R$. In addition, the noise source in these devices is shot noise instead of Johnson noise [29]. Crunching the numbers indicates that a TMR device needs more than twice the $\Delta R/R$ for its SNR to break even with a comparable GMR device. An additional difficulty is that TMR cannot be located at the ABS without creating smear shorts across the junction. Therefore a flux pipe to a remotely located device is required. This approach may achieve commercial success but it has an up hill battle ahead.

11.8 Perpendicular Recording

Thermal instability of fine grains motivates a transition to perpendicular recording in the areal density range from 50–100 Gb/in^2. The low BAR, reduced SNR and marginal thermal stability of a longitudinal system should begin to look unattractive by comparison to a perpendicular system with better performance on these parameters (at the same density). As density increases the motivation to switch to this new technology, with its debug problems, should become overwhelming (see Chap. 8 for more detail).

There are several reasons that perpendicular will be superior on these issues. The strongest one is that the use of a soft under layer (SUL, see Fig. 11.13) effectively places the media in the write gap of the head. In longitudinal recording the fringe field of the gap is used to write. Therefore writeability of longitudinal media requires that it have a lower coercivity. This translates into reduced anisotropy energy density and therefore into more voluminous thermally stable grains. The larger grain area then implies

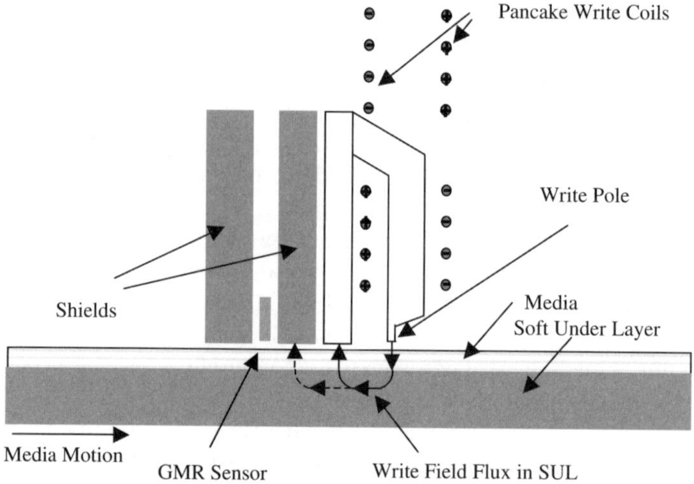

Fig. 11.13. Separate read and write structures to avoid domain noise in the perpendicular recording read back process

fewer grains per bit and worse SNR. Perpendicular gets a second advantage here in that its media can be made much thicker than that of longitudinal media whose demagnetisation field increases with thickness (see Eq. 11.26). Therefore, for a given grain volume, perpendicular grains have a smaller foot print at the media surface.

A third benefit of perpendicular is that there exist mechanisms for achieving a very high degree of orientation (e.g., $\Delta\theta_{50} < 3$ degrees) to the grains. This in principle allows a full range of switch fields (Stoner–Wohlfarth particles) of less than 13% of the minimum vs a full range for longitudinal of twice the minimum. Combining this with the fact that self-demagnetisation sharpens perpendicular transition while it blunts longitudinal ones, means that very small transition widths are possible. In that transition noise power is inverse with the square of the width (see Eq. 11.35), perpendicular can also have much better transition SNR. Therefore it will need fewer grains per bit. In fact, simulations by Roger Wood et al. [30] indicate that with a BAR = 4, as few as 10 grains per bit should suffice.

The high orientation ratio of perpendicular also allows for essentially no DC erase noise with proper media design. Thermal stability requires that the media have $Hc \gg 4\pi M_S$ (see Eq. (11.21) below). If this is achieved, along with a good $\Delta\theta_{50}$, and a small dispersion of H_K, then $S = M_r/M_S$ will be essentially one. Then, the only variation in media surface magnetic charge density will be at the non-magnetic grain boundaries. With unoriented longitudinal media the longitudinal magnetisation of the grains ranges from zero to M_S with an average value of 71% of M_S. This variation creates a lot on DC erase noise. If oriented longitudinal media could achieve a $\Delta\theta_{50}$

of 60 degrees DC erase noise would be greatly reduced. This is because the longitudinal variation from Ms is a $(1-\cos\theta)$ effect. However transition noise would not be significantly improved. This is because significant narrowing of the switching (Stoner–Wolfarth switching astroid) field distribution requires very small values of $\Delta\theta_{50}$ (see Eq. (11.17) below).

11.9 Perpendicular Recording Write Heads

A typical perpendicular recording head is shown in Fig. 11.13. Structurally it is very similar to a longitudinal head. Though media with no SUL and ring head write (Fig. 11.1) certainly works, its write field limitations are worse than those of longitudinal (the gradient is worse and with perpendicular H_C/H_K is almost twice as large). So this section assumes the use of a SUL.

The structure shown in Fig. 11.13 is a monopole writer with a bucking pancake coil driver. This coil arrangement mimics the magnetic performance of a solenoidal coil without the associated production problems (process complexity/reliability). From a resistance viewpoint it is significantly worse than a solenoid or a single unbucked pancake. However this is less of a problem with SUL perpendicular recording in that the number of required amp-turns is much less than longitudinal (e.g., <30% of longitudinal) and resistance goes with the square of the turns for a pancake. The inductance of the bucking double pancake is almost as low as that of a solenoid (the stray coil inductances buck each other). Again the required turns is low so this is a non-issue.

The motivation for using the bucking double pancake (beyond inductance reduction) is that it greatly reduces the field between the shields/upstream pole and the media SUL. A single pancake will produce large fields there (e.g., >10 percent of the write field) even in the absence of a write pole and yoke. These fields will enhance thermal decay of data under the shields. At high write currents the shields can corrupt data with only one pass if the media is marginally stable.

Though the monopole writer+SUL is not the best design, it is better than a ring head with no SUL and it is simple to produce. The spatial distribution of the horizontal and perpendicular fields of this head averaged over the media thickness (media thickness $= 2h$) are shown in Fig. 11.14. The horizontal fields are important because the switching field of a Stoner-Wohlfarth particle is:

$$H_{\text{switch}} = H_K(\sin^{2/3}(\theta) + \cos^{2/3}(\theta))^{-3/2} \,. \tag{11.17}$$

For a field parallel to the easy axis ($\theta = 0$) the switching field is H_K . But at the edge of a monopole, the head field direction deviates from normal by 7.5 degrees (at the media centre 2/3 of the way to the SUL – see Fig. 11.14). Here, its strengths is down by 10% (from maximum deep under the pole. But Eq. (11.17) predicts that at 7.5 degrees the switching field is only 72% of H_K. Therefore at this point the media will switch when the deep gap field is 80%

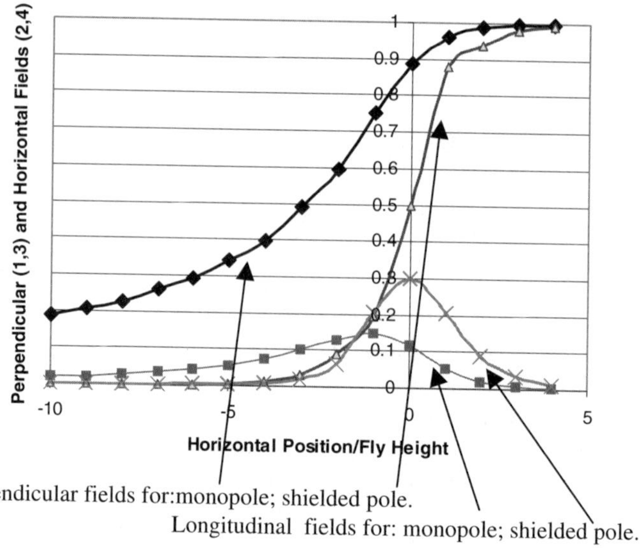

Fig. 11.14. Normalized fields (same EMF) for unsaturated monopole and shielded pole heads in arbitrary units for fly height = 1, media thickness = 2, and gap = 1.5 arvitrary spacial units

of H_K. The ratio $|H|/H_{\text{switch}}$ vs horizontal position for a monopole head with a deep gap field of H_K is shown in Fig. 11.15. We can see that it peaks at the pole tip corner. Therefore when the head field is switched, a closely spaced dibit transition pair will nucleate in this vicinity. This is illustrated in Fig. 11.16. However the increasing write field will be much less than H_K at this moment because the self demagnetising field of the media adjacent the infant dibit will add to the head field. (The strength of this field will be somewhat less than $4\pi M_S$ because of the proximity of the pole tip edge and the field of the dibit itself.) If $4\pi M_S = H_K/4$ and $h = \delta/2$ (see below), the initial boost is a little less (neglecting pole tip images) than 35% of the head field. Therefore the nucleation of the dibit will occur when the field has reached approximately 0.65 H_K for these conditions.

As the down stream transition moves away from the pole edge, the dibit spacing grows, because the upstream one, B, remains fairly stationary at the pole edge (if the rising head field is just barely sufficient to continue switching the media at that point), as the deep gap field continues to rise. When the deep gap field approaches about 75% of $M_{S,\text{poletip}}$, saturation at the corner starts to become significant (see above). However, at this point the downstream transition has already formed and moved away from the corner. Therefore limiting the deep gap field to 75% of $M_{S,\text{poletip}}$ is not necessary here provided that the head field does not rise too rapidly. In fact a partly saturated tip is desirable, because this allows the highest possible H_K for

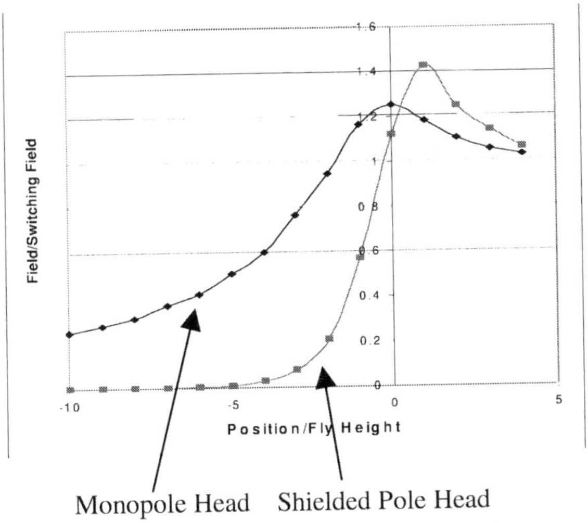

Fig. 11.15. Head field/media switching field, for the field under the pole = H_K, for a monopole and a shielded pole head (gap = half of tip to SUL spacing). The fly height = 1 (arbitrary units) and media thickness = 2

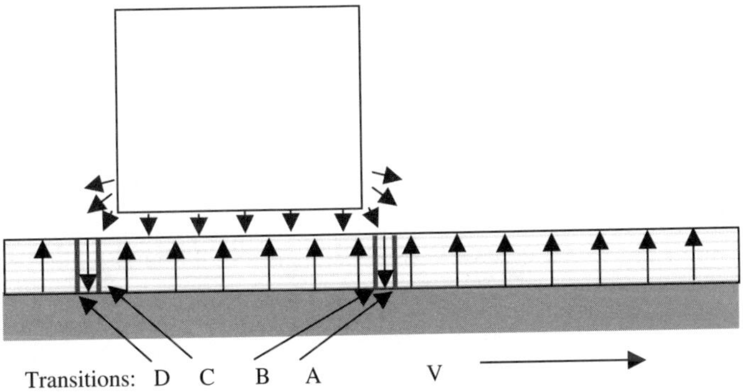

Fig. 11.16. Initiation of the monopole write process with the nucleation of a dibit

thermal stability with small grains. As the written transition, A continues to move away, the self-field at B becomes predominantly horizontal. Its value approaches $4M_S\ln(3)$. This field adds to the horizontal field of the head so that the net deviation from perpendicular approaches 16 degrees at the media centre (for $4\pi M_S = H_K/4$ and media thickness of twice the spacing). Equation (11.17) then predicts a required switching field strength of only 60% of H_K at this point. From Fig. 11.15, the deep gap field is then required to be 67% of H_K.

The situation that requires the largest deep gap field is the switching of the media between transitions C and B as the dibit spacing between them approaches zero (media motion carries C to stationary B). When the B–C spacing is nearly zero (near the pole tip edge) the media demagnetising field is somewhat larger than $4\pi M_S h/(h+\delta)$ (due to partial pole tip imaging near the edge). Then the net perpendicular field is that of the head (Fig. 11.14) minus $4\pi M_S h/(h+\delta) = 4\pi M_S/3$ (for $\delta = 2h$). This tilts the field (net) angle up to about 8.5 degrees if $4\pi M_S$ is about HK/4. From Eq. (11.17), the required net write field, H, is then 0.71 H_K. In order to overcome the demagnetising field, the head must supply $H = 0.71\ H_K + 4\pi M_S h/(h+\delta) = 0.71\ H_K + 4\pi M_S/3$. Now the media field in the poles is $4\pi M_S\ (\delta)/(h+\delta)$ (with no head field), so the flux density at the pole and SUL is:

$$B_w = 0.71 H_K + 4\pi M_S h/(\delta+h) + 4\pi M_S \delta/(\delta+h)$$
$$B_w = 0.71 H_K + 4\pi M_S < 4\pi M_{S,\text{poletip}}\ . \tag{11.18}$$

The above writability constraint can now be used to find the most thermally stable writeable value of $4\pi M_S$. The anisotropy energy barrier density (e.g., $V = 1$) in the presence of the self-demagnetising field, $H_d = 4\pi M_S$, is:

$$E_B = K_u(1 - H_d/H_K)^2 = H_K M_S (1 - 4\pi M_S/H_K)^2/2\ . \tag{11.19}$$

Substituting $(B_w - 0.71 H_K)$ for $4\pi M_S$ gives:

$$E_B = H_K(B_w - 0.71 H_K)/8\pi((1 - (B_w - 0.71 H_K)/H_K)^2\ . \tag{11.20}$$

Substituting $Z = H_K/B_w$ gives:

$$E_B = ((B_w)2/(8\pi Z))(1 - 0.71 Z)(1.71 - 1/Z)^2\ . \tag{11.21}$$

It is easy to verify that Eq. (11.21) maximises for $Z = 1.083$. Then Eq. (11.19) gives:

$$4\pi M_S = 0.21 H_K\ . \tag{11.22}$$

At this thermal stability maxima, the writeable (saturated monopole tip) energy barrier density is then (noting that $Bw = 4\pi M_{S,\text{poletip}}$ because corner saturation is OK for overwriting the C–B dibit):

$$E_B = 0.0063 B_{\text{gap}}^2 = 0.99 M_{S,\text{poletip}}^2\ . \tag{11.23}$$

In practice a larger value of $4\pi M_S$ would be used in order to get more signal flux. For example if $4\pi M_S = H_K/4$ (as assumed above), the read flux would increase by 19% while the energy density would go down by only 2%. In addition, a small amount of intergranular exchange coupling (e.g., about 2% of bulk [31]): increases the magnetisation at the optimum; reduces the coercivity; and improves stability (though at the cost of transition noise).

The switching energy density can be further improved upon by using a writer design [5] that enhances the write field gradient by more than a factor

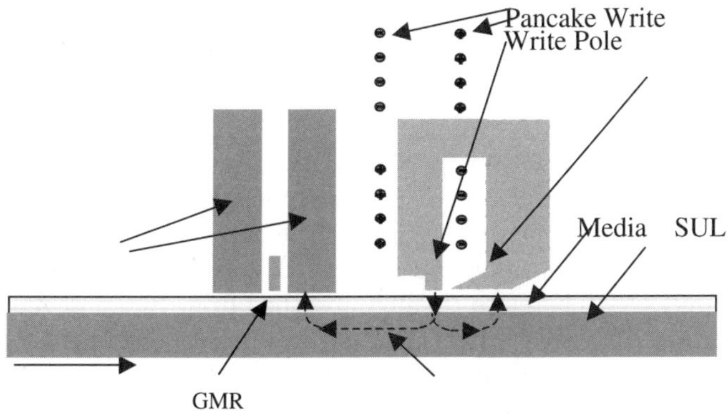

Fig. 11.17. Shielded pole writer

of two. The transition width then goes down comparably and the transition noise power will decrease by even more (inverse with the width squared). This design also doubles the horizontal field. This allows the writing of media with even higher H_K. It accomplishes these benefits by providing a downstream shield that absorbs the fringing field of the write poles. A shielded pole writer design is illustrated in Fig. 11.17. Its field distributions are also shown with that of the monopole in Fig. 11.14.

The ratio between H_{switch} and H_K is shown in Fig. 11.15 with the maximum field (under the write pole) set to H_K. It is 14% better than that of the monopole. In the worst case condition of annihilating the upstream dibit (formed by transitions B and C in Fig. 11.16 above) the net field tilt angle is enhanced by the need to buck the self-demagnetising field (as before). The net result is that the switching field is only 61% of H_K at a point upstream of the gap centre where the perpendicular field is still high (e.g., 88% of the maximum).

As before, the write field must exceed the switching field by the media self-demagnetising field, $4\pi M_S h/(h+\delta)$. So H at the write point is $0.61\ H_K + 4\pi M sh/(h+\delta)$. Also the flux density here includes the media flux density of $4\pi M s\delta/(h+\delta)$. Therefore the required flux density at the tips nets to $0.61\ H_K + 4\pi M_S$. When this substitution is made in Eq. (11.19), the maximum is at $Z = H_K/B_w = 1.23$. The energy barrier density then increases 24% (from the monopole result) to:

$$E_B = 1.23 M_{S,\text{poletip}}^2 \,. \tag{11.24}$$

11.10 Perpendicular Recording Read Heads

Longitudinal read back heads will work well for perpendicular recording with minor modification. However it should be kept in mind that the read back wave form for perpendicular looks like the integral of the longitudinal read back. Therefore it has a lot of low frequency flux and the resolution (ratio of high f to low f amplitude) will be degraded. In addition the low frequency flux will drive the sensor to saturation if the gap is too large. Therefore small read gaps are essential in a PRML channel.

Consider a system in which $H_K = 16$ kOe, $4\pi M_S = 4$ kG, and $\delta/h = 2$. The DC erase media flux density in the shields is then $4\pi M_S \delta/(h+\delta) =$ 2.67 kG. For an advanced head the free layer might be only 3 nm of $Co_{90}Fe_{10}$ with a flux carrying capacity of about 50 kG-nm (if it is optimally biased in the horizontal direction). Therefore the flux absorbed from 19 nm of ABS is sufficient to saturate it. Now the free layer gathers flux from the central 60% of the total gap, G_{ss}. Therefore if $G_{ss} > 19/0.6 = 32$ nm, saturation will occur near the ABS. This gap is about 1/3 of what is typical and would represent a severe process problem. Ceramic AF layers (e.g., γ-Fe_3O_4, NiO) with their thermal problems, may be required along with thin high resistivity shield layers adjacent the gap (e.g., X + Fe_3O_4, NiFeHfO, etc.). In addition one might chose a value of $4\pi Ms/H_K = 1/6$ instead of 1/4. This would allow for a 50% increase in the gap (to 48 nm). For the shielded pole writer, this would be a slight improvement in switching energy density.

This saturation problem will get worse as H_K and $4\pi M_S$ approach their limiting values (e.g., $H_K = 32$ kOe, $4\pi M_S = 5$ kG). For a 3 nm free layer and $\delta/h = 2$, the saturation limit is $G_{ss} < 38$ nm. One potential solution to this problem is to use a CPP design (Fig. 11.11). Then free layers of about 3 nm of CoFe alternating with spacers of about 2 nm of Cu (this is getting into the AF bias regime) have an effective moment range of ±5.1 kG when they are PM biased at 30 degrees. Of course there is still the low impedance problem mentioned above.

A perpendicular reader design that helps on the resolution problem is a shielded gradiometer design [32]. In effect, two GMR sensors are electrically and magnetostatically coupled by a fairly thick (e.g., 20 nm) resistive layer. It can be the AF pinning layer. This design is sensitive to the derivative of the media magnetisation. Therefore it produces a pulse that looks like a longitudinal pulse. Thus it does not need the high frequency channel boost required for of a conventional shielded GMR. The sense CIP current provides ±30 degrees of bias for the free layers. Despite the use of two free layers there is still a saturation problem in that this bias uses up half the dynamic range of each layer.

11.11 Perpendicular/Longitudinal Comparison

Longitudinal and perpendicular recording compete in a multidimensional parameter space. At the highest level we have performance, fabricability, and history. History is very much in favour of longitudinal. A garage full of engineers must be working on a very superior technology to compete with tens of thousands who are working on the old technology. On the dimension of fabricability, history still plays a roll in that capital has been invested in the old technology. However, the above discussion indicates that the same tool base can be applied to perpendicular as to longitudinal. This is less so for patterned media or thermally assisted recording, however. On performance, perpendicular has always had a small (e.g., +50%) advantage on BPI because of its sharp transitions. This has never been sufficient to out weight history.

At this time the primary motivation for perpendicular is that it allows areal density growth into a regime where longitudinal is too unstable; too difficult to execute (low BAR problems); or too noisy. The above discussion shows that all of these issues are coupled. Therefore a comparison between longitudinal and perpendicular should fix the parameters as much as possible. So we will assume that the two approaches have the same: BAR; thermal stability; $M_{S,\mathrm{poletip}}$; and grain number per bit. We will then attempt to answer two questions. How much higher density can be obtained from the perpendicular system? And how much better is the media SNR? In order to answer the density/stability question we need a detailed analysis of longitudinal recording. Here, the reduction in energy barrier from self-demagnetisation is expressed by:

$$E_{bl} = K_u(1 - H_d/H_{co})^{3/2} \ . \tag{11.25}$$

The use of the 3/2 power here is well-established [6]. For non-oriented and partly "oriented" thin film media H_{co} is approximately $H_K/2$. The worst case demagnetising field (two free poles with line charge per unit length of $8\pi Mr\delta$ at a distance of $S_b/2$) for a sharp dibit with spacing, S_b, is:

$$H_d = 16M_r\delta/S_b \ . \tag{11.26}$$

So, using (11.3) we get (unoriented $Mr = 0.707Ms$) the energy surface density as:

$$E_{bl} = 21/2M_r\delta H_{co}(1 - 16M_r\delta/S_bH_{co})^{3/2} \ . \tag{11.27}$$

It is easy to verify that this maximises for:

$$H_d/H_{co} = 16M_r\delta/SbH_{co} = 2/5 \ . \tag{11.28}$$

Therefore the optimal switching energy barrier density is:

$$Eb_l = 0.0164 \cdot S_b.H_{co}^2 \ . \tag{11.29}$$

In practice the media will not be thickened to the optimum predicted by Eq. (11.28) in order to reduce NLTS and percolation. This most thermally stable thickness yields a Williams Comstock [33] transition width, a, of $S_b/2.83$ for zero magnetic spacing. This slightly violates the percolation rule of thumb that the width parameter should be less than a third of the transition spacing. Zero magnetic spacing also implies negative fly heights. This is not reasonable. Therefore the media should (and is) thinned so that a $< S_b/3$ with a magnetic spacing ($d = h+\delta/2$) that is comparable to about 70% of S_b (0.1 for the Wallace spacing loss at high frequency). From [33] the transition width parameter (for an arctan(x/a) shape) is:

$$a = d(1 - S^*)/\pi Q + ([d(1 - S^*)/\pi Q]^2 + 4M_r\delta d/QH_{co})^{1/2} , \qquad (11.30)$$

where S^* is the media coercive squareness, Q is the head gradient factor (e.g., $Q = 0.8$ for $G \gg h$) and $M_r\delta$ is in emu/cm^2. For exchange decoupled media S^* is 0.71. Using the above conditions on a and d we get a value for $M_r\delta$ that is 2/3 of the most thermally stable value predicted with Eq. (11.28). The demagnetising field of the dibit is then 27% of the coercivity instead of 40%. This is also more consistent with what is commonly used in products. The loss in switching energy barrier however is only 11%. Therefore with this percolation reality constraint Eq. (11.29) becomes:

$$E_{bl} = 0.0146 S_b H_{co}^2 . \qquad (11.31)$$

Micromagnetic simulations and analysis indicate that a deep gap field of at least $2H_{co}$ times is required for writing longitudinal media well. This is consistent with a recent IBM 35 Gb/in^2 demonstration [20]. The "no significant saturation at the corner condition" requires that this be less than 75% of the saturation field of the pole tip material. So:

$$E_{bl} = 0.32 S_b M_{S,\mathrm{poletip}}^2 . \qquad (11.32)$$

Finite element simulations indicate that with a shielded pole head one can write well on perpendicular media with a fly height equal to the bit space. However the read back Wallace spacing loss criteria of 70% of the bit space is more stringent. So, with $h = \delta/2 = 0.7S_b$, use of Eqs. (11.24) and (11.32) gives the surface energy density ratio between perpendicular and longitudinal as:

$$E_{bp}/E_{bl} = 1.23 \cdot 1.4 S_b M_{S,\mathrm{poletip}}^2 / 0.32 S_b M_{S,\mathrm{poletip}}^2 = 5.4 . \qquad (11.33)$$

Now if we use this advantage to scale to higher density the media thickness must go down as the square root of the areal density. So we can get a density improvement of 5.42/3 = **3.1 density gain** from perpendicular recording, in the thermal decay limit with the same number of thermally stable grains per bit.

In addition to the density gain, for the same number of read grains per bit, perpendicular also provides improvement in SNR from a greatly reduced transition width parameter, a. This is because the perpendicular transition has no demagnetising field to broaden it. Also, in principle, it can have a very narrow switching field distribution. So for zero easy axis dispersion and zero anisotropy energy dispersion, the magnetic centres of grains as suggested by Wood [30] can define the transitions in perpendicular. Now for simplicity, assume that the grains are square with width, C, and are in upstream to down stream columns with random phase. The transition in column i has a location variance of $\sigma_i^2 = C_i^2/12$ (e.g., 1/12 = the integral from −0.5 to 0.5 of x^2). For simplicity (again) we neglect the grain size dispersion. Then the transition location is the simple average of all N columns across the read width, W_r. The variance of the perpendicular transition location is then:

$$\sigma_p^2 = C^2/(12N) = C^3/(12W_r) \,. \tag{11.34}$$

For longitudinal recording, with thick thermally stable media, the transition width, a, is large compared to the grain diameter, D. For exchange decoupled grains, the autocorrelation length, A, is half the diameter and the variance of the longitudinal transition position is given by [34]:

$$\sigma_l^2 = (\pi^4/48)a^2A/W_r = (\pi^4/96)a^2D/W_r \,. \tag{11.35}$$

So from Eqs. (11.34) and (11.35), noting that $C = D\,(\pi/4)^{1/2}$, with $a = S_b/3$, and for the same number of grains per bit, we get:

$$\sigma_l^2/\sigma_p^2 = 17.5a^2/D^2 = 1.94S_b^2/D^2 = 1.392(S_b/D)^2 \,. \tag{11.36}$$

For BAR = BPI/TPI = 4, a read width of 60% of the pitch, and 15 grains per read bit, $S_b/D = 2.2$. The predicted jitter ratio is then 3.1. This result partly explains the prediction of Wood that data can be recovered with only 6 read grains per read bit with an ideal write process [30]. Simulations indicate that the shielded pole write head can approach this ideal.

This 9.8-dB improvement in signal to noise ($\mathrm{SNR}\alpha\sigma/S_b$) can be translated into an areal density gain. Equation (11.5) gives a gain factor of **2.7**. Combining this with the energy density factor gives a net gain of 8.3. Therefore perpendicular can achieve 1 Tb/in^2 with the same SNR and BAR that longitudinal recording would use at 120 Gb/in^2. Of course the track density would be three times higher.

11.12 Other Technologies

The principle alternate contender for high-density recording is patterned media. One thermally stable and spatially defined grain per bit is better than ten marginally stable grains with random location and size. But it is important to quantify this advantage in order to know when a change over from perpendicular to patterned media might occur despite its greater complexity and cost.

With BPI/TPI = 4, Wood [30] predicts that ten Voronoi grains per bit area (1/areal density) are sufficient (corrected BER = 10^{-15} with off track margin). Equation (11.6) predicts that this could be reduced to 4.6 grains per bit if TPI = BPI. This is comparable to other predictions by Wood [30]. Now if one were to pattern the media with islands that occupy 50% of the media surface area (e.g. 30% guard bands), then areal density could be increased by a factor of 2.3. In order to achieve this gain, the media would have to have the same switching energy density of the ideal perpendicular media. Therefore the head and media for a patterned system is essentially the same as that for an ideal perpendicular system with BAR = 1. This is the shielded pole head with side shields [5].

Even though the ultimate density gain of patterned media is marginal, it will still have a place. Its fringing advantages may in fact provide the only practical way to get to a BAR of unity. However media complexity and the need for write clock servo control will delay its introduction.

Thermally assisted write has also been proposed as a way of dealing with the superparamagnetic limit. But the extremely high-density capability of perpendicular (e.g., >1 Tb/in^2) implies that it would not be needed until the track pitch is less than 50 nm. Only a near field aperture can achieve this. It is very unlikely that the energy flux at this aperture can be made compatible with a lubricated disk at a fly height <5 nm. Therefore thermally assisted write is unlikely to be a commercial player.

The optical wave length limitations of thermally assisted write also apply to optically patterned media. At 1 Tb/in^2 the islands would be 18 nm on a side. If the goal of patterning is to improve SNR, these islands must be well defined. Therefore the resolution should be about 2 nm (for 7% jitter/bit space). Cheap X-ray, E-Beam, or imprint lithography will be required.

A technology that can achieve very high TPI is based on the simultaneous writing and reading of many tracks. The principle tracking problem is the avoidance of adjacent track erasure. With a ganged writer design this is not a problem as long as enough guard band to the next set of tracks is provided. Another advantage of this approach is that the cross talk during read electronically cancelled. Also cross talk residues can be used for continuous read servo control. A final advantage is that the number of transducers multiplies the data transfer rate. A monopole write head concept for this is shown in Fig. 11.18. The trapezoidal down stream poles trim the tracks of the upstream poles even for sizeable skew angles. The low current requirements of advanced systems will allow for single turn write coils with acceptable current (e.g., 60 mA).

11.13 Conclusions

Longitudinal recording will soon run out of gas (e.g., at 50 to 100 Gb/in^2). Perpendicular recording has the capability of extending areal density growth by at least an order of magnitude beyond the thermal decay associated impediments of longitudinal. Therefore perpendicular recording is the technology of choice for areal density growth. Various options for head designs exist. Initially small modifications of longitudinal heads (e.g., monopole writers) are all that will be necessary for perpendicular products. Eventually shielded pole

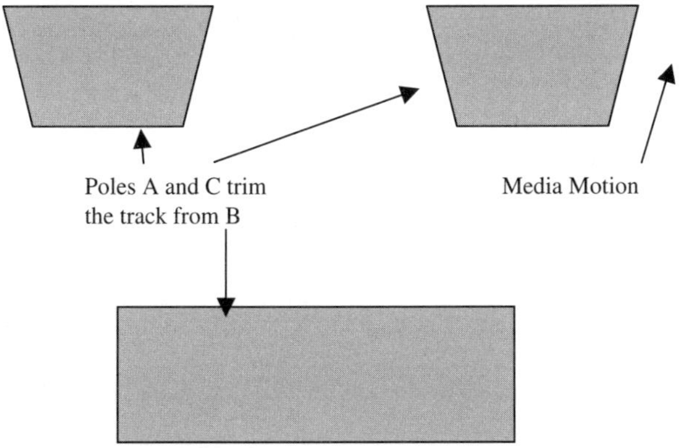

Fig. 11.18. Shielded pole perpendicular writer design

writers and improved readers (e.g., CPP) will be brought in. At this time, alternatives to perpendicular recording appear to provide only supplemental benefits. Of course, this analysis is looking at physical as opposed to practical impediments. Manufacturing 1 Tb/in^2 products with acceptable sigmas will require a mountain of art that has yet to be developed.

References

1. Mason Williams, private communication
2. S. Batra et al., Thin Film Read/Write Head for Minimising Erase Fringing and Method for Making the Same, US Pat. #5,267,112, 11/30/93; S. Batra, A. Torabi, M. Mallary, S. Ramaswamy, S. Marshall, Performance Eval. of Different Pole Geometries in Thin Film Heads, IEEE Trans Magn. Mag **30**, 3876–3879 (1994)
3. M. Mallary, Effective Field Gradient Dependence on Write Field Rise Time, IEEE Trans Magn. Mag **32**, 3527 (1996)
4. Heads sampled from Applied Magnetics Corporation
5. M. Mallary, Vertical Magnetic Recording Arrangement, US Pat#4,656,546, 4/7/87. Re #33,949 with S. Das, 6/2/92
6. M. Mallary, R. Beauregard, and A. Torabi, Switching Field Time Dependence of Thin Film Media, IEEE Trans Magn. Mag. **36**, No. 5, 2477 (2001)
7. M. Mallary, US Pat#5,801,910, Long Sat. Zone Magnetic Write Head, 9/1/98
8. C. Mao et al., J. Appl. Phys. **87**, No. 9, 5416 (5/1/00)
9. M. Mallary et al., Frequency Response of Thin Film Heads with Longitudinal and Transverse Anisotropy, IEEE Trans. Mag. **26**, (1990)
10. M. Mallary, A. Torabi, S. Batra, Three Dimensional Transmission Line Model for Flux Conduction in Thin Film Heads, J. Appl. Phys. **67(9)**, 4863 (1990)
11. M. Mallary and A. Smith, Conduction of Flux at High Frequencies by a Charge Free Magnetization Distribution, IEEE Trans. Mag. **24**, 2374 (1988)
12. M. Mallary, Conduction of Flux at High Frequencies in Permalloy Strips by Small-Angle Rotations, J. Appl. Phys. **57**, 3952 (1985)

13. J.C. Slonczewski et al., IEEE Trans. Mag. **24**, No. 3, 2045 (May 1988). For exchange energy corrections see Slonczewski and Middelhoek, Appl. Phys. Lett. **6**, No. 7, 139 (1965)
14. J.C. Slonczewski, J. Appl. Phys. **37**, 1268 (1966)
15. N. Smith et al., J. Appl. Phys. **73**, No. 10, 6013, Abstract (May 15, 1993)
16. R. Arai et al., High Frequency Characterisitcs of Multi-layer CoTaZr Cores for Thin Film Heads, IEEE Trans. Mag. **28**, No. 5, 2115 (1992); S. Narumi et al., Abstract ES-01, Digests of the InterMag Conf., Stockholm, Sweden (April 1993)
17. M. Mallary, US Pat# 5,089,334 and #5,085,935, Flux Spreading Thin Film Magn. Dev., 2/18/92
18. C. Kittel, Introduction to Solid State Physics (John Wiley & Sons, NY, 1966), p. 525
19. R. S. Tebble and D. J. Craik, Magnetic Materials, (John Wiley & Sons, 1969), p. 409
20. Mike Madison et al., J. Appl. Phys. **87**, No. 9, 4996 (5/1/00) & HT-01, Intermag 2000 Digest
21. H. Zhou, N. Bertram, A. Torabi, and M. Mallary, MMM 2001 Conference, (EA-12), January 2001
22. M. Mallary, Three Pole Magnetic Recording Head, US Pat# 4,907,113, 3/6/90
23. R. Rottmayer and J.G. Zhu, IEEE Trans. Magn. **31**, No. 6, 2597 (1995)
24. W. F. Egelhoff et al., J. Appl. Phys. **. 78**, 243 (1995)
25. S. S. P. Parkin, Z. G. Li, and J. Smith, Appl. Phys. Lett. **58**, 2710 (1991)
26. V.S. Spiriosu, etal., Spin Valves with Synthetic Ferrimagnets, Abstract AA-04, Intermag. Conf., 1996, Seatle Washington
27. K. Chahara, T. Ohnp, M. Kasai, and Y. Kozoono, Appl. Phys. Lett. **63**, 1990 (1993)
28. H. Kikuchi, M. Sato, K. Kobayashi, J. Appl. Phys. **87**, No. 9, 6055 (2000)
29. R. Coehoorn et al., IEEE Trans. Magn. **35**, No. 5, 2586 (1999)
30. R. Wood, et al., IEEE Trans. Magn. **35**, No. 5, 2253 (1999); Abstract 23aA-1, Digest of the PMRC 2000, Sendai, Japan
31. R. Victora et al., J. Appl. Phys. **87**, No. 9, 6358 (2000); Abstract 25pA-2, Digest of the PMRC 2000, Sendai, Japan
32. N. Smith et al., IEEE Trans. Magn. **28**, No. 5, 2292 (1992) is a DMR.
33. M. Williams and L. Comstock, 17th Annu. AIP Conf. Proc., vol. 5, pp. 738–742
34. X. Xing and N. Bertram, IEEE Trans. Mag. **33**, No. 5, 2959 (1997)

Index

SPRINGER SERIES IN SURFACE SCIENCES

Editors: G. Ertl, R. Gomer and D.L. Mills

Founding Editor: H.K.V. Lotsch

1 **Physisorption Kinetics**
By H. J. Kreuzer, Z. W. Gortel

2 **The Structure of Surfaces**
Editors: M. A. Van Hove, S. Y. Tong

3 **Dynamical Phenomena at Surfaces, Interfaces and Superlattices**
Editors: F. Nizzoli, K.-H. Rieder, R. F. Willis

4 **Desorption Induced by Electronic Transitions, DIET II**
Editors: W. Brenig, D. Menzel

5 **Chemistry and Physics of Solid Surfaces VI**
Editors: R. Vanselow, R. Howe

6 **Low-Energy Electron Diffraction**
Experiment, Theory and Surface Structure Determination
By M. A. Van Hove, W. H. Weinberg, C.-M. Chan

7 **Electronic Phenomena in Adsorption and Catalysis**
By V. F. Kiselev, O. V. Krylov

8 **Kinetics of Interface Reactions**
Editors: M. Grunze, H. J. Kreuzer

9 **Adsorption and Catalysis on Transition Metals and Their Oxides**
By V. F. Kiselev, O. V. Krylov

10 **Chemistry and Physics of Solid Surfaces VII**
Editors: R. Vanselow, R. Howe

11 **The Structure of Surfaces II**
Editors: J. F. van der Veen, M. A. Van Hove

12 **Diffusion at Interfaces: Microscopic Concepts**
Editors: M. Grunze, H. J. Kreuzer, J. J. Weimer

13 **Desorption Induced by Electronic Transitions, DIET III**
Editors: R. H. Stulen, M. L. Knotek

14 **Solvay Conference on Surface Science**
Editor: F. W. de Wette

15 **Surfaces and Interfaces of Solids**
By H. Lüth*)

16 **Atomic and Electronic Structure of Surfaces**
Theoretical Foundations
By M. Lannoo, P. Friedel

17 **Adhesion and Friction**
Editors: M. Grunze, H. J. Kreuzer

18 **Auger Spectroscopy and Electronic Structure**
Editors: G. Cubiotti, G. Mondio, K. Wandelt

19 **Desorption Induced by Electronic Transitions, DIET IV**
Editors: G. Betz, P. Varga

20 **Scanning Tunneling Microscopy I**
General Principles and Applications to Clean and Adsorbate-Covered Surfaces
Editors: H.-J. Güntherodt, R. Wiesendanger
2nd Edition

21 **Surface Phonons**
Editors: W. Kress, F. W. de Wette

22 **Chemistry and Physics of Solid Surfaces VIII**
Editors: R. Vanselow, R. Howe

23 **Surface Analysis Methods in Materials Science**
Editors: D. J. O'Connor, B. A. Sexton, R. St. C. Smart

24 **The Structure of Surfaces III**
Editors: S. Y. Tong, M. A. Van Hove, K. Takayanagi, X. D. Xie

25 **NEXAFS Spectroscopy**
By J. Stöhr

26 **Semiconductor Surfaces and Interfaces**
By W. Mönch
3rd Edition

27 **Helium Atom Scattering from Surfaces**
Editor: E. Hulpke

28 **Scanning Tunneling Microscopy II**
Further Applications and Related Scanning Techniques
Editors: R. Wiesendanger, H.-J. Güntherodt
2nd Edition

29 **Scanning Tunneling Microscopy III**
Theory of STM and Related Scanning Probe Methods
Editors: R. Wiesendanger, H.-J. Güntherodt
2nd Edition

30 **Concepts in Surface Physics**
By M. C. Desjonquères, D. Spanjaard*)

31 **Desorption Induced by Electronic Transitions, DIET V**
Editors: A. R. Burns, E. B. Stechel, D. R. Jennison

32 **Scanning Tunneling Microscopy and Its Applications**
By C. Bai
2nd Edition

33 **Adsorption on Ordered Surfaces of Ionic Solids and Thin Films**
Editors: H.-J. Freund, E. Umbach

34 **Surface Reactions**
Editor: R. J. Madix

35 **Applications of Synchrotron Radiation**
High-Resolution Studies of Molecules and Molecular Adsorbates on Surfaces
Editor: W. Eberhardt

36 **Kinetics of Metal-Gas Interactions at Low Temperatures: Hydriding, Oxidation, Poisoning**
By E. Fromm

37 **Magnetic Multilayers and Giant Magnetoresistance**
Fundamentals and Industrial Applications
Editor: U. Hartmann*)

*) Available as a textbook

Printing: Mercedes-Druck, Berlin
Binding: Stein+Lehmann, Berlin